WHERE TO WATCH BIRDS IN SPAIN

The 100 best sites

José Antonio Montero

IN COLLABORATION WITH EDUARDO DE JUANA AND FERNANDO BARRIO

WHERE TO WATCH BIRDS IN SPAIN

THE 100 BEST SITES

Cover photos: Ebro Delta, © Joan Gil (above); Eleonora's Falcon, Carlos Sánchez © nayadefilms.com (below left); Betancuria, © Pedro Retamar (below centre); Audouin's Gull, Carlos Sánchez © nayadefilms.com (below right).

Recommended reference:

Montero, J. A. et al. (2005). *Where to watch birds in Spain. The 100 best sites.* Lynx Edicions, Bellaterra, and SEO/BirdLife.

First edition: March 2006

© **Lynx Edicions** – Montseny, 8, 08193 Bellaterra, Barcelona
© text: SEO/BirdLife
© photographs: credited photographers.

English translation: Mike Lockwood

Printed by: Ingoprint, S.A.
Legal Deposit: B-11.889-2006
ISBN: 84-96553-04-3
ISBN: 978-84-96553-04-0

All rights reserved. No part, text or illustration, of this publication may be reproduced, stored in a retrieval system, or transmitted by any form or by any means, electronic, mechanical, photocopying, recording or otherwise, without prior written permission from Lynx Edicions.

Montero, José Antonio
 Where to watch birds in Spain : the 100 best sites / José Antonio Montero ; in collaboration with Eduardo de Juana and Fernando Barrio. – Bellaterra : Lynx Edicions and SEO/BirdLife, 2006.
 344 p. : il. col., map ; 23 cm.

 ISBN 84-96553-04-3
 ISBN 978-84-96553-04-0

 1. Birdwatchers' guides – Spain
 I. Juana, Eduardo de. II. Barrio, Fernando. III. Title

WHERE TO WATCH BIRDS IN SPAIN. THE 100 BEST SITES

CONTENTS

Prologue 9
Introduction 11
Recommended reading 17

Andalusia
1. Sierra de Andújar 23
2. Grazalema and Ronda 25
3. Sierra de Cazorla 28
4. Sierra Nevada 31
5. Laguna de Fuente de Piedra 33
6. Lagunas de Zóñar and El Rincón 36
7. Laguna de Medina 38
8. Deserts of Almería 41
9. Doñana 43
10. Sanlúcar de Barrameda 48
11. Brazo del Este 51
12. Marismas del Odiel 53
13. Bahía de Cádiz 56
14. Mouth of the Guadalhorce 58
15. Wetlands of western Almería 61
16. Cabo de Gata 63
17. Straits of Gibraltar 66

Aragon
18. Benasque 71
19. Ordesa 74
20. Sierra de Guara 77
21. San Juan de la Peña 80
22. Riglos 83
23. Monegros 85
24. Belchite steppes 88
25. Laguna de Sariñena 90
26. Galachos del Ebro 93
27. Laguna de Gallocanta 96

Asturias
28. Somiedo 100
29. Cabo Peñas 103
30. Ría de Villaviciosa 106

The Balearic Islands
31. Serra de Tramuntana 110
32. S'Albufera de Mallorca 113
33. Menorca 116
34. Cabrera 118

The Basque Country
35. Álava wetlands 123
36. Urdaibai 126
37. Txingudi 128

The Canary Islands
38. Forests of Tenerife 133
39. Southern Tenerife 136
40. La Gomera 138
41. Fuerteventura 141
42. Lanzarote 144

Cantabria
43. La Liébana 148
44. Marismas de Santoña 151

Castilla-La Mancha
45. Tablas de Daimiel 157
46. La Mancha wetlands (I) 160
47. La Mancha wetlands (II) 163
48. Laguna de Pétrola 166
49. Castrejón and Azután 168
50. Valle del Tiétar 172
51. Alto Tajo 174
52. Cabañeros 177

Castilla y León
53. La Moraña 183
54. Lagunas de Villafáfila 186
55. Laguna de La Nava 188
56. Azud de Riolobos 191
57. Hoces del Duratón 193
58. Montejo de la Vega 196
59. Picos de Europa 199
60. Hoces del Alto Ebro 202
61. Valle del Arlanza 204
62. Sierra de Gredos 207
63. La Granja and Segovia 209
64. Arribes del Duero 212

Catalonia
65. Ebro Delta 217
66. Aiguamolls de l'Empordà 221
67. Llobregat Delta 224
68. Aigüestortes 227
69. Cadí and Moixeró 230
70. Lleida Plains 232

CONTENTS

Comunitat Valenciana
71. L'Albufera 238
72. Marjal del Moro 241
73. El Hondo 243
74. Salt-pans of southern Alicante 245

Extremadura
75. Monfragüe 251
76. Sierra de San Pedro 254
77. Las Villuercas 256
78. Campo Arañuelo 259
79. Cáceres Plains 262
80. La Serena 265
81. Puerto Peña and Orellana 267
82. Vegas Altas del Guadiana 270
83. Mérida 273

Galicia
84. Ría de Ortigueira 278
85. Costa da Morte 281
86. Ría de Arousa 283
87. Islas Atlánticas 286

La Rioja
88. Sierra de La Demanda 291
89. Iregua, Leza and Cidacos 294

Madrid
90. Valle del Lozoya 298
91. Alto Manzanares 301
92. South-west Madrid 304
93. Valle del Jarama 307
94. South-eastern wetlands 309

Murcia
95. Mar Menor 314

Navarre
96. Roncesvalles and Irati 319
97. Lumbier and Arbayún 322
98. Bardenas Reales 325
99. Laguna de Pitillas 328
100. Embalse de Las Cañas 330

Species index 334

PROLOGUE

I first met José Antonio Montero in the summer of 1990 when he joined the staff of the magazine *Quercus*. I had been working there for a couple of years already and from that moment onwards we worked side-by-side on the arduous task of publishing every month one of the most rigorous, influential and committed publications on the Spanish environmental scene. I freely admit thus that my opinions regarding this book are somewhat coloured by our longstanding friendship.

I have been involved from the word go with this book, starting with the preliminary meetings in the SEO/BirdLife HQ, continuing through the routine work enlivened by animated lunchtime debates, and ending up with the final editorial decision-making. I can honestly affirm that the author has applied the same serious criteria as in his work for *Quercus* and has put his whole heart into the writing of this book. He has checked all the maps – having drawn the preliminary sketches himself – and knows in detail the majority of all the suggested routes. As well, he has worked with many of the contributors to *Quercus* in the task of checking and bringing up-to-date the book's least details. Throughout the writing of the book, he has had to sift through vast quantities of written material and websites during marathon (and unhealthily long) working days, and has spent many hours on the phone. All in all, José Antonio has done everything in his power to ensure that readers of this book will enjoy the fruits of his labours.

Nevertheless, birdwatching is not an exact science and all too often depends on the climate, the season, the time of day and, above all, on the willingness of birds to put in an appearance. Given that this is a book designed for a very general audience, it is worthwhile remarking here that birdwatching is not like visiting a zoo: birds have the habit of not always appearing on cue and so, despite all José Antonio's efforts, merely brandishing this book will not necessarily guarantee a successful day in the field. Even so, all those with a genuine empathy for wildlife will be able to enjoy the routes described in this book, above all if well provisioned with the patience, perseverance and discretion needed by all good naturalists.

Finally, I must congratulate both SEO/BirdLife and the publishers Lynx for their decision to publish two highly accurate and affordable books designed to initiate readers into the wonderful world of birds. First came *Guía de las aves de España* by Eduardo de Juana and Juan Varela, and now this 'Where to watch birds' guide. Both books are practical, easy-to-read and full of the hope that they will encourage young and old alike to take up birdwatching as a serious hobby. No one will contradict me if I say that birdwatching today is now no longer the preserve of the specialists, but rather a privilege open to all.

Rafael Serra
Director of *Quercus*

INTRODUCTION

Where to watch birds in Spain, coordinated and edited by Eduardo de Juana, was published in 1993 and contained information on a vast number of excellent birdwatching sites. It was the fruit of collaboration between SEO/Birdlife – who wrote the texts – and the publishers Lynx: now, over ten years later, the same partnership has produced a similar guide under the guidance of Eduardo de Juana and Fernando Barrio.

In Where to watch birds in Spain: the 100 best sites we have used similar criteria to those used in the first book, although to make it more manageable we have limited ourselves to just 100 sites, all magnificent places for watching birds in the wild. All important birdwatching habitats have been included – uplands, forests, wood pastures, agricultural areas, steppes, river canyons, wetlands, coastlines and even urban areas – and we have ensured that no autonomous region has been left uncovered. Even so, it is inevitable in a work of this nature that we have had to leave out some excellent sites, and we apologise in advance for having done so.

Another important factor in the choice of sites was ease of access. We have thus avoided including out-of-the-way sites, only accessible with great difficulty, and have emphasised sites with good visitor facilities. Naturally, particularly sensitive sites have been excluded and we have scrupulously followed the advice given by the over 200 ornithologists from the whole of Spain, all equipped with valuable local information, who have collaborated with this guide.

Without the data and comments provided by these top-quality collaborators, most of whom are members

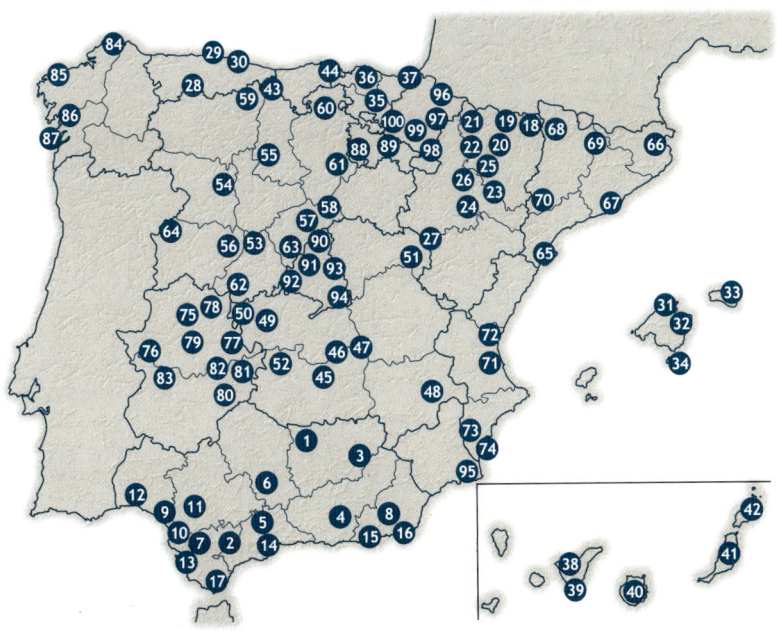

of SEO/BirdLife, this book would not have been possible. Thanks to them, the information contained in this guide is – to the best of our knowledge – practical, highly accurate and perfectly up to date.

CHAPTER CONTENTS

The sites are grouped together by autonomous community (including the Balearic and Canary Islands) and each chapter begins with a brief description of the region's natural habitats and their levels of legal protection; where we have seen fit, we also discuss the human pressure on the region's wildlife and existing conservation problems. In the following section 'Ornithological Interest' we describe the birds and groups of bird species that make the region attractive and worthy of protection.

However, it is not until the following section entitled 'Itineraries' that we provide the essential information needed by birdwatchers who wish to visit the sites. In most cases, a number of routes are suggested and descriptions include detailed explications and reference points, many of which are numbered in the text and can be cross-referenced to the maps. A lot of information is provided about the most interesting or characteristic species that can be easily seen at different points of the itineraries.

The meaning of the symbols that appear at the beginning of each itinerary is described below.

indicates that the itinerary is to be done by car, on foot or by combining the two.

indicates whether the itinerary requires a whole or just a half-day (taking into account the numerous halts that will slow overall progress).

indicates whether the itinerary is circular and takes you back to your starting point or not.

indicates the length of the itinerary (usually expressed in kilometres). In the non-circular itineraries, distances do not include the return journey.

The section 'Practical Details' provides additional practical information such as the main access roads and places to spend the night that will make visits more fruitful. Any accommodation or infrastructure especially designed for birdwatchers is always mentioned, as are facilities such as interpretation centres and hides. We also remark on the lack of such facilities as something else to be taken into account by visiting birdwatchers.

The best time of year to visit and the most interesting species of bird to be seen in each season, as well as practical access details, are also included in this section. We give the names of groups and organisations that might be able to provide birdwatchers with additional information, and also mention useful books and other publications and web sites. A good map is always useful and the number of the respective Spanish military map sheet (*Servicio Geográfico del Ejército*: SGE), normally at a scale of 1.50,000, is always given.

The final section of each site 'Most Representative Species' consists of a list of the most interesting birds, grouped according to the criteria of 'residents', 'summer visitors', 'winter visitors' and 'on migration', that may be seen during the itinerary. The commonest species found in many sites are not included; on the other hand, those species that are especially hard to see, be it due to their shyness or

INTRODUCTION

scarcity, are likewise not included so as to avoid creating false expectations. Furthermore, we have deemed it wise to avoid mentioning particularly vulnerable or sensitive species that it is best not to search for.

Other than the Doñana and the Ebro Delta, two huge areas with endless possibilities for birdwatching, all the site descriptions are more or less the same length. Each text is accompanied by a representative photograph of the site and, most importantly, one, two or three maps that have been especially drawn for this book.

The maps have been designed to make the itineraries easier to follow; the unbroken red line indicates that the itinerary follows a road, while the broken red line indicates an off-road section. Successively numbered points marked on the itinerary on the map refer to points highlighted in the text. Car-parks and information points are also marked, as are rivers, peaks, provincial and state boundaries, roads and nearby population nuclei. The following key explains these features.

ℹ	information points
—	rivers
▲	peaks
- - -	boundaries
=	roads
·······	tracks/paths
◉	villages/towns
◉	hides
⛪	religious buildings
━	boundaries of protected areas

Given that the size of the map varies with the size of the site and length of the itinerary described, the scale of the map is always provided. Wherever possible, all the itineraries in a single site have been drawn on the same map, although in some of the larger sites this has not been possible and more than one map has been included.

BIRDS COME FIRST

The information provided in *Where to watch birds in Spain: the 100 best sites* is sufficiently wide-ranging and precise for readers to be able to get the best out of the suggested itiner-

Greater Flamingos at La Laguna de Fuente de Piedra © Pedro Retamar

aries. However, we recommend readers to spend further time planning their field trips and investigating more fully the ornithological riches of the area they plan to explore. Without doubt, a little extra planning will enable readers to benefit even more from the details provided in the book.

Although each site description gives a certain amount of information regarding the best times to visit, in general it is best to avoid the best-known sites during peak holiday periods such as Easter. As well, much of Spain is too hot for comfortable birdwatching during much of the summer, although there are certain exceptions such as some mountain regions and migration watch-points.

Morning and evening excursions when birds are more active than during the heat of the day are always the most rewarding. It is best to wear appropriate, sober-coloured clothing in the field and to carry binoculars (telescope if possible), field guide and notebook for taking field notes.

We trust that the common sense of users of this guide will ensure that they do not disturb birds whilst in the field: the well-being and the conservation of the birds under observation come first. Please remember that very often the behaviour of a few inconsiderate individuals can blacken the image of a whole group of people.

As such we recommend that you do not approach nest sites too closely and be especially cautious during the breeding season. Never shout or gesticulate wildly in order to let your friends know what you have just seen; never run after birds or use CDs or tapes to attract them; and finally take care who you pass information on to about the birds you see and, above all, their nest-sites.

It is also very important not to stray from public and/or waymarked paths: respect private property and only enter if permission has been granted. As well, we trust that you will treat the local people you meet with respect and that wherever possible you will consume local products and services as a means of promoting sustainable local development. Be environmentally aware with your rubbish – as little as possible deposited in the appropriate places – and try and use your vehicles only when necessary.

A GUIDE THAT WELCOMES CHANGES

As times passes, some of the information provided by *Where to watch birds in Spain: the 100 best sites* may be rendered out of date; this will occur as much with the birds to be seen as with the access and visitor facilities on offer.

In future editions some sites may be replaced by others to ensure that the guide is the best possible up-to-date tool for birdwatching in Spain. Our hope is that the inclusion of a new site will be seen as a reward for the success of a particular institution, be it public or private, in protecting one of our country's natural areas, above all if birdwatching as a result becomes more rewarding.

We would therefore ask all readers to send us information regarding necessary updates, as well as any errors they have detected, to the following address: SEO/BirdLife, seo@seo.org.

THE MOST COMMONLY USED ABBREVIATIONS

ha: hectares
km: kilometres
m: metres
SGE: *Servicio Geográfico del Ejército* (Spanish military cartographic service)
SPA: Special Protection Area for Birds

INTRODUCTION

ACKNOWLEDGEMENTS

The authors would like to thank the following people: in **Andalusia**, David Barros, José Bayo, Francisco Chiclana, Carlos Dávila, Manuel Garrido, Jorge Garzón, Alfonso Godino, Fernando Ibáñez, Juan José Jiménez, Manuel Jiménez, Enrique L. Carrique, José Manuel López, José Manuel López-Martos, Juan Manrique, Francisco Montoya (COCN), Mariano Paracuellos, Jesús Pinilla, José Manuel Rivas, José Manuel Sayago, Antonio Tamayo and José Antonio Torres-Esquivias. In **Aragon**, Juan Carlos Albero, Javier Blasco, Alberto Bueno, Juan Carlos Cirera, Carmina Franco, David Gómez, Luis Lorente, Javier Mañas, Ramiro Muñoz, José Luis Rivas, Josele J. Saiz, José Manuel Sánchez and Eduardo Viñuales. In **Asturias**, Antonio Alba, Clemente Álvarez, César Álvarez-Laó, Ángel González-Losa, Manuel Quintana and Cristino Ruano. In **the Balearic Islands**, Juan Salvador Aguilar, Pep Amengual, Oscar G. Febrero, Santi Cachot, José Miguel González, Jordi Muntaner, Biel Perelló, Juan José Sánchez, Rafel Triay, Miquel Truyol and Pere Vicens. In **the Basque Country**, Teresa Andrés, Mikel Etxaniz, Jon Hidalgo, Luis Lobo and Jon Maguregi. In **Cantabria**, Felipe González. In **Castilla-La Mancha**, Manuel Carrasco, Miguel Ángel de la Cruz, Miguel Ángel Hernández, Pepe Jiménez, Marino López de Carrión, Alejandro del Moral, Roberto Carlos Oliveros, Antonio Paredes, Juan Picazo, Juan Francisco Sánchez, Carlos Torralvo and Ángel Vela. In **Castilla y León**, Enrique Álvarez, Ángel Blanco, Miguel Briones, Sara Brizuela, Jesús Cobo, Ángel Contreras, Paco Cosme, Fidel José Fernández, Ángel Gaspar García-Miranda, David González, Nicolás González, Enrique Gómez, Carlos Gutiérrez, Alfredo Hernández, Octavio Infante, Francisco Martín, Borja Palacios, Carlos Palma, Francisco Purroy, Juan José Ramos, José Enrique Remis, Miguel Rey, Adolfo Rodríguez, Mariano Rodríguez, Fernando Román, Raquel Romero, Francisco Sánchez-Aguado, Joaquín Sanz-Zuasti and Gabriel Sierra. In **Catalonia**, David Bigas, Gerard Bota, Lali Caballé, Jordi Canut, Joan Estrada, Miguel Ángel Franch, Jordi García, Ricard Gutiérrez, Raimon Mariné, Jordi Martí, Ramón Martínez, Jordi Palau and Ignasi Ripoll. In the **Comunitat Valenciana**, Sergio Arroyo, Nacho Díes, Mario Giménez, Elías Gomis, Miguel Ángel Pavón, Luis Fidel Sarmiento, Marcial Yuste and Antonio Zaragozi. In **Extremadura**, José Antonio Álvarez, José María Benítez, Javier Briz, Manuel Calderón, Emilio Costillo, Manuel Gómez-Calzado, Casto Iglesias, Juan Carlos Núñez, Alfredo Ortega, Víctor Pizarro, Javier Prieta, Juan Pablo Prieto, Domingo Rivera and Chema Traverso. In **Galicia**, Pablo Gutiérrez, José Luis Rabuñal, Cosme-Damián Romay, Antonio Sandoval, José Antonio Souza and César Vidal. In **La Rioja**, Álvaro Camiña, Ignacio Gámez and Luis Lopo. In **Madrid**, Fernando Ávila, Eugenio Castillejos, Emilio Escudero, Manuel Fernández, Oscar Frías, Gonzalo García, Javier Grijalbo, Miguel Juan, Blas Molina, Máximo Muñoz, Juan Carlos del Moral, Eladio G. de la Morena, Javier de la Puente and Federico Roviralta. In **Murcia**, Gustavo Ballesteros, Pedro García and Antonio Hernández. In **Navarre**, Juan Deán, Alberto Jiménez, José Antonio Martínez, Antonio Munilla, Alfonso Senosiáin, Beatriz Taracelli and Alejandro Urmeneta.

In the **Canary Islands**, thanks also to Juan Antonio Lorenzo from the SEO/BirdLife office in Santa

Cruz de Tenerife, who wrote the first draft of the five sites in the Canary Islands.

Thanks also to all the wildlife photographers who donated their pictures.

And, above all, many thanks to Cristina Vega, Rafael Serra, Ángeles de Andrés, José Manuel Reyero and all the members of the editorial staff of the magazines *Quercus* and *Turismo Rural*.

José Antonio Montero

RECOMMENDED READING

GENERAL

Alamany, O. & Vicens, E. 2003. *Parques Nacionales de España. 26 itinerarios para descubrirlos y conocerlos.* Lynx Edicions, Barcelona.
Butler, J.R. 2001. *Birdwatching on Spain's Southern coast.* Santana Books, Fuengirola (Málaga).
Clavell, J. 2002. *Catàleg dels ocells dels Països Catalans.* Lynx Edicions, Barcelona.
Crozier, J. 1998. *A Birdwatching Guide to the Pyrenees.* Arlequin Press, Chelmsford.
Finlayson, C. 1991. *Birds of the Strait of Gibraltar.* Poyser, London.
Finlayson, C. 1993. *A Birdwatchers' Guide to Southern Spain and Gibraltar.* Prion, Huntingdon.
García, E. 1995. *Dónde observar aves en la España meridional.* Omega, Barcelona.
García, E. & Paterson, A. 2001. *Where to watch birds in Southern & Western Spain.* Helm, London.
Gosney, D. 1993. *Finding Birds in Northern Spain.* Available from BirdGuides.
Gutiérrez, R. *Rare birds in Spain.* URL: http://www.rarebirdspain.net.
de Juana, E. 1993. *Dónde ver aves en España peninsular.* Lynx Edicions, Barcelona.
de Juana, E. & Varela, J.M. 2000. *Guía de las aves de España.* Lynx Edicions, Barcelona.
Martí, R. & del Moral, J.C. (eds.). 2003. *Atlas de las aves reproductoras de España.* DGCN-SEO/BirdLife. Madrid.
Palmer, M. 1994. *A Birdwatching Guide to the Costa Blanca* (second edition). Arlequin Press, Chelmsford.
Palmer, M. 1997. *A Birdwatching Guide to Southern Spain.* Arlequin Press, Chelmsford.
Palmer, M. 1999. *A Birdwatching Guide to Eastern Spain.* Arlequin Press, Chelmsford.
Rebane, M. 1999. *Where to watch birds in North and East Spain.* Helm, London.
Roche, J.C. & Chevereau, J. 2001. *Guía sonora de las aves de Europa* (10 CDs of 442 species). Lynx Edicions, Barcelona.
Rose, L. 1995. *Dónde observar aves en España y Portugal.* Helm, London.
Sterry, P. 2004. *Guía fotográfica de las aves de España y del Mediterráneo.* Lynx Edicions, Barcelona.
Svensson, L., Grant, P.J., Mullarney, K. & Zetterström, D. 2001. *Guía de aves. La guía de campo de aves de España y de Europa más completa.* Omega, Barcelona.

ANDALUSIA

Barros, D. & David, R. 1999. *Aves del Parque Natural Sierra de Grazalema.* Ornitour, La Línea (Cádiz).
Barros, D. & Ríos, D. 2002. *Guía de aves del Estrecho de Gibraltar, Parque Natural de Los Alcornocales y Comarca de La Janda.* Ornitour, La Línea (Cádiz).
Bernis, F. 1980. *La migración de las aves en el Estrecho de Gibraltar. Volumen I: aves planeadoras.* Universidad Complutense, Madrid.
Chiclana, F., Lama, J.A. & Salcedo, J. 2002. *Aves de la provincia de Sevilla.* Grupo local SEO-Sevilla, Sevilla.
Garrido, H. 1996. *Aves de las marismas del Odiel y su entorno.* Editorial Rueda, Alcorcón (Madrid).
Garrido, H. (ed.). 2000. *Anuario Ornitológico de Doñana, nº 0.* Ayuntamiento de Almonte, Almonte (Huelva).
Garrido, M., Alba, E. & González, J.M. 2002. *Las aves acuáticas y marinas en Málaga y provincia.* Diputación de Málaga, Málaga.

Garrido, M. & Alba, E. 1997. *Las aves de la provincia de Málaga*. Diputación de Málaga, Málaga.
Gómez, Mena, J. 1998. *Visita el Parque Natural de Cazorla, Segura y las Villas*. Editorial Everest, León.
Tellería, J.L. 1981. *La migración de las aves en el Estrecho de Gibraltar. Volumen II: aves no planeadoras*. Universidad Complutense, Madrid.
Torres-Esquivias, J.A. 2004. *Lagunas del sur de Córdoba*. Diputación de Córdoba, Córdoba.

ARAGON

Mañas, J. (coord.). 2001. *Guía de la naturaleza de Gallocanta*. Prames, Zaragoza.
Pedrochi, C. (coord.). 1998. *Ecología de Los Monegros*. Instituto de Estudios Altoaragoneses, Huesca.
Viñuales, E. 1999. *9 itinerarios naturalísticos por los Monegros oscenses*. Prames, Zaragoza.
Woutersen, K. & Grasa, M. 2002. *Parque Nacional de Ordesa y Monte Perdido. Atlas de las aves*. KW Publicaciones, Huesca.

ASTURIAS

Arce, L.M. 1996. *La ría de Villaviciosa. Guía de la naturaleza*. Trea, Gijón.
Noval, A. (ed.). 2000. *Guía de las aves de Asturias*. Gijón.

BALEARIC ISLANDS

Escandell, R. & Catchot, S. 1997. *Aves de Menorca*. GOB Menorca, Maó.
Gosney, D. 1993. *Finding Birds in Mallorca*. Available from BirdGuides
Hearl, G. 1996. *A Birdwatching Guide to Menorca, Ibiza and Formentera*. Arlequin Press, Chelmsford.
Hearl, G. & King, J. 1999. *A Birdwatching Guide to Mallorca* (second edition). Arlequin Press, Chelmsford.

King, J. & Hearl, G. 1999. *The Birds of the Balearic Islands*. Poyser, London.
Pons, G.X. (ed.). 2000. *Las Aves del Parque Nacional Marítimo-Terrestre del archipiélago de Cabrera*. Ministerio de Medio Ambiente-GOB, Palma de Mallorca.
Rebassa, M. & Muntaner, J. 2002. *Aus de les Illes Balears*. Perifèrics, Palma de Mallorca.

BASQUE COUNTRY

Etxaniz, M. 1998. *Txingudi*. Departamento de Ordenación del Territorio y Medio Ambiente, Vitoria.
Galarza, A. 1989. *Urdaibai. Avifauna de la ría de Gernika*. Diputación Foral de Vizcaya, Bilbao.
Hidalgo, J. & del Villar, J. 2004. *Urdaibai: guía de aves acuáticas*. Departamento de Medio Ambiente y Ordenación del Territorio del Gobierno Vasco, Vitoria.
Riofrío, J. 2000. *Avifauna de Txingudi*. Departamento de Ordenación del Territorio y Medio Ambiente, Vitoria.

CANARY ISLANDS

Lorenzo, J.A. & González, L. 1993. *Las aves de El Médano*. ATAN, Santa Cruz de Tenerife.
Martín, A. & Lorenzo, J.A. 2001. *Aves del archipiélago canario*. Francisco Lemus editor, La Laguna (Tenerife).

CANTABRIA

Pérez de Ana, J.M. 2000. *Aves marinas y acuáticas de las marismas de Santoña, Victoria, Joyel y otros humedales de Cantabria*. Fundación Marcelino Botín, Santander.
Saiz, J. & Fombellida, I. 1999. *Aves de Cantabria*. Creática, Santander.

RECOMMENDED READING

CASTILLA-LA MANCHA

Jiménez, J. 1995. *Aves de Cabañeros y su entorno.* Ecohábitat, Madrid.
Jiménez, J., del Moral, A., Morillo, C. & Sánchez, M.J. 1991. *Las aves del Parque Nacional de las Tablas de Daimiel y otros humedales manchegos.* Lynx Edicions, Barcelona.

CASTILLA Y LEÓN

Cobo, J. & Suárez, L. 2000. *Guía del Refugio de Rapaces de Montejo de la Vega.* WWF/Adena, Madrid.
Martín, J.L. & Sierra, G. 1999. *Guía de las aves de La Moraña y Tierra de Arévalo.* Asodema, Ávila.
Palacios, J. & Rodríguez, M. 1999. *Guía de fauna de la Reserva de las Lagunas de Villafáfila.* Junta de Castilla y León, Valladolid.
Román, J., Román, F., Ansola, L.M. Palma, C. & Ventosa, R. 1996. *Atlas de las aves nidificantes de la província de Burgos.* Caja de Ahorros del Círculo Católico, Burgos.
Sánchez, F. 1995. *Cañón del Duratón, el lento trabajo del río.* Edilesa, León.
Sanz-Zuasti, J. & Velasco, T. 1997. *Guía de las aves de las lagunas de Villafáfila y su entorno.* ADRI Palomares, Zamora.
Sanz-Zuasti, J. & Velasco, T. 2001. *Guía de las aves de Castilla y León.* Náyade Producciones, Medina del Campo (Valladolid).
Vicente, J.L., Palacios, J., Martínez, A. & Rodríguez, M. 2000. *Arribes del Duero: el hogar del águila perdicera y de la cigüeña negra.* Junta de Castilla y León, Valladolid.

CATALONIA

Bas, J., Curcó, A. & Orta, J. 2005. *Itineraris pels parcs naturals de Catalunya.* Lynx Edicions, Barcelona.
Estrada, J., Pedrocchi, V., Brotons, Ll. & Herrando, S. 2004. *Atles dels ocells nidificants de Catalunya 1999-2002.* Institut Català d'Ornitologia (ICO)/Lynx Edicions, Barcelona.
García Petit, J. 1997. *Fauna del Parc Natural del Cadí-Moixeró (Vertebrats).* Lynx Edicions, Barcelona.
Gutiérrez, R., Esteban, P. & Santaeufemia, F.X. 1995. *Els ocells del delta del Llobregat.* Lynx Edicions, Barcelona.
del Hoyo, J. & Sargatal, J. 1989. *On observar ocells a Catalunya.* Lynx Edicions, Barcelona.
Llobet, T. & Feliu, P. 2001. *Parc natural dels Aiguamolls de l'Empordà. Guia d'itineraris.* Edicions del Brau, Figueres (Girona).
Martínez-Vilalta, A. & Motis, A. 1991. *Els ocells del delta de l'Ebre.* Lynx Edicions, Barcelona.

COMUNITAT VALENCIANA

Català, F.J., Díes, B., Díes, J.I., García, F.J. & Oltra, C. 1999. *Las aves de l'Albufera de Valencia.* Vaersa, Valencia.
Ramos, A.J. & Sarmiento, L.F. 1999. *Las aves de los humedales del sur de Alicante y su entorno.* Editorial Club Universitario, Alicante.
Ramos, A.J., Sarmiento, L.F. & Pavón, M.A. 1999. *Las aves del Clot de Galvany.* Ayuntamiento de Elche, Elche (Alicante).

EXTREMADURA

Holgado, P. & Caldera, J. 1997. *Villuercas-Ibores. Valores y tradiciones de una comarca desconocida.* Ed. Prunus lusitanica, C.B., Madrid.
López Gallego, A. 2000. *Dónde ver aves y naturaleza en Extremadura. Guía detallada de las mejores zonas para observar aves en Extremadura, de los espacios naturales más valiosos.* Albarragena, Monesterio (Badajoz).

Muddeman, J. 1999. *A Birdwatching Guide to Extremadura*. Arlequin Press, Chelmsford.

GALICIA

Guitián, J., Munilla, I., González, M. & Arias, M. 2004. *Guía de las aves de la sierra de O Caurel*. Lynx Edicions, Barcelona.

MURCIA

Ballesteros, G.A., Casado, J., Robledano, F. & Saura, V. 2000. *Guía de aves acuáticas del Mar Menor*. Dirección General de Medio Natural, Murcia.

Cavero, L. 1998. *El Parque Regional de las Salinas y Arenales de San Pedro del Pinater*. Consejería de Medio Ambiente, Agricultura y Agua, Murcia.

WHERE TO WATCH BIRDS IN SPAIN. THE 100 BEST SITES

ANDALUSIA

1. Sierra de Andújar
2. Grazalema and Ronda
3. Sierra de Cazorla
4. Sierra Nevada
5. Laguna de Fuente de Piedra
6. Lagunas de Zóñar and El Rincón
7. Laguna de Medina
8. Deserts of Almería
9. Doñana
10. Sanlúcar de Barrameda
11. Brazo del Este
12. Marismas del Odiel
13. Bahía de Cádiz
14. Mouth of the Guadalhorce
15. Wetlands of western Almería
16. Cabo de Gata
17. Straits of Gibraltar

Andalusia occupies 17% of the total surface area of Spain and is arguably the most exciting birdwatching region in the country, as a visit to the Doñana National Park, the most important site for birds in southern Europe, will reveal. However, this particular site is merely the best known of the many wonderful natural areas to be found in Andalusia, the second-largest autonomous community in Spain (only smaller than Castilla y Leon).

The biodiversity of Andalusia is well represented in the region's network of natural and national parks (the most complete in any autonomous community). Amongst the fauna, pride of place must go to the Pardel Lynx, whose only viable world population (less than 200 animals in all) is confined to Andalusia.

This region has a wonderfully variable landscape, in which plains and uplands alternate on either side of two important mountain chains. The northern-most part of Andalusia is dominated by the Sierra Morena, a barrier of low, rounded mountains that runs for approximately 500 km along the region's frontier with Extremadura and Castilla-La Mancha. Its forests and Mediterranean wood pastures (*dehesas*) are home to Spanish Imperial Eagle (which also breeds in Doñana), Black Vulture and Black Stork. The Sierra de Andújar (1) is the main home for the Pardel

White-headed Duck © Juan Martín Simón

Lynx and a marvellous example of the range of habitats to be found in the Sierra Morena.

The Sierra Béticas, the other great Andalusian mountain range, is a larger and far more heterogeneous area. Here, the best populations of cliff-breeding raptors – above all of the endangered Egyptian Vulture and Bonelli's Eagle – in the south of the Iberian Peninsula are to be found. The most interesting zones are around Grazalema and Ronda (2) and Cazorla (3), the latter the last place the Lammergeier bred in Spain outside the Pyrenees and site today of a reintroduction programme using captive-bred birds. The Sierra Nevada (4), the highest mountain range in the Iberian Peninsula, possesses interesting high-level birds, some of which are at the southern-most points here of their European ranges.

Between these two ranges lies the Guadalquivir valley, the most densely populated part of Andalusia. Here, as in other similar spaces, agriculture dominates and the most attractive sites for birdwatchers are probably the numerous endorheic (inwardly draining lagoons) that dot the landscape. Fuente de Piedra (5) with its breeding colony of Greater Flamingos stands out, while the lagoons of Zóñar and El Rincón in Córdoba province and Medina (7) in Cádiz are all important for their breeding White-headed Ducks and Red-knobbed Coots.

Another low-lying area of interest is that of the so-called 'Deserts' of Almería (8). The Desert of Tabernas is home to the largest European mainland population of Trumpeter Finch, a bird more commonly associated with the arid climes of North Africa, Asia and the Canary Islands.

Nevertheless, the coastal sites are probably the most interesting areas for birds in Andalusia. No other region in Spain can offer such variety and without doubt the highlight is the Doñana National Park and its buffer zone (9), lying at the heart of the even today immense Guadalquivir saltmarshes. On the east bank of the Guadalquivir, the Bonanza salt-pans near the town of Sanlúcar de Barrameda (10), as well as the fluvial habitats of Brazo del Este (11) and the surrounding paddy-fields near the city of Seville, are also of exceptional ornithological importance.

This colossal wetland is the most important wintering ground in Europe for wildfowl, and the annual arrival of thousands of Greylag Geese is one of the most spectacular bird-watching sights on offer in Spain. Breeding birds include the best Spanish populations of Purple Heron, Eurasian Spoonbill, Glossy Ibis and Purple Swamp-hen, as well as populations of birds such as Marbled Duck and Spanish Imperial Eagle, both threatened on a world scale.

Beyond Doñana, a whole series of wetlands stretches along the Andalusian coastline. So close to Africa, the gentle and sandy Atlantic coasts west of Gibraltar harbour the Odiel salt-marshes (12) and the Bahía de Cádiz (13), both home to breeding colonies of Eurasian Spoonbill and of extraordinary interest for wintering birds and passage waders. The more abrupt and rocky Mediterranean coast to the west of Gibraltar is interrupted by the mouth of the Guadalhorce river (14), the wetlands of western Almería province (15) and the salt-pans of Cabo de Gata (16).

Finally, the Gibraltar area itself (17) is of special relevance as the main entry point for birds migrating to Europe from Africa. The annual passage of hundreds of thousands storks and raptors is one of the most spectacular events that our country can offer a birdwatcher.

1. SIERRA DE ANDÚJAR

This site consists of a large extension of mid-altitude mountain ridges separated by valleys and covered by superb holm, cork, Lusitanian and Pyrenean oak forests and wood pastures. This section of the Sierra Morena in the north-east of Jaén province holds the main world population of the Pardel Lynx, the most emblematic and threatened European mammal. Here too take refuge the last Iberian Wolves in Andalusia, another endangered species.

Cimbarra waterfall, Sierra de Andújar © Pedro Retamar

Almost the whole of the area is in private hands and given over to vast hunting estates, a fact that explains the abundance of game species such as Red Deer and Wild Boar. Also within the area lies the Sierra de Andújar Natural Park (68,000 ha), declared an SPA, and two National Hunting Reserves, Lugar Nuevo and Contadero-Selladores, in the extreme south-east and north-east, respectively, of the natural park.

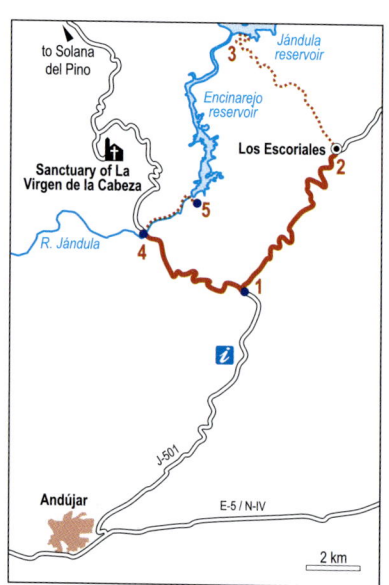

ORNITHOLOGICAL INTEREST

Andújar is undoubtedly the best place in Andalusia to see Mediterranean raptors. Top of the bill is the Black Vulture (over 500 breeding pairs), closely followed by an extraordinarily important number of Spanish Imperial Eagles (recent estimates talk of a dozen breeding pairs). Egyptian Vulture and Golden, Bonelli's, Short-toed and Booted Eagles all breed, as do a few pairs of Black Stork.

Passerines associated with Mediterranean forest and scrub habitats are abundant, both during the breeding season and in the winter, when vast numbers of fruit-eating birds arrive from northern Europe.

ITINERARIES

The road between the town of Andújar and the sanctuary of La Virgen de la Cabeza is the best route into the natural park. At km point 14.4 (1), just past Los Pinos tourist complex, a narrow road heads off right towards the valley of the river Jándula, the eastern limit of the natural park. After 8 km and just as the asphalt finishes, you reach a crossroads at the mines of Los Escoriales (2).

Here you should continue left, either by car or on foot, along a track. On your right you will see a chimney with an abandoned White Stork's nest and then 8 km further on, you reach the end of this itinerary at the Jándula reservoir (3).

Wood pasture and Mediterranean scrub dominate the whole itinerary and it is fairly easy here to observe Black Vulture, Spanish Imperial Eagle and other smaller raptors, as well as Black Stork. Amongst the other typical Mediterranean species found in these habitats, search for Wood Pigeon, Great Spotted Cuckoo, Red-rumped Swallow, Sardinian and Orphean Warblers, Southern Grey and Woodchat Shrikes, Azure-winged Magpie and Cirl and Rock Buntings. In winter Song and Mistle Thrushes and Redwings are common.

The access to this itinerary is the same as for the previous one, except that once at Los Pinos tourist com-

ANDALUSIA

PRACTICAL INFORMATION

- **Access:** at the entrance to the town of Andújar (Jaén province) from the N-IV (Autovía de Andalucía), turn right at the roundabout outside the Del Val hotel along the narrow J-501. This road will take you past the sanctuary of La Virgen de la Cabeza, through the Sierra de Andújar Natural Park from south to north and onto the village of Solana del Pino and town of Puertollano, both in the neighbouring province of Ciudad Real.
- **Accommodation:** in Andújar, La Carolina and Bailén. The rural accommodation in the Los Pinos tourist complex (tel. 953 549 079, recepcion@complejolospinos.com) at km point 14.2 on the J-501 is at the beginning of the itineraries described above.
- **Visitor facilities:** the natural park visitor centre (tel. 953 549 030, cv_vinasdepenallana@egmasa.es) is at km point 12 on the J-501. Open Friday-Sunday, morning and afternoon; Thursday and spring and summer Sundays mornings only.
- **Visiting tips:** spring and autumn visits advisable as summer is too hot. Aside from exceptions such as these itineraries, private estates with limited access are the norm in the Sierra Morena. However, road itineraries allow stops to be made to observe raptors and landscapes. Note the only petrol available is in the town of Andújar.
- **Further information:** maps 1:50,000, n°. 883 and 904 (SGE).

plex, continue along the J-501 road. In 8 km and just after crossing the river Jándula, a track (4) heads off right along the river to the Encinarejo reservoir (5).

This track can be walked and enables you to enjoy a stretch of fluvial woodland that in winter holds Great Cormorant, Grey Heron, Green Sandpiper and large numbers of passerines. Kingfishers are seen regularly throughout the whole year.

MOST REPRESENTATIVE SPECIES

Residents: Griffon and Black Vultures, Goshawk, Spanish Imperial, Golden and Bonelli's Eagles, Peregrine Falcon, Eagle Owl, Kingfisher, Robin, Blue Rock Thrush, Dartford and Sardinian Warblers, Firecrest, Crested Tit, Southern Grey Shrike, Azure-winged Magpie, Hawfinch, Cirl and Rock Buntings.
Summer visitors: Black Kite, Egyptian Vulture, Short-toed and Booted Eagles, Hobby, Great Spotted Cuckoo, Roller, Red-rumped Swallow, Black-eared Wheatear, Subalpine, Orphean and Bonelli's Warblers, Golden Oriole, Woodchat Shrike.
Winter visitors: Great Cormorant, Grey Heron, Green Sandpiper, Dunnock, Song Thrush, Redwing, Blackcap, Chiffchaff.

2. GRAZALEMA AND RONDA

The Sierra de Grazalema is a limestone massif stretching between the north-east of Cádiz and north-west of Málaga provinces at the extreme western end of the Cordillera Bética. Despite its modest height, its mountains are impressively abrupt and karstic rock landscapes abound aboveground (Garganta Verde) and underground (Hundidero-Gato and La Pileta caves). Winds off the Atlantic bring generous quantities of rain (this is in fact the wettest spot in the Iberian Peninsula) and ensure the

Barranco de Bocaleones, Sierra de Grazalema   © Pedro Retamar

survival on the shady northern slopes of the Sierra de El Pinar of a large forest of Spanish Firs, a species of tree found only in this part of Andalusia and northern Morocco.

The Serranía de Ronda is the continuation of the Cordillera Bética in Málaga province: at its eastern end rise the spectacular limestone mountains of the Sierra de las Nieves. Like Grazalema, these mountains, carved open by spectacular gorges and riddled with extensive cave systems, harbour a spectacular Spanish Fir forest (3,000 ha).

A natural park of over 50,000 ha and an SPA protect the Sierra de Grazalema; the Sierra de las Nieves encompasses a natural park of 20,000 ha and, like Grazalema, has been declared a Biosphere Reserve.

ORNITHOLOGICAL INTEREST

The best cliff-breeding raptor communities in western Andalusia are to be found in Grazalema and Ronda and include over 300 pairs of Griffon Vulture. There are as well almost 30 pairs of Egyptian Vultures, the greatest concentration in Andalusia of this worryingly declining raptor, Golden and Bonelli's Eagles, Peregrine Falcon and Eagle Owl.

Forest passerines abound and look out too for White-rumped Swifts (occupying old Red-rumped Swallow nests) and in the surrounding valleys migrant birds moving between Africa and Europe across the Straits of Gibraltar.

ITINERARIES

1.

The famous gorge in the town of Ronda (Málaga province) is a breeding site for Lesser Kestrel, Alpine Swift, Red-billed Chough and other cliff-breeding species. From Ronda take the C-339 towards Los Algodonales (Cádiz province) and Seville and after 12 km turn left (1) into the Sierra de Grazalema Natural Park on a road that leads towards Montejaque.

In Montejaque a track leads towards the *Cortijo de Líbar*, an isolated farm. It is best to head on foot into Los Llanos de Líbar (2), a series of plains straddling the border between Málaga and Cádiz provinces. Here the spectacular karstic scenery includes low-lying poljes used as pastures framed by rocky limestone ridges and Lusitanian and holm oak forests. This is a good area for Egyptian and Griffon Vultures and other cliff-breeding raptors.

2.

Two interesting itineraries begin and end in the town of Grazalema. The first passes through the villages of Benamahoma, El Bosque, Ubrique, Benaocaz and Villaluenga del Rosario. It is worth walking the paved Roman road (3) between Benaocaz and Ubrique and also through La Manga (4) and the plains of El Republicano (5). The second circuit takes you to Zahara de la Sierra and to the access points for two of the park's best known routes, the Pinsapar or Spanish fir forest (6) and La Garganta Verde (7), both of which require a permit in advance from the park offices. The return journey to Grazalema passes alongside the Zahara reservoir and then through the Gaidovar valley (8).

3.

This itinerary consists of a walk from the *cortijo* of Los Quejigales – a mountain hut open to the public – in the heart of the Sierra de las Nieves Natural Park. Take the A-376 from Ronda towards San Pedro de Alcántara (on the Costa del Sol) and after 13 km, turn left (9) along a track with a Natural Park sign and continue on to the car-park at Los Quejigales (10).

The path up Pico Torrecilla, the highest peak in Málaga province, begins here. First, you head along La Cañada del Cuerno through a forest of giant Spanish Firs, home to interesting forest birds such as Common Redstart and Coal Tit. A steep uphill section then takes you up to the saddle at Los Pilones (11) with vegetation dominated by Lusitanian oaks, brooms and junipers. This is a good point for Ring Ouzel and Fieldfare in winter and Golden and Bonelli's Ea-

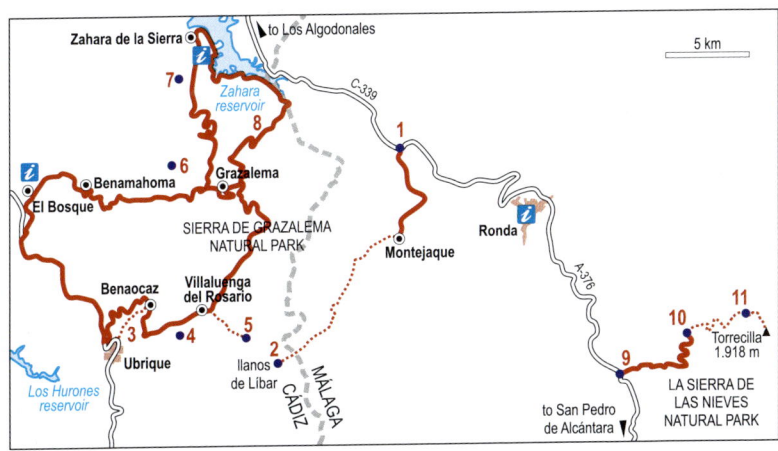

PRACTICAL INFORMATION

- **Access:** both Grazalema (Cádiz province) and Ronda (Málaga province) are good starting points.
- **Accommodation:** in the mountain villages (above all, Grazalema, Ubrique and Zahara de la Sierra) in Cádiz province or Ronda, where there is a *Parador* (smart state-run hotel)
- **Visitor facilities:** there are Sierra de Grazalema Natural Park visitor centres in El Bosque and Zahara de la Sierra. The Palacio de Mondragón in Ronda acts as an information point for the Sierra de las Nieves Natural Park.
- **Visiting tips:** Spring and early summer are best.
- **Recommendations:** the company Orni Tour (tel. 956 794 684, info@ornitour.com) organises birdwatching trips into the Sierra de Grazalema Natural Park.
- **Further information:** maps 1:50,000, n°. 1050, 1051, 1064 and 1065 (SGE). Also more information available from the natural park offices in Grazalema (tel. 956 727 029) and in El Bosque, and for the Sierra de Nieves (tel. 952 877 778) in Ronda. Useful book: *Aves del Parque Natural Sierra de Grazalema* by David Barros and David Ríos (Orni Tour, 1999).

gles all year round; then from here on, look for Northern Wheatear, Rock Thrush and, in winter, Alpine Accentor. The keenest mountaineers can tackle the peak of Torrecilla from this spot.

MOST REPRESENTATIVE SPECIES

Residents: Griffon Vulture, Golden and Bonelli's Eagles, Peregrine Falcon, Eagle Owl, Thekla Lark, Skylark, Crag Martin, Black Wheatear, Blue Rock Thrush, Crested and Coal Tits, Southern Grey Shrike, Red-billed Chough, Rock Sparrow, Common Crossbill, Rock Bunting.
Summer visitors: Egyptian Vulture, Short-toed and Booted Eagles, Lesser Kestrel, Alpine and White-rumped Swifts, Wryneck, Common Redstart, Northern and Black-eared Wheatears, Rufous-tailed Rock Thrush.
Winter visitors: Osprey, Water Pipit, Dunnock, Alpine Accentor, Ring Ouzel, Fieldfare, Redwing, Bullfinch.
On migration: Honey Buzzard, Black Kite.

3. SIERRA DE CAZORLA

These imposing limestone mountains, lying in the east of Jaén province, form part of the Cordillera Bética and are the origin of two of the great rivers of southern Spain, the Guadalquivir and the Segura. Deep gorges and sheer cliffs are omnipresent, as are extensive pine forests, many the products of past reforestations, but some still composed of the native black pine. The area's flora, which includes almost 30 endemic species of plant, is exceptional. Wild ungulates such as Spanish Ibex, Muflon, Roe and Red Deer and Wild Boar are common everywhere and the Spanish Algyroides (a lizard) is found here and nowhere else in the world.

A large part of this mountain range is included in a huge natural park (214,000 ha), the largest protected area in Spain, and has been declared a Biosphere Reserve and an SPA.

ORNITHOLOGICAL INTEREST

Outside the Pyrenees (where today its populations are stable), the Lammer-

ANDALUSIA

Sierra de Cazorla © Pedro Retamar

geier last bred in Spain in the 1980s in the Sierra de Cazorla and this vulture is currently being bred in captivity in the Guadalentín breeding centre, run by the Andalusian Government, in the Sierra de Cazorla Natural Park.

Cazorla is also important for both forest and cliff-breeding raptors, including 300 pairs of Griffon Vulture, as well as Egyptian Vulture, Golden, Bonelli's and Booted Eagles, Peregrine Falcon and Eagle Owl. Red-billed Choughs are remarkably common and this is also one of the southern-most outposts in Spain for Citril Finch, which breeds in black pine forests above 1,500 m.

ITINERARIES

The network of tracks and roads that criss-cross the Natural Park provide visitors with many options; the following are two of the most recommendable excursions.

1.

From the village of Cazorla take the road to the Natural Park as far as the village of La Iruela, where you should look on your right in the main square for the uphill road to Riogazas-El Chorro. Soon after the road turns

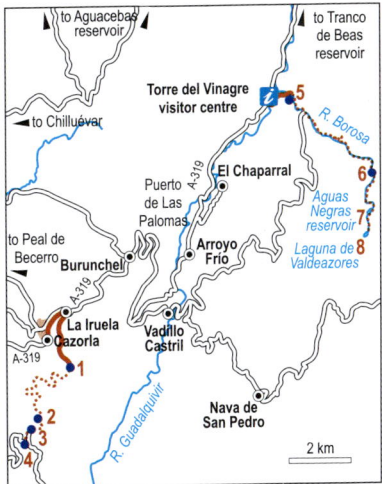

into a track, you reach a park control point at Riogazas (1), and then further on a forest hut at El Chorro (2). Continue and take the right-hand option where the track splits into two (3), which will take you to a viewpoint with an information panel about the birds of the area. From here there are good views of a vulture colony (4) on two impressive cliff-faces separated by a waterfall that surges out after rain. Habitually present here are Griffon and Egyptian Vultures, Common Kestrel, Alpine Swift, Black Wheatear and Blue Rock Thrush.

2. 12 km

From Cazorla head for the visitor centre at Torre del Vinagre, from where an asphalted road leads off to the right towards the fish farm on the river Borosa (5), home also to the *Centro de Interpretación Fluvial* (River Interpretation Centre). From here a path heads upstream providing good views of birds such as Dipper and Grey Wagtail, typical of mountain streams. The signposted path to Cerrada de Elías takes you into a river gorge with sections of the path on overhanging boardwalks.

After reaching the HEP station at Salto de los Órganos (6), the valley widens out and you are confronted by enormous cliffs, frequented by Red-billed Choughs, Griffon Vultures and the occasional Golden Eagle. From here on, the path narrows and starts to climb seriously (not suitable for non-walkers) and reaches the tunnels that take you out to the Aguas Negras reservoir (7). Cross the dam, follow the left bank of the reservoir and pick up the path to Laguna de Valdeazores (8), a lake with Little Grebe, Grey Heron and Eurasian Coot. Flocks of Common Crossbill are commonly seen in the surrounding pine woods.

MOST REPRESENTATIVE SPECIES

Residents: Griffon Vulture, Goshawk, Golden and Bonelli's Eagles, Common Kestrel, Peregrine Falcon, Eagle Owl, Kingfisher, Great Spotted

PRACTICAL INFORMATION

- **Access:** the village of Cazorla is the main access point to the Natural Park. The visitor centre, over 30 km from Cazorla village, can be reached along the A-319: cross the Las Palomas pass and then follow the river Guadalquivir downhill towards the Tranco reservoir.
- **Accommodation:** in Cazorla and in many of the villages in and around the natural park.
- **Visitor facilities:** the Natural Park visitor centre at Torre del Vinagre (tel. 953 713 017) has a botanical garden. Open morning and afternoon every day. The *Centro de Interpretación Fluvial* lies nearby next to the fish farm on the river Borosa.
- **Visiting tips:** spring and autumn are the best times to visit, although these periods can be rainy. Summer is too hot and too crowded.
- **Recommendations:** the cooperative Turisnat (tel. 953 721 351, info@turisnat.org) organises 4-wheel drive excursions and other activities.
- **Further information:** maps 1:50,000, nº. 907, 928 and 929 (SGE). Useful book: *Visita el Parque Natural de Cazorla, Segura y las Villas* by Joaquín Gómez MENA (Editorial Everest, 1998).

ANDALUSIA

Woodpecker, Crag Martin, Grey Wagtail, Dipper, Black Wheatear, Blue Rock Thrush, Mistle Thrush, Firecrest, Crested Tit, Eurasian Nuthatch, Red-billed Chough, Rock Sparrow, Citril Finch, Common Crossbill, Rock Bunting.
Summer visitors: Egyptian Vulture, Short-toed and Booted Eagles, European Nightjar, Pallid and Alpine Swifts, Northern and Black-eared Wheatears, Rufous-tailed Rock Thrush, Subalpine, Orphean and Bonelli's Warblers.
Winter visitors: Little Grebe, Grey Heron.

4. SIERRA NEVADA

The 80-km long Sierra Nevada, the highest mountain chain in both the Mediterranean and the Iberian Peninsula (summit: Mulhacén, 3,842 m), rises majestically between the fertile plains of the Vega de Granada and the sub-deserts of Almería. The holm oak forests of the lowest ridges give way to a myriad of forest types (Scots pine, juniper, maples and Pyrenean and Lusitanian oaks), today cleared in many places, which eventually fade into broom scrub, high mountain pastures and bare summits. Around 2,100 higher plants – a quarter of the total number of species for the whole Iberian Peninsula – have been recorded, of which over 60 are only found here; the Sierra Nevada is thus the most important area in Europe for endemic flora.

In 1989 170,000 ha of the Sierra Nevada were declared a natural park and in 1999 86,000 ha of this park – above all the peaks above 2,200 m – were awarded the stricter levels of protection offered by national park status. The area is also a Biosphere Reserve.

ORNITHOLOGICAL INTEREST

The Sierra Nevada is home to interesting Mediterranean montane bird

North face of Mulhacén, Sierra Nevada © Pedro Retamar

communities, with species such as Dunnock, Alpine Accentor, Chiffchaff, Hawfinch, Common White-throat and Ortolan Bunting at the southern-most points in their Iberian distributions. Golden Eagle, Red-legged Partridge, House Martin, Northern Wheatear and Black Wheatear all breed at over 3,000 m, their highest breeding grounds in Europe. Montane zones have Bonelli's Eagle, Red-billed Choughs are abundant on cliffs and Dippers frequent mountain streams. Wintering Ring Ouzels are frequent and good numbers of birds, including raptors, pass through on passage.

ITINERARIES

From the village of Güéjar-Sierra, take the road towards Barranco de San Juan down to the river Maitena. After crossing the narrow bridge (1), the road follows the river Genil along the course of the old Sierra Nevada tramway, passing through a number of tunnels before reaching restaurant Chiquito (2).

Park and continue on foot along the river Genil for Kingfisher, Dipper, Grey Wagtail and other river birds. The valleys sides are home to Crag Martin, Blue Rock Thrush and Black Wheatear. A path off to the right takes you to the confluence of the rivers San Juan and Genil (3), the starting point of one of the classic climbs in the Sierra Nevada. The path is only apt for the fittest, but provides a good chance to explore various habitats and see different bird species as you gain height.

At restaurant El Charcón on the road back to Güéjar-Sierra, a narrow winding road, only apt for small vehicles, heads up left (4). After a climb through wooded north-facing slopes, you come to a junction with a somewhat broader road (5). Just 200 m to the right lies the El Dornajo visitor centre; leftwards you start to climb towards the Prado Llano ski station. The first Rock Thrushes and Thekla Larks appear on the rocky slopes here and at Las Sabinas (6), a pass at over 2,000 m, Northern Wheatears are frequent and with luck a Golden Eagle will fly over.

The climb continues up to the area of the mountain huts and towards the Peñones de San Francisco. Between km points 32 and 33 (7), a rough asphalted track passing through a Scots pine plantation heads off left and can be explored on foot. Look for Firecrest, Coal Tit, Common Crossbill and other forest birds. Back on the main road, continue as far as the military hostal Hoya de la Mora (8) at 2,500 m, a magnificent viewpoint. On the rock

ANDALUSIA

PRACTICAL INFORMATION

- **Access:** leave Granada for Güéjar-Sierra on the old road to the Sierra Nevada via Cenes de la Vega.
- **Accommodation:** in Granada, Pinos Genil and Güéjar-Sierra, as well as around the Prado Llano ski station. Once in the National Park there are two mountain huts with wardens – Poqueira at the foot of Mulhacén and Postero Alto at the foot of Picón de Jeres.
- **Visitor facilities:** the National Park has no visitor infrastructure as yet, although the surrounding Natural Park has two visitor centres. The main centre is El Dornajo (tel. 958 340 625), at km point 23 of the road to the Sierra Nevada. It is open every day, morning and afternoon.
- **Visiting tips:** best in spring and at the beginning of summer. In high mountain areas avoid exposed ridges (danger of lightning); take care when crossing snow patches, frozen terrain or wet rocks; check the weather information before setting out; always wear sun cream, sun-glasses and hat.
- **Recommendations:** the National Park (tel. 958 026 300) provides two guided routes in mini-buses, one from Hoya de la Mora on the north side of the mountains, and the other from the village of Capileira in the south. Sierra Nevada Natural (tel. 958 489 759, info@serranevadanatural.com) organises birdwatching activities.
- **Further information:** map 1:50,000, n°. 1027 (SGE). Highly recommendable book: *Parque Nacional de Sierra Nevada* by Rafael Delgado et al. (Cansecoa, 2001). Also check www.mma.es/parques/lared/s_nevada.

outcrops towards Alto de San Francisco and above the university hostal, Rock Thrush and Alpine Accentor (in winter) can be seen.

MOST REPRESENTATIVE SPECIES

Residents: Golden Eagle, Peregrine Falcon, Thekla Lark, Skylark, Dipper, Dunnock, Alpine Accentor, Black Redstart, Black Wheatear, Blue Rock Thrush, Chiffchaff, Firecrest, Coal Tit, Red-billed Chough, Rock Sparrow, Common Crossbill, Hawfinch, Rock Bunting.
Summer visitors: Booted Eagle, Pallid Swift, Tawny Pipit, Northern Wheatear, Rufous-tailed Rock Thrush, Common Whitethroat, Subalpine and Bonelli's Warblers, Woodchat Shrike, Ortolan Bunting.
Winter visitors: Hen Harrier, Ring Ouzel, Redwing, Brambling, Siskin.
On migration: Common and Honey Buzzards, Black Kite, Alpine Swift, Water and Meadow Pipits.

5. LAGUNA DE FUENTE DE PIEDRA

In the north of Málaga province, amidst olive groves, cereal fields and row upon row of sunflowers, lies the large endorheic (inwardly draining) lagoon of Fuente de Piedra (over 6 km long and with a total surface area of almost 1,400 ha), the largest inland wetland in Andalusia and, along with the lagoon of Gallocanta, one of the largest lakes in Spain. Its name is taken from that of a popular nearby spring, whose waters have been used for centuries for curing gallstones and kidney stones.

Fuente de Piedra is a shallow saline lagoon that often dries up in summer, although in some years water levels are maintained artificially by pumping to safeguard the lagoon's bird colonies. The walls of the former salt-pans form lineal islands that are used by thousands of Greater Flamingos, terns, gulls and waders as nesting sites. Owing to this ornithological richness, the lagoon has been declared a *reserva natural*, a Ramsar site and SPA.

ORNITHOLOGICAL INTEREST

Fuente de Piedra harbours the largest breeding colony of Greater Flamingo in Spain and the western Mediterranean. In some years half of all the region's flamingos concentrate here: record figures of almost 20,000 pairs were recorded in 1998 and 50,000 adults and young in summer 2000. The Greater Flamingo also breeds *en masse* in the Ebro Delta and Doñana.

Fuente de Piedra is also home to important colonies of Black-winged Stilt, Avocet, Kentish Plover, Black-headed and Yellow-legged Gulls and Gull-billed Tern; Slender-billed Gulls have not bred since 2000. More sporadically, Shelduck, Red-crested Pochard, White-headed Duck, Purple Swamp-hen and Collared Pratincole also breed. During both passage periods winter duck, wader and gull numbers increase (including a recent rise in wintering Lesser Black-backed Gulls), and groups of Black-necked Grebe and Common Crane turn up.

ITINERARIES

1.

Fuente de Piedra is totally surrounded by roads and it is fairly straightforward to stop wherever you want to on the road around the lagoon; it is however strictly forbidden to cross the perimeter fence or enter the restricted-use agricultural tracks.

Built upon the traces of an old roadway, the observation point of Las Vicarias (1) can be reached by following the road around the north-west

Laguna de Fuente de Piedra © Pedro Retamar

ANDALUSIA

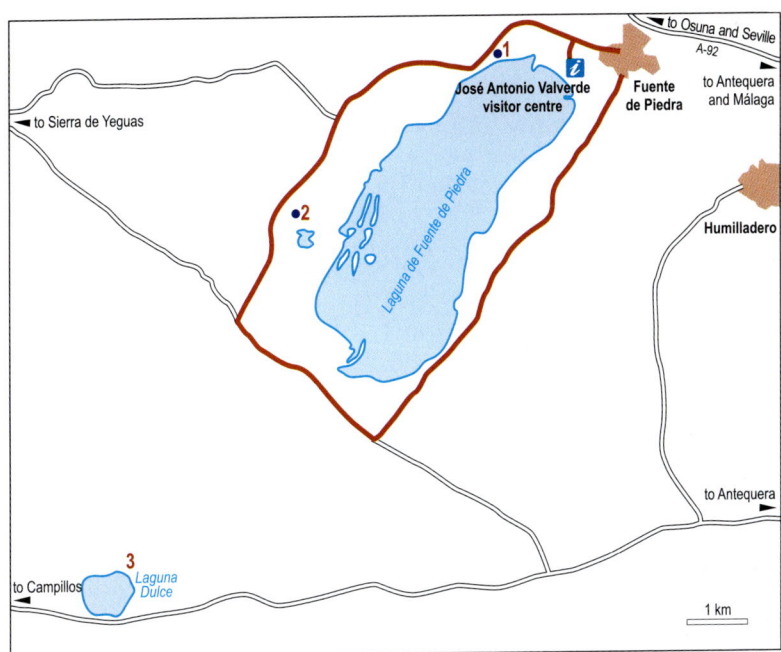

bank of the lagoon to where a sharp bend (after 2 km) takes the road close up to the lagoon shore. Park here and walk towards the lagoon for good views of Greater Flamingo, ducks and waders, as well as of numerous passerines in the lagoon's fringing halophytic vegetation. Just 5 km further on you reach the area of La Madriguera (2), another good vantage point over the lagoon.

2.

The much smaller lagoon of Laguna Dulce (3) lies on your right less than 3 km before the village of Campillos on the road from the village of Fuente de Piedra. From next to an information panel, a short path leads to a hide where, depending on water levels, you can expect to see Black-necked Grebe, Greater Flamingo, White-headed and other ducks and waders, as well as Marsh Harrier and Purple Swamp-hen.

The surrounding fields and fallows are the best areas in Málaga province for steppe birds: search for Little Bustard and Stone Curlew all year around, with Montagu's Harrier and Lesser Kestrel in spring and summer. Winter sees the arrival of Common Crane, Hen Harrier and Merlin.

MOST REPRESENTATIVE SPECIES

Residents: Greater Flamingo, White-headed Duck, Purple Swamp-hen, Little Bustard, Stone Curlew, Kentish Plover, Black-headed Gull, Calandra Lark, Sardinian Warbler, Southern Grey Shrike.

Summer visitors: Little Bittern, Montagu's Harrier, Lesser Kestrel, Black-winged Stilt, Avocet, Collared Pratincole, Slender-billed Gull, Gull-billed Tern, Short-toed Lark, Yellow Wagtail, Black-eared Wheatear, Reed, Great Reed and Spectacled Warblers.

PRACTICAL INFORMATION

- **Access:** leave the A-92 Seville to Málaga motorway at junction 132 for the village of Fuente de Piedra. From here take the local road towards Sierra de Yeguas and, 100 m after crossing over the railway, turn left towards the lagoon and its visitor centre.
- **Accommodation:** a camp site near the village Fuente de Piedra; more accommodation on offer in the nearby town of Antequera.
- **Visitor facilities:** The José Antonio Valverde visitor centre (tel. 952 111 715) is open Tuesday-Sunday, morning and afternoon. Nearby two hides have been built: El Cerro del Palo, with good views of waders and Greater Flamingos in the north-east of the lagoon, and El Laguneto, overlooking a small pool with plenty of vegetation and many interesting water birds.
- **Visiting tips:** in light of the local climate and water levels, the best time to see large numbers of Greater Flamingos is March, April and May, above all early in the morning.
- **Recommendations:** the year's flamingo chicks are rung during a work camp organised for volunteers by the Andalusian Regional Ministry of the Environment. The visitor centre will provide information about this camp, as well as details of guided routes entering restricted areas run by a private company, Villa di Fonte.
- **Further information:** map 1:50,000, n°. 1023 (SGE).

Winter visitors: Black-necked Grebe, Grey Heron, Shelduck, Eurasian Wigeon, Common Teal, Northern Shoveler, Red-crested Pochard, Marsh and Hen Harriers, Merlin, Common Crane, Little Stint, Dunlin, Black-tailed Godwit and other waders, Lesser Black-backed Gull.

6. LAGUNAS DE ZÓÑAR AND EL RINCÓN

In the south of Córdoba province, surrounded by cereal fields, vineyards and olive groves, lie a group of small, semi-saline lagoons – Zóñar, El Rincón, Amarga, Tíscar, Los Jarales and El Salobral – of which some such as Zóñar (near the town of Aguilar de la Frontera) are permanent and relatively deep. This lagoon, almost 1 km long and with a surface area of almost 40 ha, is the largest of the six; El Rincón and Amarga, both covering around 10 ha, are also permanent.

Together, these wetlands are part of a *reserva natural* that has also been declared an SPA and, in the case of the three permanent lagoons, a Ramsar Site.

ORNITHOLOGICAL INTEREST

These lagoons are ornithologically famous as a breeding and wintering site for the White-headed Duck, one of the most threatened ducks in Spain. By 1977 the entire European population of this singular duck had been reduced to around 20 birds concentrated in Zóñar. Nevertheless, this small nucleus proved sufficient for a spectacular recovery in the fortunes of the Iberian populations of the White-headed Duck.

Once the lagoons were legally protected and hunting banned, the White-headed Duck began to spread to the other lagoons in the south of Córdoba and then eventually to wetlands in other areas of Andalusia, Valencia and Castilla-La Mancha. By 2000, the total number of birds in Spain as a whole was calculated at 4,500.

ANDALUSIA

Laguna de Zóñar © José Manuel Reyero

Today, around a dozen female White-headed Ducks breed in these lagoons, mostly at El Rincón. In 2002 the Red-knobbed Coot, another threatened water bird, was introduced into this group of interesting wetlands.

ITINERARIES

1. 1 km

Zóñar can be reached along the A-309 between Aguilar de la Frontera and Puente Genil. Around 4 km south of Aguilar, a 500 m-long track closed to vehicles heads off right to the lagoon visitor centre (1). Here there is a hide, the only permitted viewing point of the lagoon.

In recent years Zóñar has become less interesting for birdwatchers, above all in comparison with other, more varied Andalusian wetlands. Marsh Harriers are occasional visitors and in winter there is a large (5,000+) Lesser Black-backed Gull roost and groups of other gulls, Great Cormorant, duck and coot can be observed. As well, there are also large Cattle Egret and Common Starling roosts.

2. 1,5 km

El Rincón is the best place to observe White-headed Ducks close-up. From the road between Aguilar de la

PRACTICAL INFORMATION

- **Access:** Aguilar de la Frontera lies 60 km from Córdoba towards Málaga and is the best base for visiting these lagoons.
- **Accommodation:** in Córdoba, Antequera and Montilla (the nearest).
- **Visitor facilities:** the Laguna de Zóñar visitor centre is next to the lagoon of the same name. It is wise to seek information here if you plan to visit these lagoons. Opening times vary according to the time of year.
- **Visiting tips:** White-headed Ducks can be seen all year round here, although spring is best for seeing the species' spectacular breeding display; greatest numbers are found in winter.
- **Recommendations:** In the 1970s, the NGO 'Amigos de la Malvasía' (oxyura@teleline.es) raised the alarm regarding the possible extinction of the White-head Duck in Spain and launched a highly successful campaign to save it. They publish a magazine *Oxyura*, dedicated to the species.
- **Further information:** map 1:50,000, n°. 1023 (SGE). The book *Las malvasías cordobesas, veinticinco años después* ('The White-headed Ducks of Córdoba, 20 years on'; published by the Diputación de Córdoba in 2003) is a description of the recovery of the species written by José Antonio Torres-Esquivias, one of the founders of Amigos de la Malvasía.

Frontera and the village of Moriles, turn right along a rough track (2) (not apt for vehicles) and walk towards the lagoon. This lagoon is fenced off and there are no recommended routes or hides; it is best thus to employ the services of a local guide from the visitor centre at Zóñar.

The small population of White-headed Duck is the main attraction at Rincón, although some of the recently reintroduced Red-knobbed Coots may also put in an appearance. Despite the small size of the lagoon, there are considerable numbers of easily observable wintering wildfowl, including Black-necked Grebe, Red-crested Pochard and Tufted Duck.

MOST REPRESENTATIVE SPECIES

Residents: Little and Great-crested Grebes, Mallard, Common Pochard, White-headed Duck, Marsh Harrier, Moorhen, Purple Swamp-hen, Eurasian and Red-knobbed Coots.

Summer visitors: Booted Eagle, Short-toed Lark, Common and Great Reed Warblers.

Winter visitors: Black-necked Grebe, Great Cormorant, Cattle Egret, Grey Heron, Eurasian Wigeon, Gadwall, Common Teal, Marbled and Tufted Ducks, Pintail, Northern Shoveler, Red-crested Pochard, Northern Lapwing, Black-headed and Lesser Black-backed Gulls, Common Starling.

7. LAGUNA DE MEDINA

Since the draining in the 1970s of the wetland of La Janda, the Laguna de Medina has become the largest lagoon in the province of Cádiz and in Andalusia is only smaller than Fuente de Piedra in Málaga province. Today it covers some 120 ha and is the main point of interest in the group of lagoons that are scattered around the centre and west of the province.

The Laguna de Medina, surrounded by gently rolling hills exploited for agriculture and stock-raising, has a

ANDALUSIA

Laguna de Medina © Fernando Barrio

sizeable fringe of aquatic vegetation and is seasonal and slightly saline. Nevertheless, unlike the other lagoons in Cádiz, it rarely dries up completely and as such acts as emergency refuge for many birds whenever the famous Doñana salt-marshes dry out. Medina is a *reserva natural*, Ramsar Site and an SPA.

ORNITHOLOGICAL INTEREST

Medina is an excellent refuge for birds from other nearby wetlands such as the Guadalquivir salt-marshes, from where, for example, arrive the Eurasian Coots in summer that can build up into flocks of over 20,000 birds. There are also important flocks of White-headed Duck – almost 800 birds – and this threatened duck also breeds here on occasions.

This is also one of the most significant sites for Red-knobbed Coot in Spain and there are at least 10 breeding pairs. Other breeders include Black-necked Grebe, Purple Heron, Red-crested Pochard, Marsh Harrier and Purple Swamp-hen. Hundreds of Greater Flamingos from the breeding colony at Fuente de Piedra come here to feed. Spectacular numbers of hirundines pass through in autumn – the lagoon's reed-beds play home to roosts of tens of thousands of Barn Swallows – and winter sees the arrival of a great diversity and abundance of duck.

ITINERARY

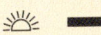

2 km

The path passing along the southern side of the lagoon that coincides with a drover's road is the only possible access to the site. The best panorama of the lagoon is from very the start of the walk.

The path passes through a strip of Mediterranean vegetation dominated by large wild olive trees and lentiscs that separates the surrounding fields from the belt of aquatic vegetation (reeds, tamarisks, bulrushes and sedges) girding the lagoon. These two belts of vegetation provide good cover for many birds, but do rather inhibit observation of the lagoon itself: therefore, it is best to stop to view the lagoon wherever possible.

Common Pochard. Rather less common are Greylag Goose, Shelduck, Red-crested Pochard and Tufted Duck.

The fringing vegetation holds various warblers and other passerines, including Zitting Cisticola and Great Reed Warblers, while Marsh Harrier, Greater Flamingo and various species of heron and gull are all regularly seen.

MOST REPRESENTATIVE SPECIES

Residents: Little and Great-crested Grebes, Cattle and Little Egrets, Greater Flamingo, Red-crested Pochard, Marbled and White-headed Ducks, Marsh Harrier, Purple Swamphen, Common and Red-knobbed Coots, Southern Grey Shrike.
Summer visitors: Black-necked Grebe, Little Bittern, Purple Heron, Black-winged Stilt, Avocet, Collared Pratincole, Kentish Plover, Whiskered Tern, Rufous Bush Robin, Black-eared Wheatear, Reed and Great Reed Warblers.
Winter visitors: Greylag Goose, Shelduck, Ferruginous, Tufted and other ducks, Bluethroat, Penduline Tit, Reed Bunting.

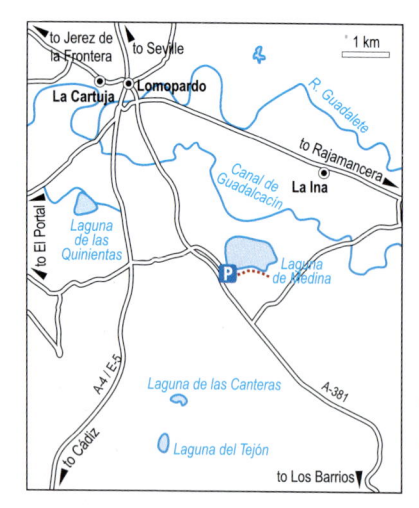

The Red-knobbed Coot, an endangered species that in Europe is only found in a few Spanish localities, is the faunistic jewel in Medina's crown. Nevertheless, the large size of the lagoon and this coot's relative scarcity compared to the similar Eurasian Coot makes observation of this singular species difficult.

Little and Black-necked Grebes breed in years with good levels of water: there are also post-nuptial concentrations of White-headed Duck and winter brings important numbers of Mallard, Gadwall, Common Teal, Northern Shoveler and

PRACTICAL INFORMATION

- **Access:** the lagoon is signposted off the A-381 dual-carriageway (between Jerez de la Frontera and Los Barrios) after 10 km, just after a quarry and a cement factory. At the car-park there is a small control point that regulates access to the lagoon.
- **Accommodation:** in Jerez de la Frontera.
- **Visitor facilities:** no visitor centre or infrastructure for birdwatchers.
- **Visiting tips:** all year is good except summer, when there are fewer birds, unbearable heat and the chance the lagoon will be dry. Spring and winter are ideal and the use of a telescope is highly recommended given the dimensions (1 km x 1.5 km) of the lagoon. As a *reserva natural*, leaving the marked path is strictly forbidden.
- **Further information:** map 1:50,000, n°. 1062 (SGE). The lagoon is managed by the office of the Bahía de Cádiz Natural Park (tel. 956 590 405 and 956 590 971, pn.bahiadecadiz.cma@juntadeandalucia.es) in San Fernando.

On migration: Squacco Heron, Black Stork, Garganey, Roller, Barn Swallow, Whinchat.

8. DESERTS OF ALMERÍA

Tabernas, the Sierra Alhamilla and the karstic scenery at Sorbas are the three surprising components of the site known conjointly as the Deserts of Almería, in which animal and plant communities, unique to Europe, eke out a living in a desolate landscape of ecological extremes.

The desert of Tabernas lies to the north of the city of Almería and is dominated by a succession of dry gullies and *ramblas* (seasonal river beds) jumbled together in a singular, lunar-like landscape lying in the rain shadows of the Sierra de Los Filabres to the north and the Sierra Alhamilla to the south. The relatively smooth slopes of this area of mid-altitude mountains are only interrupted by a few rocky outcrops and numerous deeply incised seasonally dry gullies. Almost 12,000 ha of the Tabernas desert and 8,500 ha of Sierra Alhamilla have been classified as a *paraje natural* (site of natural interest) and an SPA.

ORNITHOLOGICAL INTEREST

These deserts are home to an important community of steppe birds and, above all, to the largest European mainland population of the scarce Trumpeter Finch. In 2001 Cream-coloured Coursers bred here for the first time. Other species of interest include Little Bustard, Stone Curlew, Black-bellied Sandgrouse and, amongst the passerines, Short-toed and Thekla larks and Black and Black-eared Wheatears.

Cliff-nesting raptors include the increasingly endangered Bonelli's Eagle, Peregrine Falcon, Common Kestrel and Eagle Owl.

Rambla de Lanújar, Deserts of Almería © Pedro Retamar

ITINERARIES

1. ⟨2 km⟩ 🚶 ☀ ➡

This walk begins at the petrol station located at the Tabernas turn-off of the A-92 Almería-Granada motorway. From here, pick up an access road that will take you down into the dry bed of the Rambla de Tabernas (1) and walk either way along this *rambla* (with just a few trickles of water in places) and adjoining ones in search of the Trumpeter Finch, a finch of North African origin.

Other typical birds of these semiarid environments include Little Ringed Plover, European Bee-eater, Alpine Swift, Thekla Lark, Olivaceous Warbler and Rock Sparrow.

Remember that breeding colonies of Trumpeter Finch are very mobile and you are mostly likely to come across then in flight. From one year to another they often change their breeding areas and during the year continually move from one feeding site to another. Thus, despite being relatively common, it is difficult to guarantee observations in any one particular area.

2. ⟨35 km⟩ 🚗 ☀ ➡

From the same petrol station as before, continue along the A-370 towards the town of Tabernas; after just 500 m, turn right along a metalled road (2) that passes outside the gates to a tourist complex (a Wild West village and a mini-zoo, pompously referred to as a 'reserve'). This narrow and sinuous road climbs the northern flanks of Sierra Alhamilla (take care!) up to the television repeater station on top (3). Here the tarmac ends and a track continues along the top of the main ridge and then down to the village of Turrillas, from where it is easy to return to the road between Tabernas and Sorbas).

The interest in this route lies in the splendid views and the chance to move from a semi-desert environment, through areas of scrub and up to the patches of pine and holm oak

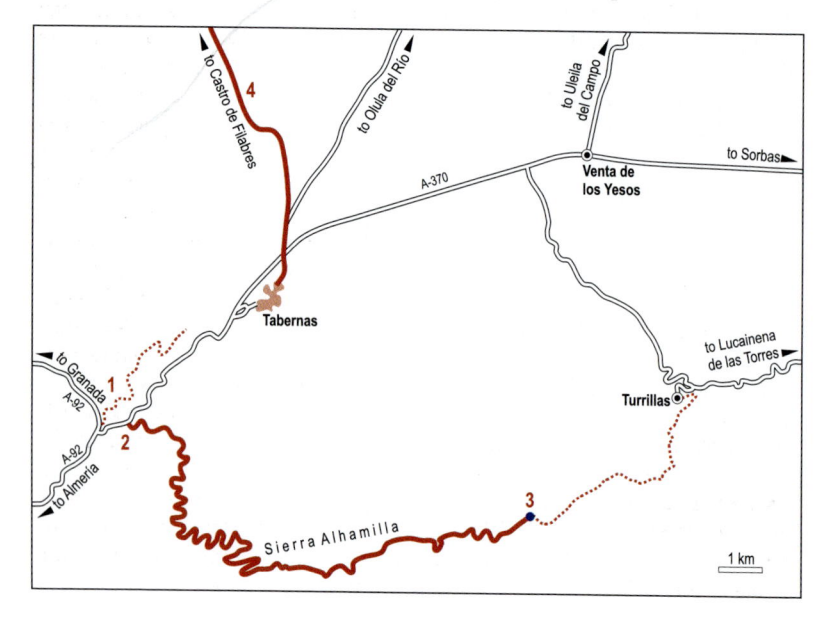

ANDALUSIA

PRACTICAL INFORMATION

- **Access:** the Tabernas turn-off of the A-92 Almería-Granada motorway connects to the A-37 towards Sorbas.
- **Accommodation:** in Almería just 20 km away.
- **Visitor facilities:** no visitor centre or infrastructure for birdwatchers.
- **Visiting tips:** The best time of year is spring, the worst summer (extreme heat). It is always best to go out early in the morning or in the evening. Take care in case of sudden heavy rainstorms, which can provoke dangerous flash floods in these *ramblas*. Check your petrol before undertaking any of these itineraries by car since there are few petrol stations in the area.
- **Further information:** maps 1:50,000, n°. 1030 and 1045 (SGE). The offices of the Cabo de Gata-Níjar Natural Park in Rodalquilar will also provide information about the Tabernas and Sierra Alhamilla area (tel. 950 389 742, pn.cabodegata.cma@juntadeandalucia.es).

woodland that surprisingly cover the higher slopes. This variation in habitats brings with it the possibility of observing a greater number of species, including forest birds and Bonelli's Eagle and other raptors.

Alpine and Pallid Swifts, European Bee-eater, Short-toed Lark, Red-rumped Swallow, Rufous Bush Robin, Black-eared Wheatear, Olivaceous and Spectacled Warblers, Woodchat Shrike.

3. 20 km

9. DOÑANA

From Tabernas take the road (4) north towards Castro de Filabres through an easily explored area of dusty plains scored by numerous gullies that is good for Stone Curlew, Black-bellied Sandgrouse, other steppe-loving species and, above all, for Short-toed and other Larks.

MOST REPRESENTATIVE SPECIES

Residents: Bonelli's Eagle, Common Kestrel, Peregrine Falcon, Stone Curlew, Black-bellied Sandgrouse, Rock Dove, Eagle and Little Owls, Calandra and Thekla Larks, Black Wheatear, Blue Rock Thrush, Sardinian Warbler, Southern Grey Shrike, Rock Sparrow, Trumpeter Finch.
Summer visitors: Little Ringed Plover, Turtle Dove, Great Spotted Cuckoo, Red-necked Nightjar, Roller,

Doñana, one of the greatest natural landscapes in Western Europe, consists essentially of a vast area of salt-marshes at the mouth of the river Guadalquivir, of which today a large part is given over to paddy fields and other types of agriculture. The untouched salt-marshes dry out in summer and then only revive with the arrival of the autumn and winter rains. Bordering on the salt-marshes are the fixed dunes, covered in vast areas of scrub (*cotos*) with a few patches of cork oak and pine woodland. Beyond here, the dunes become gradually more mobile and stands of stone pine and patches of scrub, as well as numerous freshwater pools, survive in the inter-dune slacks (*corrales*). Finally, the dunes give out to a vast unspoilt beach, over 30-km long. Elsewhere, there are narrow strips of grazing land (*veras*) separating these habitats from all others, as well as a few

Doñana National Park © Pedro Retamar

rivers and streams with important patches of gallery woodland.

This mosaic of ecosystems plays home to exceptional levels of biodiversity, although of greatest import in a world context are the few remaining Pardel Lynxes. The National Park (50,000 ha) was set up in 1969 and contains two biological reserves: Doñana, almost 7.000 ha, and Guadiamar, over 3,000 ha. Five discrete areas in the National Park's buffer zone have been designated a natural park (54,000 ha). As a whole, the area also enjoys triple protection as a Biosphere Reserve, a Ramsar Site and an SPA.

ORNITHOLOGICAL INTEREST

Doñana is the great birdwatcher's paradise in southern Europe, both during the breeding season and during migration. Highly endangered species such as Spanish Imperial Eagle (max. 10 pairs), Marbled Duck, whose most important Iberian populations are shared between Doñana and southern Alicante province, and Red-knobbed Coot, all breed here. A recent re-coloniser is the Glossy Ibis, whose numbers now include over 1,000 breeding pairs; furthermore, the enigmatic Small Button-quail (formerly the Andalusian Hemipode) has been heard calling in the area in recent years.

All the Iberian Peninsula's breeding herons are found here, including over 500 pairs of Purple Heron, a few hundred of Squacco Heron, the rare Great Bittern and over 1,000 pairs of Eurasian Spoonbill, the largest breeding colony in Spain. As well, Doñana harbours nationally significant breeding populations of Greater Flamingo, Pintail, Marsh Harrier, Black Kite, Booted Eagle, Purple Swamp-hen, Black-winged Stilt, Avocet, Collared Pratincole, Kentish Plover, Slender-billed Gull and Gull-billed and Whiskered Terns.

Doñana is also a superb place for winter birdwatching, with between 40,000 and 80,000 Greylag Geese, Osprey, Common Crane, spectacular

ANDALUSIA

numbers of duck – above all of Shelduck, Red-crested Pochard, Ferruginous and White-headed Ducks – and waders crowding into the area. Finally, during migration hundreds of thousands of birds (above all, birds of prey, waders and passerines) pass through Doñana: of special interest are the numbers of Black and White Storks and Audouin's Gull, three species that also winter in the area.

ITINERARIES

The following itineraries investigate the right-bank of the river Guadalquivir (see the sites of Sanlúcar de Barrameda and Brazo del Este for the left-bank) and, given the size of the area, are grouped into two: southern Doñana (south-east Huelva province) and northern Doñana (south-west Seville province).

SOUTHERN DOÑANA

1. 4 km

A visit to the natural salt-marshes lying opposite the village of El Rocío provides the birdwatcher with a remarkable range of species of bird. The first autumn rains flood the salt-marshes just as the first passage and wintering wildfowl – above all Greylag Geese and ducks – begin to arrive. Waders crowd together on islands and mudflats and in spring Little Bittern, Night and Purple Herons, Collared Pratincole and Whiskered Tern are present. All year round you can expect to find here (as long there is some water) Glossy Ibis, Eurasian Spoonbill and Greater Flamingo, and Spanish Imperial Eagle is also a distinct possibility.

Many of these species can be seen from the walk that departs from the church of El Rocío (1A) and runs along the edge of the salt-marches. It is worth stopping at the bird observatory run by the SEO/BirdLife (2A), whose terrace was one of the best places in the whole of Europe for watching birds before it was unfortunately destroyed by fire in 2002. It is currently being rebuilt.

Carrying alongside the salt-marshes, the path takes you past a small sewage works and onto a track that ends up at Boca del Lobo at the fence surrounding the National Park (3A). Check the large cork oaks in the distance for perched raptors.

2. 30 km

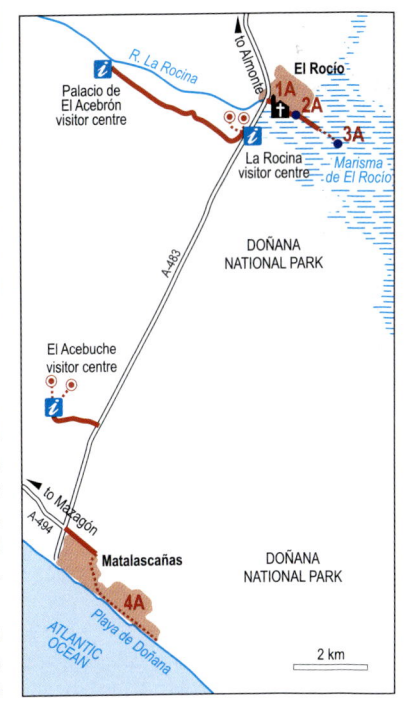

The National Park has various visitor centres that are free and open all year. The centre at La Rocina is on your right at km point 2 on the road to Matalascañas and from here a 3 km-boardwalk runs alongside a stream – Arroyo de La Rocina – to a number of hides. Most of the herons that breed in Spain can be found here, along with Eurasian Spoonbill, Greater Flamingo, Red-crested Pochard, Marsh Harrier, Purple Swamp-hen and many waders. During much of the year a Glossy Ibis roost forms here.

A small metalled road takes you from La Rocina to the visitor centre at Palacio del Acebrón, from where a short trail passes through an area of scrub and the leafy woodland surrounding another stretch of the Arroyo de la Rocina. This is a good area for many forest and riverine passerines.

The main park visitor centre is El Acebuche, signposted off the road to Matalascañas at km point 13. There are a number of itineraries and a dozen hides and this is also the departure point for the famous four-wheel drive buses that enter into the heart of the National Park every morning and afternoon; excursions last four hours and can be booked on 959 430 432.

3. 2 km 🚶 ☀️ ➡️

Other than in the summer, when large numbers of tourists flock to the area, the beach at Matalascañas (4A) is worth a visit. On foot from the town, in winter and during migration periods it is easy to observe seabirds, including Cory's and Balearic Shearwaters, Northern Gannet, Common Scoter, Audouin's Gull and Razorbill, as well as many waders and other species of gull.

NORTHERN DOÑANA

1. 12 km 🚗 ☀️ ➡️

Around 8 km from the village of La Puebla del Río (Seville province) heading towards Isla Mayor (formerly called Villafranco del Guadalquivir) lies La Cañada de los Pájaros (1B), an environmental education centre built around a restored gravel pit. Home to a large heronry, La Cañada is also a captive breeding centre for endangered species such as Marbled Duck and Red-knobbed Coot and both species can be seen in the centre's installations. Open every day (including lunchtime): more information, tel. 955 772 184, canadadelospajaros@canadadelosparajos.com.

Continue along the road from La Puebla del Río and then, just over the roundabout at La Venta del Cruce (2B), turn left at the junction towards La Dehesa de Abajo (3B), an estate owned by the local town council that is of great interest for birdwatchers. The entrance is signposted and just beyond there is a car-park, from where marked itineraries lead to a number of hides. An interpretation centre is planned for the future.

La Dehesa de Abajo boasts the largest known colony of White Storks built on trees (wild olives), in all hundreds of nests. Other breeders in the area include Black Kite and Booted Eagle; Red-knobbed Coot have bred on the nearby temporary La Rianzuela lagoon, where Glossy Ibis, Eurasian Spoonbill, Greater Flamingo and Marbled Duck are frequently seen.

2. 50 km 🚗 ☀️ ➡️

The same road as leads to Dehesa de Abajo will also take you into the northern sector of the Doñana salt-marshes. After crossing the current bed of the river Guadiamar at the ford at Don Simón (4B), turn left along the road that connects the villages of Villamanrique de la Condesa and Isla Major for a couple of kilometres until just before the road

ANDALUSIA

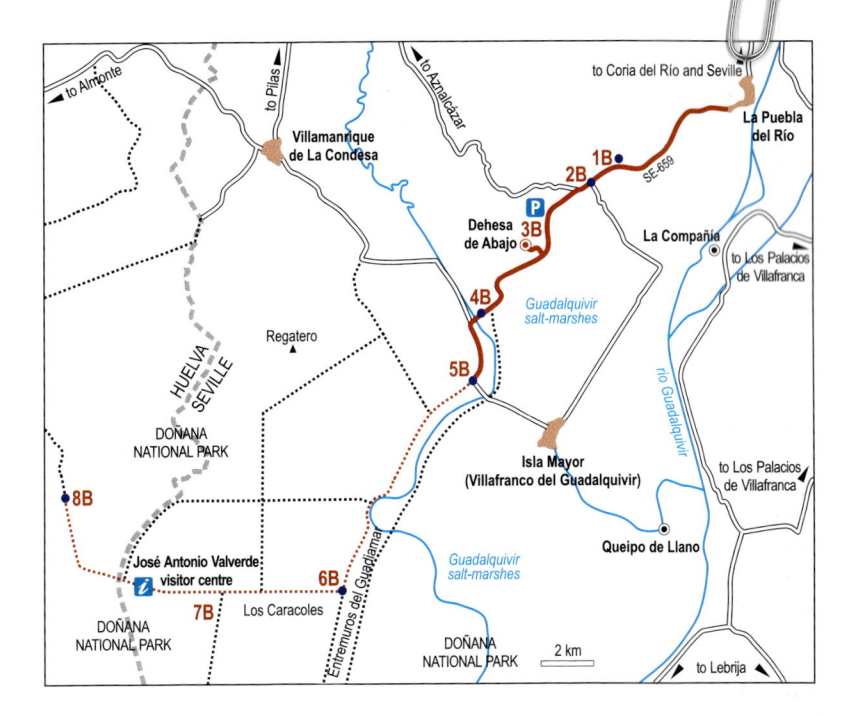

crosses the Guadiamar at the Los Vaqueros bridge.

Here, keep to the western (right) bank of the river, along the raised track on the top of the embankment that marks the western limit of this broad river, here known as 'Entremuros' (literally, 'between walls') given that the eastern (left) bank is likewise marked by another wall. After 10 km, turn right (westwards) (6B) and continue along the northern limit of the Caracoles estate.

Soon on your left you will see the Lucio del Lobo (7B) (a *lucio* is a natural lagoon that maintains its water even when the salt-marshes dry out), a breeding site for Red-knobbed Coot and a roosting site in winter for Common Crane and Greylag Goose. Further on you reach the José Antonio Valverde visitor centre, also known as Cerrado Garrido, named after the pioneering zoologist who was one of the instigators of the Doñana National Park. From here there are wonderful views of the salt-marshes and the largest colony of Glossy Ibis in Europe, here with nests built in stands of bulrushes. You can even continue along the FAO wall as far as the gate at Las Escupideras (8B), the limit of the National Park.

Depending of the time of year, the species on show along this itinerary are myriad and include most of Doñana's star birds. This is largely due to the fact that the itinerary passes through the transition area between the natural and reclaimed salt-marsh, the latter largely given over to paddy fields and where many birds take refuge when the former dries out.

MOST REPRESENTATIVE SPECIES

Residents: Black-necked Grebe, Night Heron, White Stork, Glossy Ibis, Greater Flamingo, Shelduck,

PRACTICAL INFORMATION

- **Access:** from Seville along the A-49 motorway towards Huelva. Turn off at Bollullos Par del Condado-La Palma del Condado along the A-483, which will take you through Almonte and on to El Rocío (southern Doñana). The northern part of Doñana is reached via Puebla del Río from the outskirts of the city of Seville itself.
- **Accommodation:** in El Rocío for southern Doñana: Hotel Toruño (tel. 959 442 323) or the Cortijo de los Mimbrales (tel. 959 442 237). There are also lots of hotels in Matalascañas: more information, www.turismodedonana.com. For northern Doñana, try Seville and nearby towns.
- **Visitor facilities:** the four visitor centres are mentioned in the text of the itineraries; a fifth centre, El Fábrica de Hielo, is described in the site text for Sanlúcar de Barrameda.
- **Visiting tips:** any time of the year, although October-May, when the salt-marshes are flooded and fewer tourists are around, is probably the best. In summer birds take refuge in the *lucios* and paddy fields. Take care on tracks and fording points after heavy rain.
- **Recommendations:** a number of companies organise visits into the National Park buffer zone on foot, on horseback or in 4-wheel drive vehicles: Discovering Doñana (tel. 959 442 466), Doñana Tour (tel. 959 442 468) and Doñana Ecuestre (tel. 959 442 474). The magnificent birdlife of the Veta La Palma estate (tel. 954 589 237) can be appreciated on guided walks, bus tours or trips in horse and cart.
- **Further information:** maps 1:50,000, nº. 1001, 1002, 1017, 1018 and 1033 (SGE). Censuses and other data relating to birdlife from the Doñana biological station can be consulted at www.ebd.csic.es. It is also worth getting touch with the Doñana delegation of SEO/BirdLife (seo.donana@seo.org) and visiting the National Park web page www.mma.es/parques/lared/donana.

Pintail, Marbled Duck, Red-crested Pochard, Red Kite, Marsh Harrier, Spanish Imperial and Booted Eagles, Peregrine Falcon, Purple Swamphen, Red-knobbed Coot, Black-winged Stilt, Avocet, Stone Curlew, Kentish Plover, Northern Lapwing, Common Redshank.

Summer visitors: Squacco and Purple Herons, Eurasian Spoonbill, Black Kite, Egyptian Vulture, Short-toed Eagle, Hobby, Collared Pratincole, Slender-billed Gull, Gull-billed and Whiskered Terns.

Winter visitors and on migration: Cory's and Balearic Shearwaters, European Storm-petrel, Northern Gannet, Great Egret, Black Stork, Greylag, White-fronted and Barnacle Geese, Eurasian Wigeon, Common Teal, White-head, Tufted and Ferruginous Ducks, Common Scoter, Osprey, Common Crane, waders, Little and Audouin's Gulls, Caspian, Sandwich, Black and White-winged Black Terns, Short-eared Owl, Bluethroat, Penduline Tit, Reed Bunting.

10. SANLÚCAR DE BARRAMEDA

North of the town of Sanlúcar de Barrameda (Cádiz province) on the left-bank of the river Guadalquivir (opposite Doñana) stretch the Bonanza salt-marshes, most of which have been drained and transformed into grazing pastures or commercial salt-pans. The natural interest of the area is increased by the presence of La Algaida, a stone pine forest of

ANDALUSIA

Salinas de Bonanza, Sanlúcar de Barrameda © Fernando Barrio

700 ha standing on a relic dune (the most south-easterly in Doñana).

All of this area is included in the Doñana Natural Park, the protected area that safeguards some of the most valuable areas surrounding (but outside) the National Park. The Sanlúcar area (around 3,000 ha) is the most southerly part of this Natural Park.

ORNITHOLOGICAL INTEREST

Despite having lost much of its natural habitats, the Sanlúcar area is still of great interest since the large commercial salt-pans are flooded all year and provide refuge in summer for many birds when the Doñana salt-marshes dry out. A great variety of herons, ducks, gulls and waders breed or flock here once the breeding season is over; in recent years Shelduck, Little Tern, Slender-billed Gull and Marbled Duck have all bred. Greater Flamingos from breeding colonies in other parts of the western Mediterranean, including Fuente de Piedra (Málaga province), are found all year round (up to 25,000 birds).

La Algaida is home to many raptors, mainly Black Kites, a few Red Kites and a number of other species, and the wetland and surrounding areas are hunted over by Spanish Imperial Eagle and Peregrine Falcon. In some years White-headed Duck breed on Laguna de Tarelo in the extreme south-east corner of the pine forest.

ITINERARIES

1. 12 km

From Sanlúcar de Barrameda, a dirt track with a few patches of asphalt runs through La Algaida from north to south (1) and is a good way of seeing Black Kites. This track joins a road (2), where you should turn left towards the river Guadalquivir and the Bonanza salt-pans. Pass through a drained salt-marsh where Marsh Harriers hunt along the many canals; in spring look for Collared Pratincole and in winter for groups of Common Crane.

Once at the river (3), leave the road by turning left (south-west) along the raised embankment parallel to the

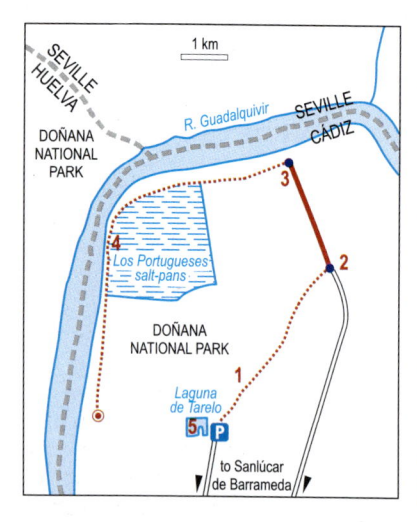

2. `100 m` 🚶 ☀️ ➡️

Just as you enter La Algaida there is a car-park, from where a short path (100 m) off to the left takes you to a hide overlooking the Laguna de Tarelo (5), an abandoned gravel pit. Look for threatened species such as Ferruginous, Marbled and, above all, White-headed Ducks, as well as Purple Swamp-hen and herons roosting in the nearby poplars.

MOST REPRESENTATIVE SPECIES

river running through a recently restored salt-marsh (good for waders). A little further on, on your left you will pass by the salt-pans of Los Portugueses (4), where many birds gather, including Greater Flamingos, Eurasian Spoonbill, Shelduck, Slender-billed Gull and, occasionally, rarer species such as Great Egret.

Eventually, the track begins to worsen and after rain becomes impassable by car. End the itinerary at a fence that acts as a hide for observing the birds on the salt-marshes.

Residents: Eurasian Spoonbill, Greater Flamingo, Shelduck, Marbled and White-headed Ducks, Red Kite, Marsh Harrier, Spanish Imperial Eagle, Purple Swamp-hen, Black-winged Stilt, Avocet, Kentish Plover, Slender-billed Gull, Caspian Tern, Lesser Short-toed Lark.

Summer visitors: Little Bittern, Squacco and Purple Herons, Black Kite, Montagu's Harrier, Short-toed and Booted Eagles, Collared Pratincole, Gull-billed, Sandwich and Little Terns, Spectacled Warbler.

PRACTICAL INFORMATION

- **Access:** from Sanlúcar de Barrameda a minor road leads to the port of Bonanza and then onto La Algaida.
- **Accommodation:** in Sanlúcar de Barrameda.
- **Visitor facilities:** the visitor centre at La Fábrica de Hielo (tel. 956 381 635) is linked to the Doñana National Park. Open every day (including lunchtimes). Book trips here on *Real Fernando*, a boat that travels 13 km upstream along the Guadalquivir. Reservations tel. 956 363 813.
- **Visiting tips:** any time of year is good, although winter and both passage periods are especially recommended. Avoid midday in summer. After heavy rain, driving on dirt tracks can be problematical.
- **Further information:** map 1:50,000, n°. 1033 (SGE). The Doñana Natural Park information centre (tel. 956 380 922, cv_bajodeguia@egmasa.es) is in Bajo de Guía, the port district of Sanlúcar de Barrameda. Also, the National Park offices (tel. 959 450 159, pn.donana.cma@juntadeandalucia.es) are in Almonte (Huelva province).

ANDALUSIA

Winter visitors: Black-necked Grebe, Common Crane, Great Egret, Black Stork, Greylag Goose, Ferruginous Duck, Osprey, Peregrine Falcon, European Golden Plover, Short-eared Owl, Bluethroat, Reed Bunting.
On migration: waders.

11. BRAZO DEL ESTE

Some 20 km to the south of the city of Seville extends Brazo del Este, a dead arm of the river Guadalquivir and one of the most important natural sites in the area surrounding Doñana.

This site consists of 14 km of broadly meandering river with dense belts of fringing vegetation surrounded by rice-paddies created by draining what was once a large area of salt-marsh. Today the area is protected as a *paraje natural* (site of natural interest) of just over 1,300 ha.

ORNITHOLOGICAL INTEREST

Brazo del Este has water all year round and so acts as a refuge for astonishing quantities of birds (in sheer numbers and in numbers of species) from Doñana when the great Guadalquivir salt-marshes dry out. Here you will find what is probably Europe's largest Purple Swamp-hen colony (thousands of pairs), as well as important colonies of Purple Heron, Little Bittern, Marbled Duck, Black-winged Stilt, Collared Pratincole and Whiskered Tern. In some years Glossy Ibis and Red-knobbed Coot (thanks to a re-introduction programme) breed.

Waders are abundant on passage and there are large post-nuptial concentrations of White and Black Storks (both of which winter), herons (including Squacco Heron and Great Egret), ducks and gulls, as well as the largest winter concentrations of Glossy Ibis in Spain (over 500 birds).

ITINERARIES

1. 9 km

Take the N-IV road from Seville towards Cádiz and at km point 549, just past the Virgen de Valme hospital, turn right along the minor road signposted 'Isla Menor' that follows the

Brazo del Este — Carlos Sánchez © nayadefilms.com

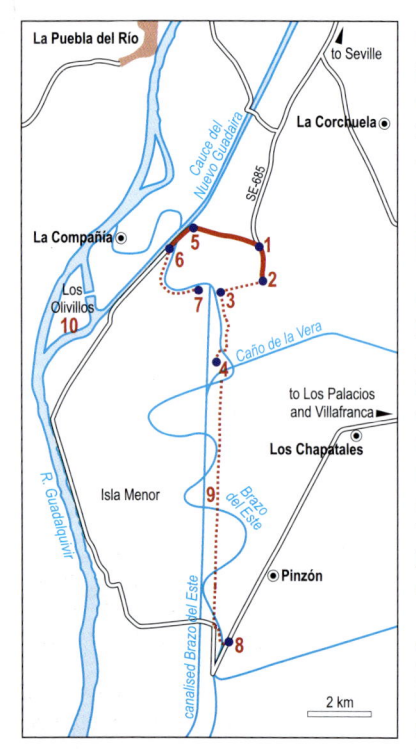

canalised course of the river El Nuevo Guadaira. At around 20 km from Seville you reach the warehouses of Mediterráneo Algodón (1). When the road forks, keep left towards Poblado Adriano and as far as a sign reading 'Finca Caño Navarro', where you should pick up a track heading right (2). In approximately 1.5 km you reach Brazo del Este (3) just where a path heads alongside the river as far as a canal called 'Caño de la Vera' (4).

Another option is to turn right at Mediterráneo Algodón (1) on a road that leads to El Nuevo Guadaira (5). Here, follow the canalised river as far as Brazo del Este (6), at which point a path continues along the right-bank of the river (7).

These routes are ideal for observing herons (Little Bittern and Squacco and Purple Herons), ducks (Marbled Duck, Garganey and Red-crested Pochard), waders (Avocet, Kentish Plover and Common Redshank) and many wetland passerines. This is also one of the few places in Spain with regular Temminck's Stint on passage and Wood Sandpiper in winter.

2. 5 km

Coming from Seville along the N-IV, at Los Palacios y Villafranca (between km points 568 and 569) take the minor road towards the villages of Los Chapatales and Pinzón. Once through these two *poblados*, you reach a tall rice silo (8), where you should turn right along a track to an embankment known as 'Muro de los Portugueses'. This track heads due north, cutting through various meanders of Brazo del Este (9).

This is an excellent site for all types of water birds, including Purple Heron, Marbled Duck, Marsh Harrier, Purple Swamp-hen, Black-winged Stilt, Avocet, Collared Pratincole and Whiskered Tern.

3. 12 km

Los Olivillos (10) is an ox-bow lake situated at the southern end of the spit of land separating El Nuevo Guadaira from the river Guadalquivir. The island of Los Olivillos boasts one of the largest heronries in Spain, with excellent numbers of Squacco Heron and a few pairs of Eurasian Spoonbill. Access is from Dos Hermanas or Los Palacios y Villafranca by crossing El Nuevo Guadaira or from Coria del Río via the river ferry.

MOST REPRESENTATIVE SPECIES

Residents: Night Heron, Red-crested Pochard, Marbled Duck, Marsh Har-

ANDALUSIA

PRACTICAL INFORMATION

- **Access:** from Seville along the N-IV to the town of Los Palacios y Villafranca (towards Cádiz).
- **Accommodation:** in Seville or nearby. Ideal places to stay are Dos Hermanas or, even better, Los Palacios y Villafranca.
- **Visitor facilities:** despite being a protected area, there are no visitor infrastructures or even hides. Nevertheless, the many raised embankments (with good dirt tracks on top) that cross the area make birdwatching easy.
- **Visiting tips:** a visit is worthwhile at any time of year. Beware of getting lost in the maze of embankments or stuck in the mud after heavy rain.
- **Recommendations:** the expert naturalist Francisco Chiclana (tel. 639 454 251, 954 616 052, pchiclana@terra.es) knows this site well and provides guiding services.
- **Further information:** maps 1:50,000, n°. 1002 and 1019 (SGE). The local Seville group of SEO/BirdLife has published a book (2002) entitled *Aves de la provincia de Sevilla*.

rier, Red-knobbed Coot, Purple Swamp-hen, Penduline Tit.
Summer visitors: Little Bittern, Squacco and Purple Herons, Eurasian Spoonbill, Black-winged Stilt, Avocet, Collared Pratincole, Kentish Plover, Gull-billed, Little, Whiskered and Black Terns, Short-toed Lark, Yellow Wagtail, Savi's and Olivaceous Warblers.
Winter visitors: Great Bittern, Great Egret, White and Black Storks, Glossy Ibis, Greater Flamingo, Eurasian Wigeon, Pintail and other ducks, Hen Harrier, European Golden Plover, Black-tailed Godwit, Lesser Black-backed Gull, Short-eared Owl, Bluethroat, Reed Bunting.
On migration: White-headed Duck, Temminck's Stint and other waders and Sedge Warbler.

12. MARISMAS DEL ODIEL

The intertidal salt-marshes at the mouth of the river Odiel near the provincial capital of Huelva have largely been transformed into commercial salt-pans. Unbroken by any relief features, these *marismas* stretch away seawards towards a long sand-spit (*El Espigón*) across a maze of islands separated by tidal channels (*caños* and *esteros*). Lagoons, dunes and beaches complete the site's diversity of habitats.

Aside from the triple protection of Biosphere Reserve, Ramsar Site and SPA, the Odiel marshes are protected as a *paraje natural* (site of natural interest) (7,000 ha) that includes two *reservas naturales*, El Burro and La Isla de Enmedio (500 ha each), the two best remaining areas of intertidal salt-marsh.

ORNITHOLOGICAL INTEREST

The island of Enmedio and, to a lesser extent El Burro, are home to various colonies of Eurasian Spoonbill that in total reach over 400 pairs, the largest breeding nucleus in Spain (along with those in Doñana). Nests are generally placed low down in the *Salicornia* scrub and many are lost during the breeding season if high tides coincide with heavy rain.

Other important colonial breeders include Grey and Purple Herons, Black-winged Stilt, Kentish Plover and Little Tern, here in one of its largest European breeding colonies (1,000+

Marismas del Odiel © José Manuel Reyero

pairs). Shelduck and Red-knobbed Coot also breed and there is a sizeable post-nuptial build up of Greater Flamingos.

Winter brings a diversity of seabirds and other aquatic birds such as Black-necked Grebe, Mediterranean Shearwater, Northern Gannet, Great Cormorant, Common Scoter and Osprey to the area, all complemented by a wonderful variety of ducks, waders, gulls and terns (also during passage periods).

ITINERARIES

1. 25 km 🚗 ☀ →

From the city of Huelva, take the dual carriageway towards the beach at Punta Umbría. Immediately after the bridge over the river Odiel (1), take the road towards the islands and sand-spit into the *paraje natural* and then on to the Juan Carlos I breakwater (*dique*).

The first habitat to appear are the commercial salt-pans 'Aragonesas' (2) on your right, which, apart from usually holding Greater Flamingo, duck, waders and gulls, are one of the main sites for wintering Black-necked Grebes in Spain (5,000+). Within 2 km you reach the excellent Calatilla visitor centre; from here the road continues through an area of well-established salt-pans – Isla de Bacuta (3) – and then some natural salt-marshes (4), frequented by Eurasian Spoonbill, Avocet, Black-tailed Godwit, Common Redshank, Eurasian Curlew and Whimbrel.

It is always best to visit at low tide when waders will feed close by, for example, at the coastal lagoon (5) on your right further on towards the sand-spit. This lagoon fills up with waders, gulls and terns at low tide and is one of the best sites for Eurasian Spoonbill, Red-breasted Merganser and Osprey.

From here on, the road enters the sand-spit that forms the right-bank

ANDALUSIA

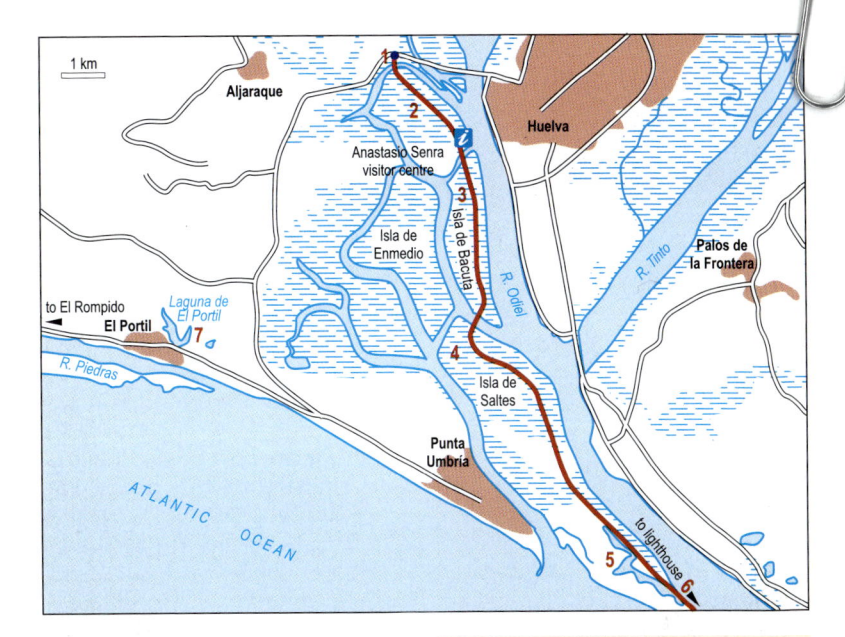

of the Odiel estuary: check the long sandy beach on your right for Oystercatcher, Sanderling, Audouin's Gull and Sandwich Tern. The final 10 km of the itinerary follow the breakwater and end at the lighthouse (*faro*) (6), from where (taking care in winter in rough seas) Common Scoter, Razorbill and other seabirds can be seen.

2. 0 km 🚗 ☀ ▬

From the road between Punta Umbría and El Rompido, good views can be had of El Portil, a freshwater lagoon situated next to the holiday complex of the same name. Little Bittern, Squacco and Purple Herons and Purple Swamp-hen breed and there are regular records of

PRACTICAL INFORMATION

- **Access:** easy from the city of Huelva. The road to the islands and El Espigón is the only unrestricted access to the *paraje natural*. The rest of the area is either private or closed off as a protected area.
- **Accommodation:** en Huelva and Punta Umbría.
- **Visitor facilities:** The Anastasio Senra visitor centre (tel. 959 509 011), known locally as 'Calatilla', is open Wednesday to Sunday, morning and afternoon.
- **Visiting tips:** any time of year, including summer (despite the tourists attracted by the beach at Punta Umbría). At low tide more birds can be seen.
- **Recommendations:** The programme 'Aves del Litoral' (tel. 959 011 500) organised by the Andalusian government welcomes volunteers for bird-related work. The company Erebea (tel. 660 414 920, erebea@terra.es) organises birdwatching walks, trips in 4-wheel drive vehicles and boat journeys around the Spoonbill colony on Isla de Enmedio.
- **Further information:** map 1:50,000, n°. 999 (SGE). Useful book: *Aves de las marismas del Odiel y su entorno* by Héctor Garrido (Editorial Rueda, 1996).

Ferruginous Duck, White-headed Duck and Red-knobbed Coot. Winter brings large flocks of duck.

MOST REPRESENTATIVE SPECIES

Residents: Little Egret, Grey Heron, Eurasian Spoonbill, Greater Flamingo, Shelduck, Marsh Harrier, Purple Swamp-hen, Eurasian and Red-knobbed Coots, Black-winged Stilt, Avocet, Stone Curlew, Kentish Plover.

Summer visitors: Little Bittern, Purple and Night Herons, Little Tern.

Winter visitors: Black-necked Grebe, Mediterranean Shearwater, Northern Gannet, Great Cormorant, Great Egret, Common Scoter, Red-breasted Merganser, Osprey, Peregrine Falcon, waders, Great Skua, Little and Audouin's Gulls, Caspian Tern, Razorbill, Short-eared Owl, Bluethroat, Penduline Tit, Reed Bunting.

On migration: ducks, waders, Mediterranean Gull and terns.

13. BAHÍA DE CÁDIZ

Despite being situated in midst one of the most densely populated stretches of the Andalusian coastline (almost half a million people), the Bahía de Cádiz is one of the most important wetlands in Spain. The site consists of a wide bay bordering the city of Cádiz to the west and includes parts of the towns of San Fernando, Puerto de Santa María, Puerto Real and Chiclana. A singular mosaic of habitats decorates the area, with broad intertidal flats alternating with dunes and beaches criss-crossed by a complex of natural (*caños*) and artificial (*esteros*) channels used for the production of salt and harvesting of shellfish.

Most of the natural salt-marshes have been transformed into commercial salt-pans, although many are now abandoned or used for shellfish production. Two of best remaining areas are the island of Trocadero and the estuary of Sancti-Petri, both of which are part of a *paraje natural* (site

Bahía de Cádiz © Pedro Retamar

ANDALUSIA

of natural interest). The whole of the Bahía de Cádiz is protected by a Natural Park (10,000 ha) and is also a Ramsar Site and an SPA.

ORNITHOLOGICAL INTEREST

Along with Doñana, the Bahía de Cádiz is the most important site for wintering waders in Spain (30,000 birds) and, in all, around 60,000 water birds winter here. Furthermore, its strategic position between the great Guadalquivir salt-marshes and the Straits of Gibraltar mean that this is an excellent site for many birds migrating between Africa and Europe. Important breeding colonies of Black-winged Stilt and Avocet exist alongside good numbers of Kentish Plover and Little Tern. In recent years a colony of around 50 pairs of Eurasian Spoonbill has established itself in the area.

ITINERARIES

The Bay can be conveniently divided into two parts: the northern section to the south of the town of Puerto de Santa María and the southern section south of Puerto Real. The following itineraries concentrate on the salt-pans and intertidal habitats of the southern section.

1. 6 km

From the car-park halfway between San Fernando and the beach at Camposanto, walk west along the bank of the former salt-pan of Tres Amigos, recently restored by the of the Andalusian Regional Ministry of the Environment with birds in mind. The path splits (1): turn right along the bank of the tidal Río Arillo, which at low tide becomes open mudflats and at high tide a shallow lagoon.

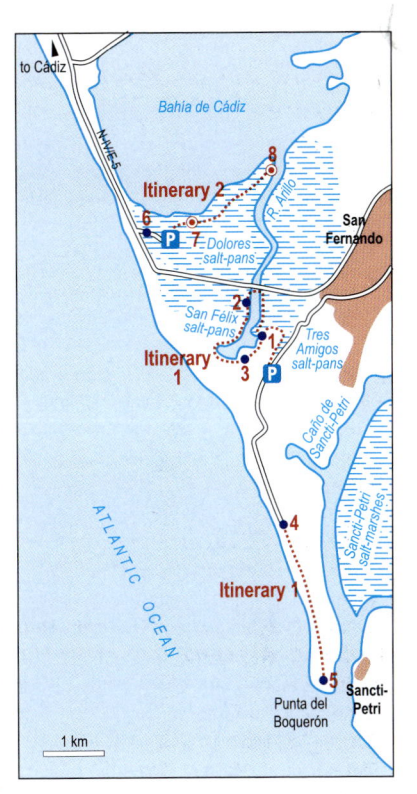

After following the path right around the river and passing by the San Félix salt-pan (2), you then come to a couple of hides (3) before rejoining the original path.

This is an excellent itinerary in winter and during migration periods for finding waders such as Black-winged Stilt, Avocet and numerous others. Other species to look out for include Greater Flamingo and Black-headed, Slender-billed, Audouin's, Lesser Black-backed and Yellow-legged Gulls, as well as various species of tern.

Back at the car-park, continue by car to the beach and on to the starting point of a linear walk (4) to La Punta del Boquerón (5) that bisects the dunes and beaches to the west and the magnificent salt-marshes of Sancti-Petri – full of feeding waders – to the east.

PRACTICAL INFORMATION

- **Access:** the Cádiz-San Fernando dual carriageway passes between the areas covered by the two itineraries and close to their starting points.
- **Accommodation:** in Cádiz, San Fernando and other nearby towns.
- **Visitor facilities:** both itineraries pass by information panels and hides, although no visitor centre as such exists.
- **Visiting tips:** good all year, but best during migration and in winter. In summer avoid the hot midday sun and carry drinking water.
- **Recommendations:** The University of Cádiz runs a summer volunteer programme on the region's best beach – Playa de Levante – to protect nesting Kentish Plovers and Little Terns (tel. 956 016 593, voluntariado.ambiental@uca.es).
- **Further information:** map 1:50,000, n°. 1068 (SGE). Also, contact the offices of the Bahía de Cádiz Natural Park in San Fernando (tel. 956 590 405 and 956 590 971, pn.bahiadecadiz.cma@juntadeandalucia.es).

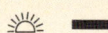

2. 2,5 km

From the dual carriageway between Cádiz and San Fernando, take the fly-over (6) towards Santibáñez and a sewage works and park halfway along at a car-park that marks the beginning of this itinerary. The path passes between the old Dolores salt-pan and the open waters of the bay, where patches of natural salt-marsh and intertidal flats can be viewed from two hides (7 and 8).

This is another good itinerary for waders, with many sandpipers and godwits on display. As well, Eurasian Spoonbill wander through from their nearby colony, and gulls and terns are common; in winter grebes, Great Cormorants, herons, and the occasional Osprey frequent the area. On the walls of the old salt-pans, the few remaining patches of halophytic vegetation hold Lesser Short-toed Lark and various species of warbler.

MOST REPRESENTATIVE SPECIES

Residents: Little Egret, Grey Heron, Eurasian Spoonbill, Greater Flamingo, Black-winged Stilt, Avocet, Kentish Plover, Yellow-legged Gull, Lesser Short-toed Lark.
Summer visitors: Little Tern.
Winter visitors: Black-necked Grebe, Great Cormorant, Shelduck, Osprey, Oystercatcher, Ringed and Grey Plovers, Dunlin, Black-tailed and Bar-tailed Godwits, Eurasian Curlew, Whimbrel, Common Redshank, Black-headed, Slender-billed and Lesser Black-backed Gulls, Caspian and Sandwich Terns.
On migration: seabirds, Great Crested Grebe, Knot, Sanderling, Little Stint, Curlew Sandpiper, Greenshank, Turnstone and other waders, Audouin's Gull.

14. MOUTH OF THE GUADALHORCE

In the heart of the Costa del Sol, the final reaches of the river Guadalhorce form a small deltaic plain south-east of the city of Málaga. Protected as a *paraje natural* (site of natural interest), this small reserve (less than 70 ha) lies between two arms of the river Guadalhorce – the current course and an older one known as Río Viejo – and the beach. It is today the most

ANDALUSIA

Mouth of the Guadalhorce © José Antonio Cortés

important of the wetlands that were once found along the coast of Málaga and consists of salt-marshes that flood in winter, old gravel pits, fluvial woodland and areas of scrub.

However, the recent canalisation of the final eight kilometres of the river has destroyed much of the natural habitat in the area, including all the fluvial woodland. As a form of compensation, the Confederación Hidrográfica del Sur has carried out some habitat improvement in the lagoon area.

ORNITHOLOGICAL INTEREST

Despite being so near a city as big as Málaga (and all that implies in terms of the artificiality of the site), the strategic situation of the mouth of the Guadalhorce river – near the Straits of Gibraltar and on the Bahía de Málaga – ensures that many birds pass through this small wetland on migration. There is good passage of Cory's and Balearic Shearwaters, Northern Gannet, Black Stork, Glossy Ibis, Eurasian Spoonbill, Greater Flamingo, waders, Audouin's Gull and Gull-billed Tern. To add to the site's attractions, vagrant species are regularly recorded.

Duck, waders, seabirds and many wetland passerines winter in and around the site, while Little Bittern, Purple Heron, Black-winged Stilt, Kentish Plover and, since 2003, White-headed Duck all breed.

ITINERARY

3 km

From the Guadalmar tourist complex head for the beach and park as near to the river mouth as possible. At the time of writing, the canalisation of the river was still unfinished and it was possible to cross the river on foot via the sand-bar at the mouth (1) and in this way gain access to the *paraje* (2). However, higher water levels can make this crossing point impracticable and in this case it is best to cross the river upstream (3) under the A-7 motorway bridge, accessible from the tourist complex, until a new crossing point has been built.

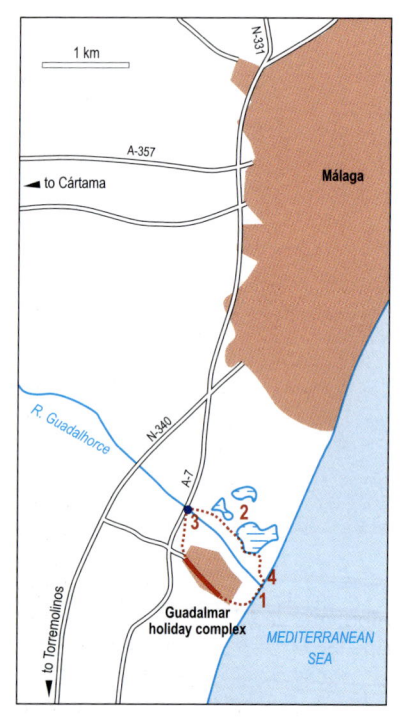

White-headed and other ducks, Osprey and Mediterranean Gull. Audouin's Gull is frequent is summer, although it does not breed.

Booted Eagles – essentially summer visitors elsewhere in Spain – winter here as in other southern coastal sites and the Peregrines that breed in Málaga cathedral come here to hunt. Amongst the passerines, Bluethroat are one of the most interesting winter visitors.

A second possibility is to wander along the beach (4) lying within the protected area (1 km). Beach-loving waders such as Sanderling and Kentish Plover are easy to observe, along with migrant or wintering seabirds such as Shearwaters, Northern Gannet, Scoters, Skuas, Terns and Razorbills.

MOST REPRESENTATIVE SPECIES

The site has a short itinerary marked with coloured wooden stakes that takes you around the old gravel pits. Winter and migration periods are the best times to see Great Cormorant, Grey and Purple Herons,

Residents: Night Heron, Little Egret, Common Pochard, White-headed Duck, Peregrine Falcon, Kingfisher.
Summer visitors: Little Bittern, Purple Heron, Black-winged Stilt, Kent-

PRACTICAL INFORMATION

- **Access:** From Málaga, take the A-7 motorway towards Algeciras and just after crossing the river Guadalhorce (here already canalised), turn off towards the Guadalmar tourist complex.
- **Accommodation:** all along the Costa del Sol between Málaga and Torremolinos. There is a 4-star hotel in the Guadalmar complex.
- **Visitor facilities:** a visitor centre and number of hides are planned.
- **Visiting tips:** the best times are during post-breeding migration between the end of summer and beginning of October and in winter.
- **Recommendations:** the Málaga SEO/BirdLife group (tel.952 625 129, seo-malaga@seo.org) provides information about this and other natural areas in the province and also organises guided visits.
- **Further information:** maps 1:50,000, n°. 1053 and 1067 (SGE). The book *Las aves acuáticas y marinas en Málaga y provincia*, by M. Garrido, E. Alba and J. Mª González (Diputación de Málaga, 2002) describes in detail the majority of species found at the mouth of the Guadalhorce.

ANDALUSIA

ish Plover, Audouin's Gull, Short-toed Lark, Yellow Wagtail, Reed Warbler.
Winter visitors: Black-necked Grebe, Great Cormorant, Northern Gannet, Marsh Harrier, Booted Eagle, Osprey, European Golden Plover, Common and Jack Snipes, Arctic and Great Skuas, Mediterranean Gull, Kentish Plover, Razorbill, Bluethroat, Penduline Tit, Reed Bunting.
On migration: Cory's and Balearic Shearwaters, Squacco Heron, Black Stork, Glossy Ibis, Eurasian Spoonbill, Greater Flamingo, Garganey, waders, Common, Little, Gull-billed, Whiskered and Black Terns.

15. WETLANDS OF WESTERN ALMERÍA

The south-west coast of Almería between the towns of El Ejido and Roquetas is characterised by a line of sandy beaches backing onto a dune cordon fixed by splendid Mediterranean scrub formations dominated by lentisc (*lentisco* or *entina*) and Phoenician juniper (*sabina*), the origin of the names of this coastline's only two headlands – Punta Entinas and Punta Sabinar.

Behind the dunes lie a number of salt-pans (Salinas de Cerrillos) and a series of shallow lagoons (Charcones de Entinas), today surrounded by acres of greenhouse cultivation that have occupied the once steppe-like coastal plain. Further inland, lie the lakes at Las Norias de Daza, in the town of La Cañada de Las Norias, the third of the sites that form the wetlands of western Almería, to which we could add the salt-pans of Guardias Viejas, drained in 1998 to build a holiday complex, and the Albuferas de Adra, a number of small brackish coastal lagoons that once formed part of a river delta.

The *paraje natural* of Punta Entinas-Sabinar (almost 2,000 ha) is an SPA and includes the first two

Punta Entinas-Sabinar, Western Almería © José Bayo

wetlands mentioned above, both or which are also *reservas naturales*. The Albuferas de Adra (outside the *paraje natural*) are *reservas naturales* and a Ramsar Site.

ORNITHOLOGICAL INTEREST

This site, the largest and best wetlands in Almería province, boasts a number of diverse habitats and is, as a result, home to a highly varied wetland avifauna. It is also a vital area for the White-headed Duck, whose colonies in the Albuferas de Adra and La Cañada de Las Norias are two of the most important in Europe.

Likewise, of equal import are the site's colonies of gulls, terns and waders that include species such as Black-winged Stilt, Kentish Plover, Little Tern and, to a somewhat lesser extent, Collared Pratincole, Avocet, Common Tern and Black-headed Gull. Squacco Herons and other herons breed in La Cañada de Las Norias, as do Marbled Duck, Red-crested Pochard and Purple Swamphen. Passage sees many ducks, waders, gulls and terns flock to the area.

ITINERARIES

1. ⬜3 km 🚗 ☀ ➡

La Cañada de Las Norias (1) consists of a group of flooded clay pits, some still being exploited, that were dug to provide soil for the intensive greenhouse-based cultivation that dominates the area. Despite being completely surrounded by acres of plastic, buildings and roads, this wetland is probably the easiest to visit and most rewarding (from a point of view of observing water birds) of all the wetlands in western Almería.

Access is from the town of Las Norias de Daza along the road between El Ejido and La Mojonera that skirts the southern side of the site. From this road and the other smaller ones that circumnavigate the lakes excellent birdwatching is guaranteed all year and, above all, between April and July when the all breeding species mentioned above and passage waders are present.

Despite its growing ornithological importance, La Cañada de Las Norias is legally unprotected, although

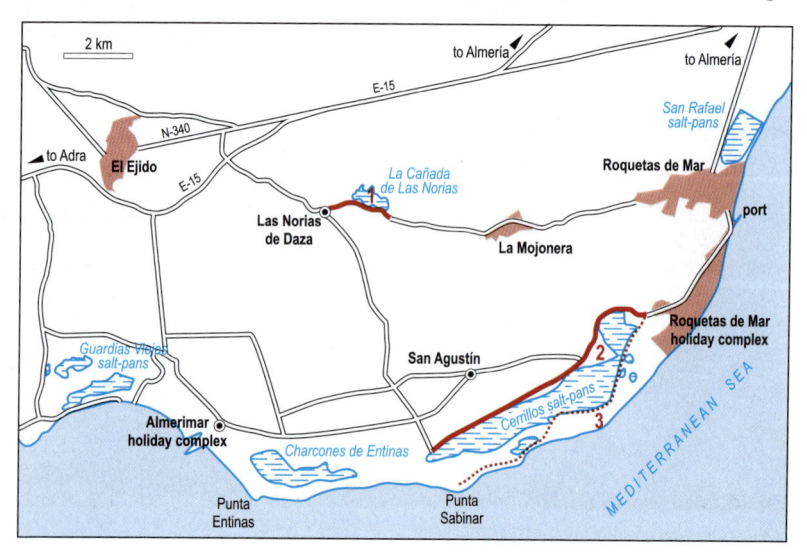

ANDALUSIA

PRACTICAL INFORMATION

- **Access:** the junctions on the E-15 motorway between Adra and Almería to the south-east of El Ejido will take you to Las Norias de Daza (itinerary 1) and Roquetas de Mar (itinerary 2).
- **Accommodation:** in El Ejido and Roquetas de Mar.
- **Visitor facilities:** no visitor centre or infrastructure for birdwatchers.
- **Visiting tips:** best in spring (for breeding birds) and autumn (for migrants). In summer, early morning and evening are the best times of day. It is important to be equipped with mosquito repellent. Telescope also very useful.
- **Further information:** map 1:50,000, n°. 1058 (SGE). More information from the Almería delegation of the Andalusian Regional Ministry of the Environment (tel. 950 011 144).

thanks to an agreement signed with the town council of El Ejido, SEO/BirdLife do in fact manage a reserve in the area.

2. 9 km

The local road between Roquetas de Mar and the Almerimar holiday complex runs along the northern side of the today abandoned Cerrillos salt-pans (2). From this road it is very easy gain access to the wetland on foot. As well, on the southern side a track (3) that can be walked also allows good access to the basins of the former salt-pans.

This area is perfect for watching waders, above all during passage periods, and is also an important site for Black-necked Grebe. Greater Flamingo and Audouin's Gulls are present almost the whole year and surrounding habitats have Lesser Short-toed Lark and Stone Curlew. From the coast, for example at Punta Sabinar, sea-watching can be rewarding.

MOST REPRESENTATIVE SPECIES

Residents: Greater Flamingo, Marbled and White-headed Ducks, Red-crested Pochard, Purple Swamp-hen, Black-winged Stilt, Avocet, Stone Curlew, Kentish Plover, Black-headed and Audouin's Gulls, Lesser Short-toed Lark.

Summer visitors: Little Bittern, Squacco Heron, Collared Pratincole, Common and Little Terns, Roller, Short-toed Lark, Reed and Great Reed Warblers.

Winter visitors: Black-necked Grebe, Purple Heron, Tufted Duck, Oystercatcher, Black-tailed Godwit, Mediterranean and Slender-billed Gulls, Sandwich Tern, Bluethroat, Penduline Tit, Reed Bunting.

On migration: Night Heron, Marsh Harrier, Osprey, waders, Whiskered and Black Terns.

16. CABO DE GATA

Cabo de Gata lies in the south of Almería province and is one of the most beautiful and best conserved stretches of coast in the whole of Mediterranean Spain. Beaches, small coves and towering sea-cliffs are complemented inland by the desert-like terrain of one of Europe's most important volcanic landscapes. Well-preserved and extraordinarily rich submarine habitats and an area of working salt-pans complete the panorama of the area.

Cabo de Gata © Eduardo de Juana

The region of Cabo de Gata is the most arid in the whole of western Europe. The Natural Park, the first in Andalusia, was declared in 1987 and includes 38,000 ha of land and 12,000 ha of sea. It is also a Biosphere Reserve, a Ramsar Site and an SPA.

ORNITHOLOGICAL INTEREST

The Cabo de Gata salt-pans (300 ha) are the best site in the Natural Park for passage and wintering ducks, waders and other aquatic birds. Audouin's Gull and Greater Flamingo are present all year but do not breed; on the other hand, Avocet, Black-winged Stilt, Kentish Plover and Common and Little Tern do breed.

The surrounding steppe areas are home to birds as scarce in Europe as Trumpeter Finch and Dupont's Lark, although the latter has suffered – along with Little Bustard and Black-bellied Sandgrouse – greatly from the loss of habitat caused by the extension of forced cultivation under plastic. Stone Curlew and Lesser Short-toed and Thekla Larks, however, are still common.

Within the mountains a few pairs of Bonelli's Eagles breed, along with Peregrine Falcon and Eagle Owl. Black Wheatear and Blue Rock Thrush are common.

ITINERARIES

1. 12 km

The narrow coastal road that links the village of San Miguel de Cabo de Gata and the lighthouse passes between the beach and the salt-pans. Just before the settlement of La Almadraba de Monteleva, a track off to the left (1) takes you to a car-park from where a short (100 m) footpath leads to a hide. From here, it is not difficult to see Greater Flamingo, Black-necked Grebe, Black-winged Stilt, Avocet and Black-tailed Godwit. The saline pools further south are a good area for Audouin's Gull, especially during post-breeding passage (October).

Continuing on towards the lighthouse, the areas around La Almadraba de Monteleva and just where the road begins to climb up into the mountains (2) should be searched for Trumpeter Finch, above all in autumn and winter, although it is a difficult species to detect and observa-

ANDALUSIA

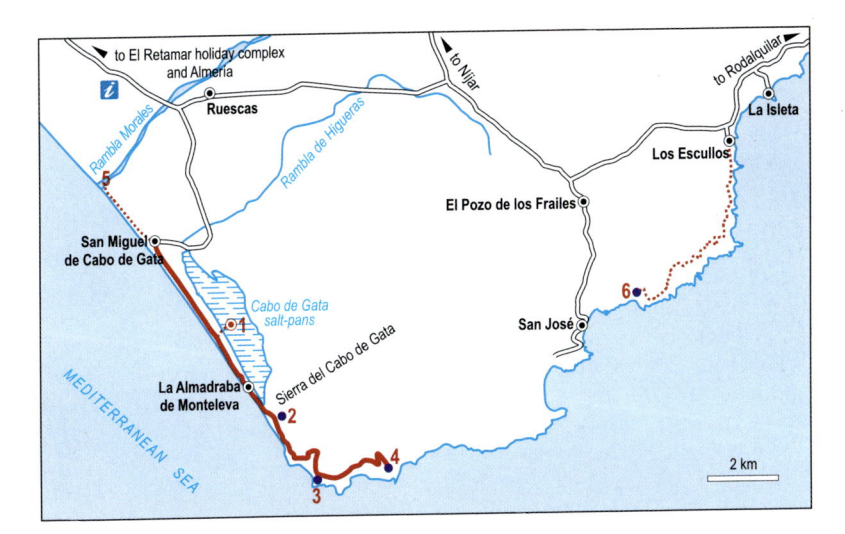

tions are not guaranteed. However, this is a good area for Thekla Lark, Crag Martin and Black Wheatear and from the lighthouse (*faro*) (3) and the tower of Vela Blanca (4), sea-watching for Balearic Shearwater and Northern Gannet can be profitable at the end of summer and in autumn and winter.

2. 2 km

From San Miguel de Cabo de Gata the mouth of the small stream of Rambla Morales (5), known as El Charco, is easily accessible along the track that is the continuation of the village's coastal promenade. Greater

PRACTICAL INFORMATION

- **Access:** from Almería take the airport dual-carriageway or the E-15 dual-carriageway that skirts the city to the north on its way from Málaga to Murcia. Take the turn-off towards San José and, when halfway there, turn south towards the village of San Miguel de Cabo de Gata.
- **Accommodation:** in Almería or Níjar or, better still, in one of the villages in the mountains of Cabo de Gata such as San José (the best equipped).
- **Visitor facilities:** the park reception centre - Las Amoladeras (tel. 950 160 435) - is at km 7 on the road to Cabo de Gata, between the holiday complex of El Retamar and the village of Ruescas. Open mornings, although between July and September it also opens until 21.30. From here there are a number of marked walks through the area of steppe known as Las Amoladeras.
- **Visiting tips:** all year, but above all during migration periods. In summer avoid midday. Because of the position of the sun, the salt-pans are best visited in the afternoon.
- **Recommendations:** the group J-126 (tel. 950 380 299, info@cabodegata-nijar.com) organise guided nature excursions on foot and in 4-wheel drive vehicles.
- **Further information:** maps 1:50,000, n°. 1059 and 1060 (SGE). The offices of the Natural Park (tel. 950 389 742, pn.cabodegata.cma@juntadeandalucia.es) are in Rodalquilar.

Flamingo, waders and other aquatic birds, which can here be very confiding, can be watched from a car or from nearby. Especially interesting in post-breeding passage.

3. 6 km 🚗 🚶 ☀️ ➡️

From the group of houses known as Los Escullos near the villages of El Pozo de los Frailes and Rodalquilar, you can drive the first part of a dirt track running parallel to the sea. Make occasional stops to search for Trumpeter Finch, above all in spring. Cala Higuera (6) offers the chance for more sea-watching.

MOST REPRESENTATIVE SPECIES

Residents: Shag, Greater Flamingo, Bonelli's Eagle, Peregrine Falcon, Black-winged Stilt, Avocet, Stone Curlew, Kentish Plover, Audouin's Gull, Black-bellied Sandgrouse, Eagle Owl, Dupont's, Lesser Short-toed and Thekla Larks, Black Wheatear, Blue Rock Thrush, Spectacled Warbler, Trumpeter Finch.
Summer visitors: Common and Little Terns, Alpine Swift, Roller, Short-toed Lark, Rufous Bush Robin, Black-eared Wheatear.
Winter visitors: Black-necked Grebe, Shelduck, Oystercatcher, Black-tailed Godwit, Slender-billed Gull, Sandwich Tern.
On migration: Balearic Shearwater, Northern Gannet, Purple Heron, Black Stork, Marsh Harrier, Osprey, waders, Whiskered and Black Terns.

17. STRAITS OF GIBRALTAR

This site consists of the extreme southern tip of Spain overlooking the Straits of Gibraltar around the towns of Algeciras and Tarifa. Inland from the long sandy beach of Los Lances the landscape is dominated by a series of low rounded ridges covered in cork oak forest, the most important such forest in Spain, that are part of the Alcornocales Natural Park (170,000 ha). In 2003 the Estrecho Natural Park

Straits of Gibraltar

Carlos Sánchez © nayadefilms.com

ANDALUSIA

(20,000 ha), covering the whole of this coastline, was declared. Part of both natural parks is included in an SPA.

ORNITHOLOGICAL INTEREST

This is the most important site for birds migrating between Africa and western Europe. Huge numbers of soaring birds – over 100,000 storks and even more raptors – funnel together and pass over the Straits in flocks in both spring and autumn. The most remarkable sights are the groups of White and Black Storks and 20 species of raptor, including Honey Buzzard, Black Kite, Egyptian Vulture, Short-toed and Booted Eagles and Montagu's Harrier.

The proximity of the two continents – a mere 14 km apart – and the variety of habitats that allow birds to feed and rest whilst they wait for favourable winds make this area one of the great places in the world for observing bird migration. Inland extends the Alcornocales Natural Park with good raptor and passerine populations, as well as breeding White-rumped Swift.

ITINERARY

[25 km]

The best views of migrating birds can be had from a series of observation points, some of which have been set up by the Andalusian Government; the best spot at any moment in time depends largely on the direction and speed of the wind. The N-340 provides access to all these points, which can be visited in a single day by travelling from Tarifa towards Algeciras and stopping as follows:

Cazalla (1): turn off left to a former wind farm at km 87 of the N-340 (a rather difficult turn, complicated by a sharp bend). A metalled track leads to the site, ideal for observing all the large soaring birds, above all those flying around the town of Tarifa. Good in both spring and autumn, especially with the wind coming from the east.

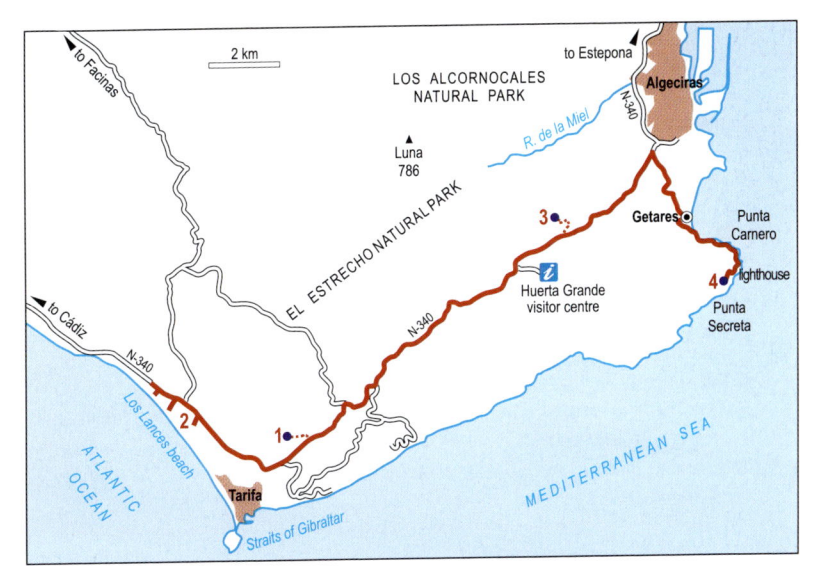

It is also worth visiting the beach of Los Lances (2), which can be reached via the football pitch at Tarifa and from signposted accesses at km points 79.9, 80.4 and 82.1 on the N-340. Kentish Plovers breed, gulls including Audouin's are to be found here all year and there is also a good passage of passerines. Waders frequent the temporary pools at the mouths of the rivers Jara and De La Vega and in winter after storms sea-watching is rewarding.

El Algarrobo (3): set up as a bird observatory, this point gives good views over La Bahía de Algeciras and the Rock of Gibraltar. The best way to approach is to go past the entrance at km point 99.1 on the N-340 and to then turn round so as to approach the signposted track (now off to your right) as if you were coming from Algeciras. Ideal in autumn with winds from the west for storks, raptors and non-soaring birds such as swifts, Barn Swallows and European Bee-eaters.

Punto Carnero (4): At a set of traffic lights on the N-340 just before km point 102, turn right towards the village of Getares and the Carnero lighthouse. Both the lighthouse car-park or 100 m further on towards Punta Secreta are excellent observation points in spring when winds are from the west. Being next to the coasts, low-flying birds are also easy to see here.

MOST REPRESENTATIVE SPECIES

On migration: Cory's Shearwater, Northern Gannet, Black and White Storks, European Bee-eater, Common and Honey Buzzards, Black and Red Kites, Egyptian and Griffon Vultures, Short-toed and Booted Eagles, Marsh, Hen and Montagu's Harriers, Osprey, Common and Lesser Kestrels, Hobby, Sparrowhawk, Common Crane, waders, Audouin's Gull, Razorbill, passerines.

PRACTICAL INFORMATION

- **Access:** from a 20 km-stretch of the N-340 in southern Cádiz province between Tarifa and Algeciras.
- **Accommodation:** try the rural accommodation in Huerta Grande (tel. 956 679 700, www.huertagrande.com), next to the visitor centre of the same name. Many hotels and camp-sites in Tarifa.
- **Visitor facilities:** The visitor centre at Huerta Grande (tel. 956 679 161) at km point 96 of the N-340 is part of the Alcornocales Natural Park, but also provides information about the birds of the Gibraltar area. Open Thursday morning, Friday afternoon and morning and afternoon at the weekend.
- **Visiting tips:** migrating birds can be seen almost all year, although the best time is undoubtedly during post-breeding migration in September and October, when large groups of migrants use well-known routes. Telescope advisable.
- **Recommendations:** SEO/BirdLife with the support of the Andalusian Government have set up the programme Migres (tel. 954 644 294, migres@seo.org) to monitor scientifically bird migration across the Straits. Volunteers welcome.
- **Further information:** maps 1:50,000, nº. 1077 and 1078 (SGE). The classic works on the subject are the two volumes of *La migración de aves en el Estrecho de Gibraltar*, the volume on soaring birds written by Francisco Bernís and the one on non-soaring birds by José Luis Tellería. Another interesting book is *Guía de aves del Estrecho de Gibraltar, Parque Natural de los Alcornocales y Comarca de la Janda* by David Barros and David Ríos (Ornitour, 2002).

ARAGON

- 18. Benasque
- 19. Ordesa
- 20. Sierra de Guara
- 21. San Juan de la Peña
- 22. Riglos
- 23. Monegros
- 24. Belchite steppes
- 25. Laguna de Sariñena
- 26. Galachos del Ebro
- 27. Laguna de Gallocanta

From the eternal snows of the high Pyrenean glaciers in the north to the semi-deserts in the south, Aragon is a region of grandiose and largely well-preserved landscapes of great biogeographical diversity. In between lie a rich succession of forests, mountains, pastures, deep river canyons, fluvial woodland, lakes and agricultural systems – in all, an invitation no birdwatcher should resist.

Travelling from north to south, the first great natural features – and one that marks the natural border between Spain and France – are the mountains of the Pyrenees. With their highest peaks crowded around the beautiful Benasque valley (18), all but on the border with Catalonia, these peaks are true islands of alpine habitat and stand guard over the most important glacial landscape in the Iberian Peninsula. Today the Pyrenees provide refuge for high mountain species such as Ptarmigan, Alpine Accentor, Wallcreeper and Snow Finch, while the magnificent mountain pine forests on lower slopes are home to two authentic jewels, Capercaillie and Tengmalm's Owl, both all but at the southern limit of their distributions.

The stunning Ordesa and Monte Perdido National Park (19) offers birdwatchers the chance to discover all these species in a magnificent setting that also provides plenty of opportunities for observing Lammergeier. The world's most important populations of this threatened vulture are found in Aragon and provide a

Lammergeier © Juan Martín Simón

source of raw material for reintroduction plans in areas such as the Alps, Picos de Europa and Cordillera Bética where the Lammergeier once flew.

Southwards the Pyrenees fade out into a jumble of abrupt mid-altitude mountain ranges known as the pre-Pyrenees, where Eurosiberian and Mediterranean influences converge and Black Woodpecker and Black Wheatear, for example, breed in relative proximity. Cliff-loving species are especially well represented here and the Lammergeier again takes pride of place in the area's strong raptor populations. The most interesting site for cliff-breeding raptors is the Sierra de Guara (20), closely followed by the nearby massifs of San Juan de la Peña (21) and Riglos (22).

The valley of the river Ebro forms a broad depression in the arid central part of Aragon. Rolling plains, given over to dry agriculture and extensive stock-raising (although irrigation is steadily gaining ground) and interrupted by just a few patches of natural vegetation, predominate and the area is a paradise for steppe birds. Some of the best sites in Europe for Pin-tailed Sandgrouse, Dupont's Lark and Lesser Short-toed Lark are to be found in the Monegros (23) and nearby areas such as Belchite (24).

These plains are interrupted by a mosaic of low ridges, dotted with a few scattered patches of pine and juniper woodland, that often exhibit splendid tabular relief forms. Rock-loving and forest birds including raptors are abundant, although the area's overall avian richness is considerably boosted by the presence of a number of natural and artificial lagoons that are frequented by wildfowl and waders in winter and on migration. The largest of these wetlands is the Laguna de Sariñena (25), which can also boast as a bonus an interesting colony of herons.

The extreme south of Aragon merges into the mountains of the Sistema Ibérico and constitutes the region's third main geographical unit.

ARAGON

Low or mid-altitude ridges alternate with broad plains, often lying at over 1,000 m above sea-level. In the heart of one of these semi-arid *parameras* lies the Laguna de Gallocanta (27), internationally famous as the main stop-over site for thousands of Common Crane on route to their wintering sites in the south of the peninsula.

These cranes are not the only birds to use Aragon as a migration corridor. The mountain passes of the Pyrenees and lowland wetlands, rivers and forests are important as refuelling sites for long-distant migrants *en route* between wintering sites in Africa and their European breeding grounds.

18. BENASQUE

This site consists of the high peaks – including many *ibones* (glacial lakes) and Aneto (3,404), the highest mountain in the Pyrenees – of north-east Huesca province that lie close to the border with France and Catalonia. Most of the area is within the Posets-Maladeta Natural Park (33,000+ ha), which has been declared an SPA. The mountain chain, here sliced open from north-east to south-west by the valley of the river Ésera, harbours the largest concentration of glaciers in Spain, which are also the most southerly in Europe. The largest of all is the Aneto glacier and the majority have been classified as a National Monument.

A significant part of the site lies above 3,000 m and is dominated by alpine pastures and bare granite rock covered by snow for most of the year. At around 2,000 m the mountain pine forests – the best in Aragon – are at their most splendid and they are occasionally visited by the European Brown Bears reintroduced onto the French side of the Pyrenees in the 1990s. Benasque is an important tourist centre and lies close by the Cerler ski station.

Estós Valley, Benasque © Pedro Retamar

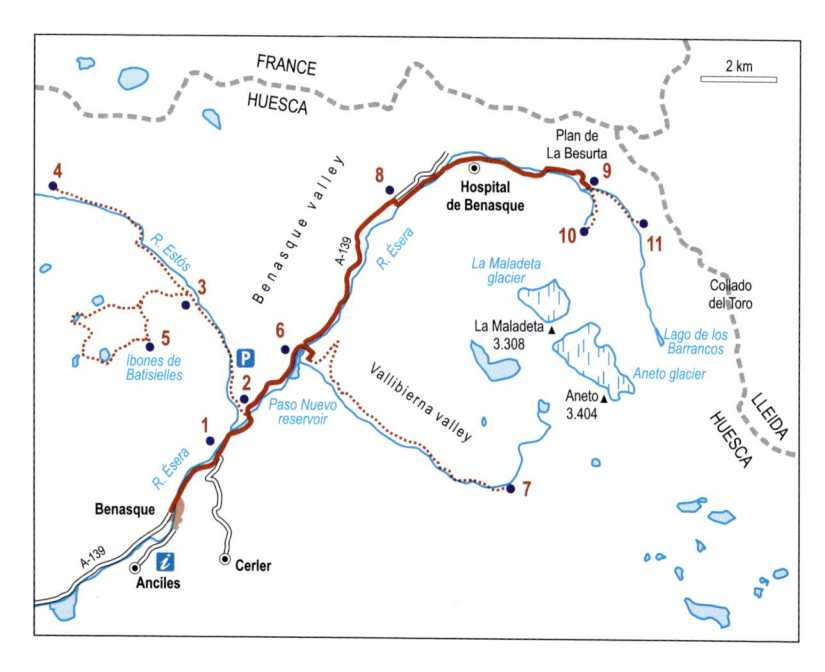

ORNITHOLOGICAL INTEREST

The valley is home to a small Lammergeier population and, above all on north-facing slopes above 2,000 m, a few Ptarmigan. Also up high, you should look out for Alpine Accentor, Ring Ouzel, Wallcreeper, Alpine Chough and Snow Finch, while the forests – in particular the subalpine mountain pine forests – are one of the main redoubts in Aragon of Black Woodpecker, Capercaillie and Tengmalm's Owl.

ITINERARIES

1. 14 km

Head up the valley from Benasque along the A-139 and at km point 99, check the cliffs on the left (1) for Lammergeier, Golden Eagle and Peregrine Falcon. Just after crossing the river Ésera at km point 100, take a track on your left (2) that skirts a camp-site and enters the Estós valley. Park after 500 m and begin the walk up the valley.

The track heads up the left-bank of the river Estós, passes by a small reservoir and reaches the Santa Ana mountain hut (3). The next stretch of the track reaches the Tormo hut and ends up at the permanently staffed Estós hut (4). Another possibility is, once past the Santa Ana hut, the path on the left that climbs steeply and with some difficult sections up to the Ibones de Batisielles (5).

This route passes through dense mountain pine forest inhabited by Capercaillie and Tengmalm's Owl, although both are exceedingly difficult to find. Easier to see are Woodcock, Black Woodpecker, Ring Ouzel, Marsh Tit, Citril Finch, Common Crossbill and Bullfinch.

2. 12 km

Once past the Paso Nuevo reservoir on the A-139 from Benasque, turn right (6) over the river Ésera into the

ARAGON

small flat area known as Plan de Senarta, where you can park. Continue along the track, first through a European silver fir forest with Black Woodpecker, and then up into the majestic Vallibierna valley. After around 4 hours of walking you reach Puente de Coronas, a bridge and a fisherman's refuge (7). This is a good site for Tengmalm's Owl, detectable above all calling at night, and Capercaillie.

3. 5 km

Continuing up the A-139, 11 km from Benasque take an asphalted track off to the right (8) that will take you past Hospital de Benasque (an old traveller's hostal converted into a hotel) and on to the plains of La Besurta (9). In summer this track is closed off and to reach La Besurta either walk along the path or take the special bus service. Fitter people can climb up from La Besurta to the staffed hut at La Renclusa (10), the classic starting point for climbing Aneto. Another option at La Besurta is take a path more to the left to the waterfall of Aigualluts (11) and then from here on to Collado del Toro and Lago de Los Barrancos.

These climbs will take you up to alpine habitats over 2,000 m and above the tree line, where typical bird species include Grey Partridge, Alpine Accentor, Alpine Chough, Snow Finch and, with luck, Ptarmigan.

MOST REPRESENTATIVE SPECIES

Residents: Lammergeier, Griffon Vulture, Goshawk, Golden Eagle, Peregrine Falcon, Ptarmigan, Capercaillie, Grey Partridge, Tengmalm's Owl, Black Woodpecker, Dipper, Alpine Accentor, Northern Wheatear, Marsh and Crested Tits, Alpine and Red-billed Choughs, Raven, Snow and Citril Finches, Common Crossbill, Bullfinch, Yellowhammer.

Summer visitors: European Bee-eater, Woodcock, Alpine Swift, Water Pipit, Whinchat, Ring Ouzel, Red-backed Shrike.

PRACTICAL INFORMATION

- **Access:** from Zaragoza, Huesca or Lleida take the N-240 to Barbastro. Continue up the Ésera valley through Graus, past Castejón de Sos and on to Benasque.
- **Accommodation:** in Benasque and Cerler. Also a number of camp-sites and mountain huts.
- **Visitor facilities:** 1 km out of Benasque on the way to Anciles, take the first left to the Natural Park interpretation centre (tel. 974 552 066). In summer, open every morning and afternoon. During the rest of the year, only weekends and public holidays. There are also interpretation centres in the villages of Aneto and Eriste.
- **Visiting tips:** Best at the end of spring and beginning of summer. Between November and May, rain, snow and ice make walking the valleys difficult and climbing towards the peaks dangerous.
- **Recommendations:** In Eresue, 10 km from Benasque, Casa Mariano (tel. 974 553 034, casamariano@imaginapuntocom.com; rural accommodation for rent) works with SEO/BirdLife and organises activities for naturalists.
- **Further information:** maps 1:50,000, n°. 148 and 180 (SGE).

19. ORDESA

High up in the north of Huesca province rises Monte Perdido (3,355 m), the highest limestone mountain in western Europe. Southwards from the summits run two grandiose glaciated valleys in the form of two arms – Ordesa and Pineta – and two deep rivers gorges – Añisclo and Escuaín. Elsewhere there is a profusion of caves, caverns, swallow-holes, gullies and cliffs with vertical faces of over 300 m, as well as a succession of habitats that change with altitude: holm and Lusitanian oaks give way to mixed deciduous and Scots pine woodland, then to mountain pine and finally to alpine pastures and bare rocks

In 2000 the last *bucardo* (the Pyrenean subspecies of the Spanish Ibex) died, thereby dashing any hopes of saving the subspecies for which the Ordesa y Monte Perdido National Park had been created in 1918. Today the park covers 15,000 ha and is also a Biosphere Reserve and an SPA.

ORNITHOLOGICAL INTEREST

Alpine Accentor, Wallcreeper, Snow Finch and other such species at home above the tree line are abundant in Ordesa; unfortunately the same cannot be said about the Ptarmigan, declining and with no breeding records in recent years. Lower down, a great variety of species including Tengmalm's Owl and the here abundant Black Woodpecker live in the forests; Capercaillies, on the other hand, are scarce and do not actually breed in the park. Lammergeiers, the Pyrenean bird *par excellence*, are frequently seen throughout.

ITINERARIES

1. 20 km

This itinerary up the Ordesa canyon consists of a longish walk of around 7 hours that is best begun early in the day. From the car-park (1), climb the steep and tiring Senda de Los Cazadores (2) through beeches and to

Ordesa and Monte Perdido National Park © Pedro Retamar

ARAGON

the viewpoint at Calcilarruego (3), from where the path contours smoothly along La Faja de Pelay to the Circo de Soaso and the famous waterfall of Cola de Caballo (4). Return via Las Gradas de Soaso and the thick beech forests of the valley bottom.

In the forests and above all in the mixed stands of beech and fir Black Woodpeckers are abundant, while in the pure mountain pine forests of La Faja de Pelay, Ring Ouzel, Eurasian Treecreeper, Citril Finch and Common Crossbill are common. Around the cliffs above the tree line, look for Alpine Swift, both Alpine and Redbilled Choughs, Golden Eagle and Lammergeier. Other alpine species such as Alpine Accentor, Wallcreeper and Snow Finch are possible if you decide to continue the walk as far as the Góriz mountain hut (6) and spend the night there.

2. 2 km

This short stroll takes you to a number of stunning viewpoints overlooking the vertical cliffs of the Escuaín gorge and the river Yaga. Begin at the curve just before the semi-abandoned village of Revilla and walk as far as a spectacular cliff (7) at the eastern limit of the National Park, one of the best places in Europe at any time of the year to see Lammergeier.

3. 10 km

This itinerary is difficult and not recommendable for those without experience of walking in mountain areas. It consists of a whole day's walking with a climb of 1,200 m up to alpine habitats with Alpine Accentor, Wallcreeper, Alpine Chough and Snow Finch, although by camping at Balcón de Pineta it can be turned into a less strenuous two-day excursion.

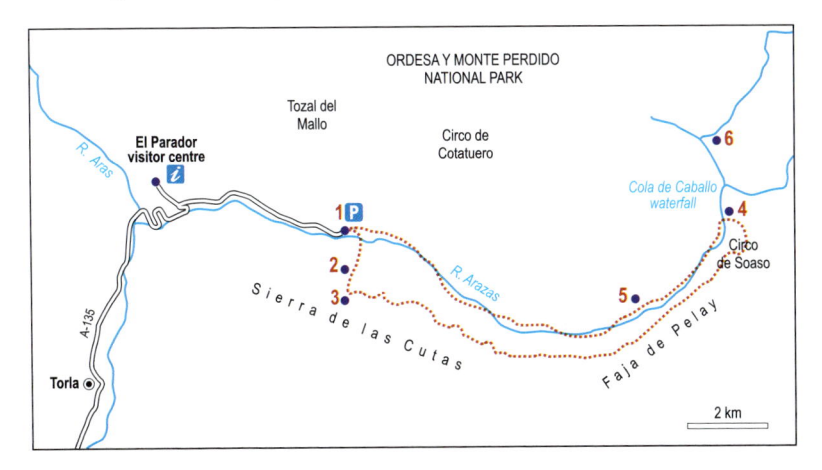

the valley. In the beechwoods at the beginning watch out for Black Woodpecker. After crossing the river, turn left (9) on to the long steep path up to Circo de Pineta. From here climb Las Cascadas de Cinca (10) and on to Balcón de Pileta at 2,500 m in a true alpine landscape. The last part of the climb is easier and ends at the frozen lake of Tucarroya or Marboré (11).

MOST REPRESENTATIVE SPECIES

From Bielsa take the road along the glaciated valley of Pineta towards the Parador of Monte Perdido. Start the walk just before the sanctuary of Nuestra Señora de Pineta (8) along the track that heads towards the head of

Residents: Lammergeier, Griffon Vulture, Goshawk, Golden Eagle, Common Kestrel, Peregrine Falcon, Grey Partridge, Black Woodpecker, Grey Wagtail, Dipper, Alpine Accentor, Dunnock, Black Redstart, Northern Wheatear, Mistle Thrush, Goldcrest, Marsh Tit, Wallcreeper, Nuthatch, Eurasian Treecreeper, Alpine and Red-billed Choughs, Raven,

PRACTICAL INFORMATION

- **Access:** from Sabiñánigo and then Biescas along the N-260 to Torla. From here it is 8 km by road to the National Park car-park (itinerary 1); at Easter and in summer this road is closed and access is by bus. From Ainsa, the A-138 follows the river Cinca. After passing Hospital de Tella, turn left to Tella on a steep narrow road to reach Revilla (itinerary 2). Continue along the A-138 for Bielsa (itinerary 3).
- **Accommodation:** in Torla and nearby towns. Within the National Park the only place to spend the night is the Góriz mountain hut (tel. 974 341 201; reservation essential).
- **Visitor facilities:** National Park information centre in Torla (tel. 974 486 472. pnomp.torla@terra.es). Open all year from Monday to Friday in the morning. The visitor centre at El Parador lies 5 km from Torla towards the Ordesa valley up a turning off to the left; open all day, morning and afternoon from Easter to the end of October.
- **Visiting tips:** best in June and July. The number of tourists during the rest of the summer makes birdwatching difficult. From November to May the climate is adverse and prevents easy access to higher habitats.
- **Further information:** maps 1:50,000, n°. 146, 178 and 179 (SGE). The book *Parque Nacional de Ordesa y Monte Perdido. Atlas de las aves*, by Kees Woutersen and Manolo Grasa (KW, 2002) is highly recommendable. Also check www.mma.es/parques/ordesa.

ARAGON

Barranco de Mascan, Sierra de Guara © Anna Motis

Snow and Citril Finches, Common Crossbill, Bullfinch, Rock Bunting, Yellowhammer.
Summer visitors: European Bee-eater, Egyptian Vulture, Short-toed Eagle, Alpine Swift, Crag Martin, Water Pipit, Rufous-tailed Rock Thrush, Ring Ouzel, Red-backed Shrike.

20. SIERRA DE GUARA

North-east of the city of Huesca lies the Sierra de Guara, the highest of all the pre-Pyrenean massifs. Its limestone ridges have been carved asunder by powerful rivers and Guara boasts one of Europe's most spectacular groups of sheer cliffs and deep river gorges.

This area of abrupt mid-altitude mountains, set between the central Pyrenees and the semi-arid Ebro Depression, is remarkable for the sharp contrasts to be found between its north- and south-facing slopes. On the former, Scots and, to a lesser extent, mountain pine forests abound, along with a little Silver Fir near its southern-most limit of its European distribution, while the latter's warmer slopes are covered by holm oak and scrub. Almost 50,000 ha are covered by a natural park and an SPA.

ORNITHOLOGICAL INTEREST

Guara holds an important concentration of around 10 pairs of Lammergeiers and over 200 pairs of Griffon Vultures, the largest population in Aragon. Other raptors include Egyptian Vulture, Golden and Bonelli's Eagle (just 1 pair of the latter), Peregrine Falcon and Eagle Owl. Rock-loving species – for example, Black Wheatear at its northern-most site in the Iberian Peninsula – abound and there are also intrusions of more Eurosiberian species such as Black Woodpecker into the site's forests. High mountain species such as Alpine Accentor and Wallcreeper take refuge in the area in winter.

ITINERARIES

1. 12 km 🚗 🚶 ☀ ➡

Salto de Roldán in the west of the Sierra de Guara consists of two great conglomerate cliffs, San Miguel and Amán, which both plunge vertically for over 400 m down to the river Flumen. Take the local road from Huesca towards Apiés and after 5 km reach the turn-off to the village of Fornillos (1), a good place for looking for the open-country species such as Black and Red Kites and Montagu's Harrier that hunt the scrub, olive groves and cereal fields around here.

Between Apiés and Sabayés, take a track off right (2) and after 4.5 km

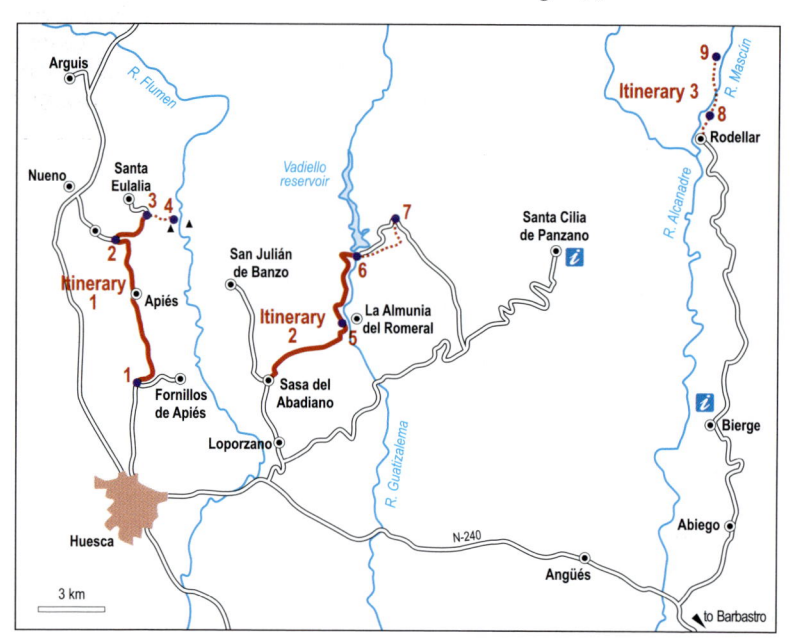

ARAGON

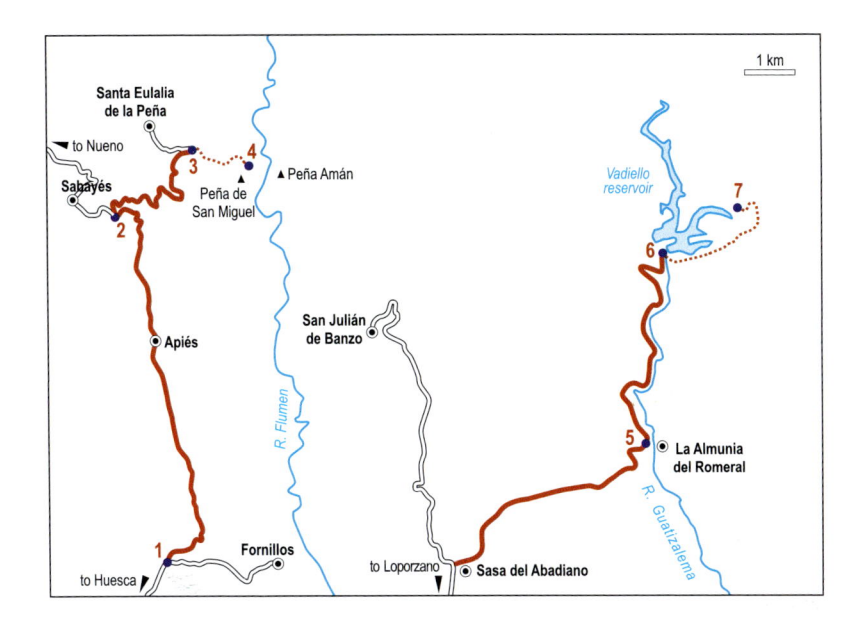

(3), instead of continuing towards Santa Eulalia de la Peña, park and continue on foot to an open area (4). From here you can approach the edge of the cliffs or take a path that climbs to the flat summit of San Miguel, taking care on the section that ascends via a metal ladder.

This is an unbeatable site for viewing all the local raptors, including Lammergeier and with luck the scarce Bonelli's Eagle. Also here look out for Black Wheatear, Rufous-tailed and Blue Rock Thrushes and Red-billed Chough.

2. 14 km

Further east in the valley of the river Guatizalema head for the reservoir of Vadiello from the city of Huesca via the village of Loporzano. After turning right in Sasa del Abadiano it is worth stopping at La Almunia del Romeral (5) just before entering the river gorge to look for raptors, other rock-loving species, warblers and buntings.

After crossing the dam (6), park and begin to walk along a forest track to the left that runs around the whole eastern side of the reservoir as far as the Conjuradero de la Cruz Cubierta; from here continue along a path to the sanctuary of Sant Cosme and San Damián (7). As from the dam, there are superb views here of the cliffs frequented by Lammergeier, Egyptian and Griffon Vultures, Golden Eagle,

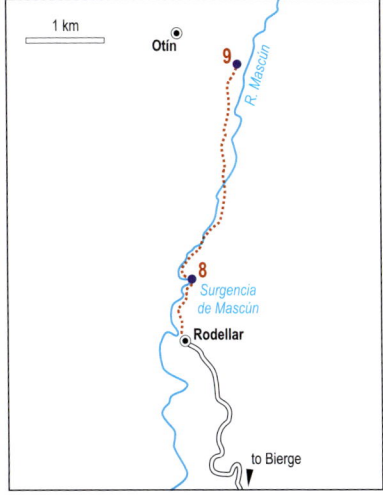

PRACTICAL INFORMATION

- **Access:** for itinerary 1, in Huesca head for Apiés. For the other two itineraries, take the N-240 from Huesca towards Barbastro. After 7 km, turn left to Loporzano for itinerary 2 or turn left after 29 km towards Abiego, Bierge and Rodellar for itinerary 3.
- **Accommodation:** in Huesca, Barbastro and in some of the smaller villages such as Alquézar.
- **Visitor facilities:** the Sierra de Guara Natural Park has a visitor centre in Bierge (tel. 974 318 238). Open every day, morning and afternoon from mid-June to September; the rest of the year only weekends and public holidays. La Casa de los Buitres (tel. 974 373 217, sargantana@sargantana.info) in the upper part of Santa Cilia de Panzano is near an observation point situated in a church belltower, from where there are views of a Lammergeier feeding station.
- **Visiting tips:** spring is best. Summer is too hot and busy. Winters are less cold than in the nearby Pyrenees and a number of alpine species spend the winter in Guara.
- **Recommendations:** just 10 km from Huesca in the village of Loporzano is Casa Boletas (tel. 974 262 027), jjsv@boletas.org), a rural guesthouse designed with ornithologists in mind and much frequented by European birdwatchers.
- **Further information:** maps 1:50,000, n°. 248 and 249 (SGE).

Peregrine Falcon, Eagle Owl, Alpine Swift and Crag Martin. Halfway along this walk look for forest birds in an interesting patch of oaks.

3. 5 km

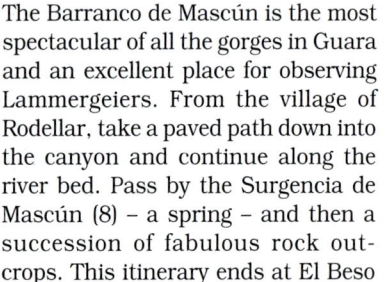

The Barranco de Mascún is the most spectacular of all the gorges in Guara and an excellent place for observing Lammergeiers. From the village of Rodellar, take a paved path down into the canyon and continue along the river bed. Pass by the Surgencia de Mascún (8) – a spring – and then a succession of fabulous rock outcrops. This itinerary ends at El Beso (9), a natural rock bridge beyond which access along the river bed becomes difficult.

MOST REPRESENTATIVE SPECIES

Residents: Red Kite, Lammergeier, Griffon Vulture, Golden and Bonelli's Eagles, Montagu's Harrier, Peregrine Falcon, Eagle Owl, Crag Martin, Dipper, Black Wheatear, Blue Rock Thrush, Dartford and Sardinian Warblers, Red-billed Chough.

Summer visitors: European Bee-eater, Black Kite, Egyptian Vulture, Short-toed and Booted Eagles, Great Spotted Cuckoo, Alpine Swift, Wryneck, Red-rumped Swallow, Black-eared Wheatear, Rufous-tailed Rock Thrush, Subalpine, Orphean and Bonelli's Warblers, Golden Oriole, Red-backed Shrike.

Winter visitors: Merlin, Alpine Accentor, Wallcreeper, Citril Finch, Bullfinch, Common Crossbill, Hawfinch.

On migration: Osprey, Ring Ouzel.

21. SAN JUAN DE LA PEÑA

The pre-Pyrenean mountain of San Juan de la Peña lies in the north-west of Huesca province near the town of Jaca. It consists of a series of high conglomerate cliffs overlooking thickly wooded slopes (in the main, Scots

ARAGON

pine) that highlight the transition between Mediterranean ecosystems on the south face and more Atlantic systems on the north face, where beech, European silver fir and even mountain pine at altitude all thrive.

The mountain takes its name from the Cistercian monastery of San Juan de la Peña built in the twelfth century under an overhanging cliff. Subsequently, a much larger second monastery was built nearby in the seventeenth century. The whole area, historically very closely linked to the birth of the Kingdom of Aragon, was declared a Natural Site of National Interest in 1920 and then reclassified as a Natural Monument in 1998 (264 ha). Along with the nearby Peña Oroel, it is also an SPA.

ORNITHOLOGICAL INTEREST

The combination of lowland Mediterranean and more high-level habitats ensures that San Juan is home to a great variety of species. Perhaps the best known is the Lammergeier, which breeds here, although there are also Egyptian and Griffon Vultures, Golden Eagle and Peregrine Falcon. The forests support Honey Buzzard and Short-toed and Booted Eagles, although possibly the most significant species in these habitats is the Black Woodpecker, here present in good numbers. Rufous-tailed Rock Thrush breed in rocky areas and wintering birds include Alpine Accentor, Ring Ouzel and Wallcreeper.

ITINERARIES

1. 5 km

This walk, with only a little uphill, skirts the southern cliffs of San Juan de la Peña in part outside the limits of the Natural Monument. Start walking in the meadows of San Indalecio (1) next to the Monasterio Nuevo (new monastery) along the narrow tarmac road that leads towards the Monasterio Viejo (old monastery). Turn left almost immediately along another narrow road closed to non-authorised traffic that takes you as far as a television repeater station (2). From here walk along the cliff edge as far as the little chapel of Tozal de San Salvador (3), from where

San Juan de la Peña © José María Cereza

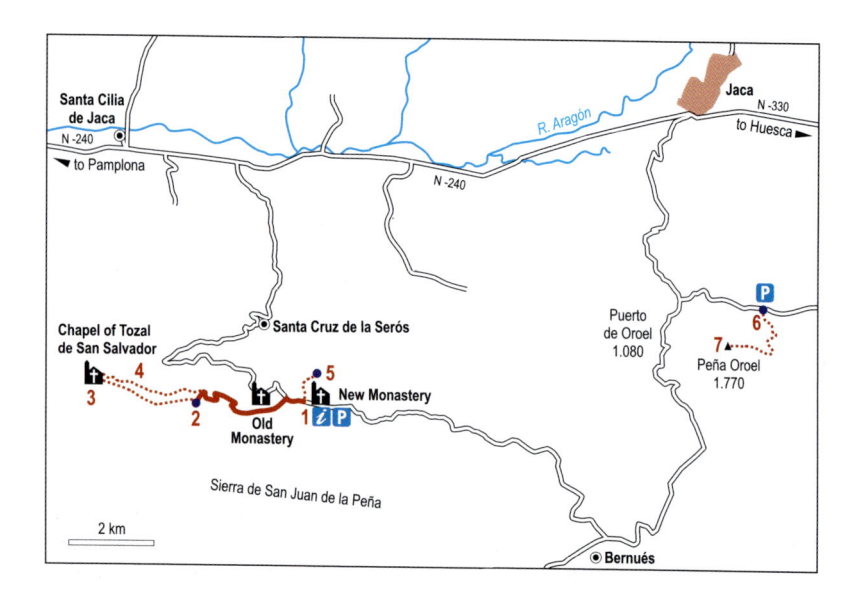

there are excellent views of the western Pyrenees.

Around the cliffs look for Alpine Swift, Black Redstart, Rufous-tailed Rock Thrush and Red-billed Chough, as well as Lammergeier and other raptors.

Return along a path that enters the fir forest (4), a good area for Black Woodpecker and other forest species such as Bonelli's Warbler, Firecrest, Goldcrest, Coal and Crested Tits, Eurasian Nuthatch, Citril Finch, Common Crossbill and Bullfinch.

2. 0,5 km

The Balcón de los Pirineos (5) is a spectacular viewpoint on the north face of the mountain, accessible along a short signposted path heading north from the new monastery. Pass through a Scots pine forest with the same forest birds as itinerary 1 (Black Woodpecker equally abundant, as far as the viewpoint, where Griffon Vultures are common and Lammergeier a possibility.

PRACTICAL INFORMATION

- **Access:** from Huesca take the N-330 towards Jaca and then the N-240 towards Pamplona. After 9 km, turn left towards the village of Santa Cruz de la Serós, from where a road leads up to the monasteries of San Juan de la Peña (itineraries 1 and 2). Return to Jaca via Puerto de Oroel, the beginning of itinerary 3.
- **Accommodation:** in Jaca or a hostal in Santa Cruz de la Serós.
- **Visitor facilities:** the interpretation centre for San Juan (tel. 974 361 476) is in the meadows at San Indalecio next to the Monasterio Nuevo. Open every day in summer, morning and afternoon. The rest of the year, only weekends and public holidays.
- **Visiting tips:** best in late spring. Many tourists arrive at weekends and in summer and so it is best to begin the routes early in the morning.
- **Further information:** map 1:50,000, n°. 176 (SGE).

ARAGON

3. 6 km 🚶 ☀️ ➡️

Once over the Puerto de Oroel (a mountain pass) on the way to Jaca, a road on the right takes you to the viewpoint at Oroel (6), with bar, restaurant and car-park. From here a path ascends to the summit of Peña Oroel through the humid, north-facing Scots pine and eventually silver fir forests. Once above the trees and out on the main ridge, a path heads west to the huge iron cross on the summit (7). Similar forest birds as around San Juan can be found in the forests here, whilst Griffon Vulture and possibly Golden Eagle and Lammergeier can be seen from the summit.

MOST REPRESENTATIVE SPECIES

Residents: Lammergeier, Griffon Vulture, Goshawk, Sparrowhawk, Golden Eagle, Peregrine Falcon, Black Woodpecker, Crag Martin, Goldcrest, Firecrest, Coal and Crested Tits, Eurasian Nuthatch, Red-billed Chough, Citril Finch, Common Crossbill, Bullfinch.
Summer visitors: European Bee-eater, Egyptian Vulture, Short-toed and Booted Eagles, Alpine Swift, Rufous-tailed Rock Thrush, Bonelli's Warbler.
Winter visitors: Alpine Accentor, Ring Ouzel, Wallcreeper.

22. RIGLOS

A series of enormous orange-coloured conglomerate pinnacles rise vertically skywards in the west of the province of Huesca. Known locally as *riglos*, these cliffs – one of the most photographed landscapes in Aragon – are an outlier of the pre-Pyrenees and overlook the beginning of the Ebro Depression. The most spectacular of all the *riglos* are found just behind the village of Riglos, although others also exist

Mallos de Riglos © Fernando Barrio

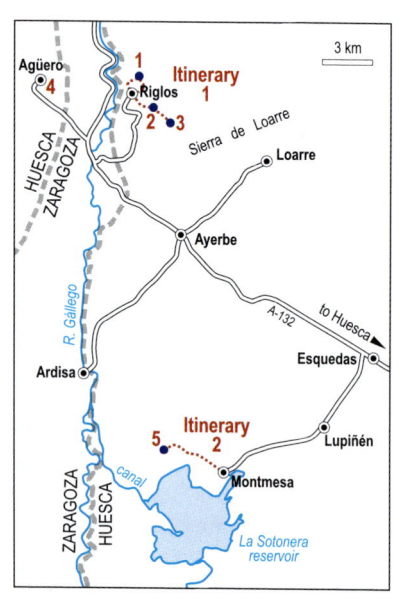

above the village of Agüero and in other areas of the area of Hoya de Huesca.

The vegetation at the base of the *riglos* is dominated by low holm and holly oak scrub and, at slightly greater altitude, by Scots pine and holm oak woodland. Nearby rivers, reservoirs and agriculture provide a touch of variety. This is one of the most popular sites in Aragon with climbers; the *riglos* lack any legal protection other than that provided by the Sierra de Santo Domingo-Caballera and Río Onsella SPA.

ORNITHOLOGICAL INTEREST

This is a good site for rock-loving birds; one pair of Lammergeiers breeds in the area, as do Egyptian and Griffon Vultures, Peregrine Falcon, Alpine Swift, Crag Martin, Blue Rock Thrush and Rock Thrush. Somewhat scarcer is the Black Wheatear, a typical bird in the south and east of the Iberian Peninsula, which here is all but at the northern limit of its restricted world range. In winter Alpine Accentor and Wall-creeper arrive from the high Pyrenees; the scrub at the base of the cliffs is home to a number of species of warbler and bunting.

ITINERARIES

1. 6 km

Leave the village of Riglos on a path heading west towards the southern and most abrupt part of the cliffs. Enter a rock amphitheatre (1), a good place for the area's cliff species and for Lammergeier. With patience and by scanning the whole area of cliffs, look for Black Wheatear and in winter Wallcreeper.

From here, head right along the base of the cliffs towards the village church, from where a track takes you into an area of allotments (2) and then up to the base of Las Buitreras (3), a huge cliff left untouched by climbers where both Griffon and Egyptian Vultures breed. Lammergeiers also appear here from time to time. The scrub here is frequented in winter by Redwing and sometimes Ring Ouzel and Fieldfare. Other, less well-known cliffs are accessible from the nearby village of Agüero (4) and harbour the same birds as Riglos.

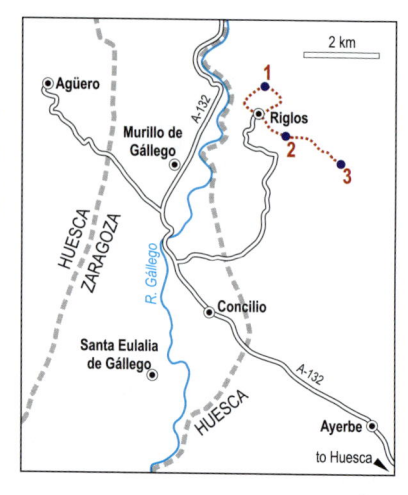

ARAGON

> **PRACTICAL INFORMATION**
>
> - **Access:** from Huesca take the A-132 towards Pamplona and after passing through Ayerbe, turn right (between km points 247 and 248) to the village of Riglos or, 2 km further on, left to Agüero. Returning towards Huesca, after Ayerbe a road on the right leads to Lupiñén and Montmesa and then to La Sotonera.
> - **Accommodation:** in Huesca, Agüero, Ayerbe or Murillo de Gállego.
> - **Visitor facilities:** currently no visitor centre or infrastructure for birdwatchers, although a centre for the observation of cliff-breeding birds with cameras placed in the vulture colonies is planned for Riglos. On the banks of La Sotonera reservoir there is a hide.
> - **Visiting tips:** April and May are the best months to visit Riglos. Winter is also interesting for Pyrenean species wintering at lower altitudes. In summer and at weekends, many climbers visit the area.
> - **Further information:** maps 1:50,000, n°. 209, 247 and 285 (SGE).

2. 2 km

Between the end of February and the beginning of March it is worth visiting the nearby Sotonera reservoir in the Hoya de Huesca, just 40 km from Riglos. It is fed by a canal connected to the river Gállego and, aside from ducks and other common wintering birds, this is a good place for spectacular viewing of large groups of Common Crane on pre-breeding migration.

Take the A-132 towards Huesca and, once through Ayerbe, turn right towards Lupiñén and Montmesa. From the second of these villages, a track heads north to an observation point (5) that is ideal for viewing the reservoir.

MOST REPRESENTATIVE SPECIES

Residents: Red Kite, Lammergeier, Griffon Vulture, Goshawk, Sparrowhawk, Golden Eagle, Peregrine Falcon, Rock Dove, Calandra Lark, Black Redstart, Black Wheatear, Blue Rock Thrush, Sardinian Warbler, Red-billed Chough, Raven, Rock Sparrow, Rock and Cirl Buntings.
Summer visitors: Black Kite, Egyptian Vulture, Short-toed and Booted Eagles, Alpine Swift, Crag Martin, Black-eared Wheatear, Dartford, Subalpine and Orphean Warblers, Woodchat Shrike.
Winter visitors: Alpine Accentor, Ring Ouzel, Fieldfare, Redwing, Wallcreeper.
On migration: Common Crane.

23. MONEGROS

The Monegros, one of the best-known areas of steppes in western Europe, lie halfway between the cities of Huesca and Zaragoza in the central and most arid sector of the great Ebro Depression. In the north the Sierra de Alcubierre divides the mosaic of habitats – plains, low hills, gullies carved out of gypsum soils, cereal fields, fallow, scrub and grasslands – into two sectors, north and south. To add further interest, there are relict patches of juniper forest and groups of shallow saline lagoons known as *saladas*.

Over the last few years the continual expansion of irrigation schemes has radically transformed the area, although an important part of the ornithologically most valuable sites (over 100,000 ha) has been saved by the protection afforded by four SPAs.

ORNITHOLOGICAL INTEREST

One of most important European sites for sandgrouse and larks, the Monegros boast some of the best Spanish populations of Pin-tailed Sandgrouse and Dupont's and Lesser Short-toed Larks, as well as good numbers of Stone Curlew, Black-bellied Sandgrouse and Little Bustard. A few Great Bustards still survive and there are breeding colonies of Lesser Kestrel. Golden Eagles visit to hunt and many other raptors such as Egyptian Vulture breed in the Sierra de Alcubierre, also home to other cliff-loving and forest bird species. When the *saladas* flood in spring and winter, there can be good numbers of passage waders and wintering duck.

ITINERARIES

1. 80 km

The Saso de Osera in the western sector of Los Monegros is the best conserved area of steppe habitat in the area and home to good numbers of Dupont's Lark. Much of the area is private and entry is forbidden, although you can walk up to the church of San Martín (1) along a track on the left, 3 km out of Osera on the road to Monegrillo.

This road crosses a surprising landscape of plains and low, flat-topped hills, cereal fields, juniper woods and patches of steppe grassland on the gypsum soils, where you should look for Stone Curlew, sandgrouse, Lesser Short-toed Lark and raptors such as Montagu's Harrier and Golden Eagle.

This itinerary continues on to San Caprasio (2), the summit of the Sierra de Alcubierre (811 m). From Monegrillo take the road heading

Monegros © Pedro Retamar

ARAGON

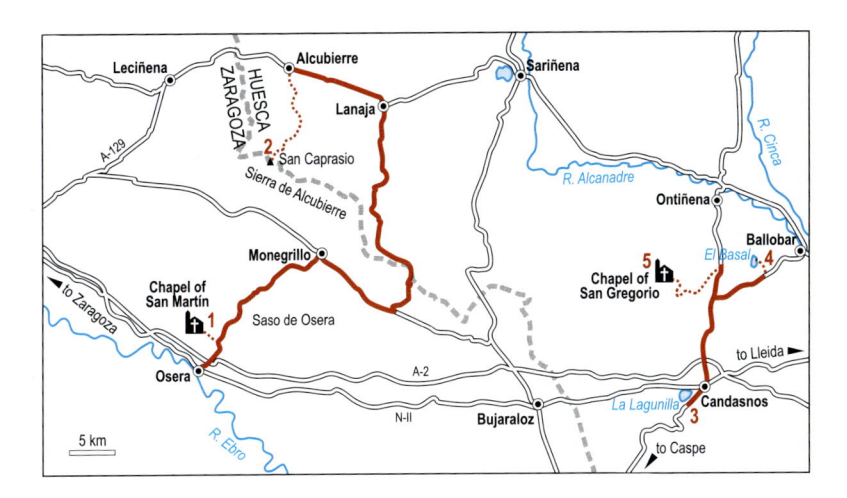

south-east and circumnavigate the whole ridge, passing through Lanaja and ending up at the village of Alcubierre itself. From here a track of 17 km heads up to San Caprasio through the north face of the ridge in an area rich in raptors: Egyptian Vulture, Golden Eagle (here breeding in trees), Booted Eagle and the here rare Eagle Owl. On the cliffs of the ridge, expect Black Wheatear, Red-billed Chough and forest birds in the pine forests.

2. `35 km`

The magnificent cereal steppes of the eastern Monegros can be tackled from the town of Candasnos. First, however, it is worth having a quick look on the outskirts of the town at La Lagunilla (3), a former *salada* transformed into a freshwater lagoon and fed by the run-off from local irrigation projects. Black-necked Grebe and Marsh Harrier breed here and in winter there is a good variety of duck.

Just 9 km out of Candasnos towards Ontiñena, turn right towards Ballobar. Then, after 6.5 km and opposite a cement cereal silo, turn left along a track that will take you in 1.5 km to El Basal (4), a normally dry lagoon covered in natural steppe vegetation with Lesser Short-toed and Dupont's Larks (very rare). If it has rained, you might find waders in spring and autumn and ducks in winter.

Return to the Candasnos-Ontiñena road and around 3 km on towards Ontiñena a left turn will take you to the church of San Gregorio (5) in the eastern part of the Sierra de Alcubierre. Along the 7 km of track, you pass through an interesting cereal steppe dotted with olive and almond groves and vines where Montagu's Harrier, Lesser Kestrel, Little Bustard, Stone Curlew, both sandgrouse and Roller breed.

MOST REPRESENTATIVE SPECIES

Residents: Black-necked Grebe, Red Kite, Marsh Harrier, Golden Eagle, Peregrine Falcon, Little Bustard, Stone Curlew, Pin-tailed and Black-bellied Sandgrouse, Eagle and Little Owls, Dupont's, Calandra, Lesser Short-toed and Thekla Larks, Black Wheatear, Mistle Thrush, Dartford Warbler, Red-billed Chough, Com-

PRACTICAL INFORMATION

- **Access:** from Zaragoza along the motorway (A-2) towards Lleida. Leave at Alfajarín and take the N-II to Osera (itinerary 1) and then to Ballobar (itinerary 2), in all 90 km from Zaragoza.
- **Accommodation:** in Zaragoza, but also in Bujaraloz, Sariñena, Fraga and other nearby towns.
- **Visitor facilities:** no visitor centre or infrastructure for birdwatchers.
- **Visiting tips:** spring is ideal for steppe birds. Begin the itineraries early in the morning; after heavy rain some of the tracks can become very slippery. Do not leave the public tracks so as to avoid conflicts with owners and farmers; few petrol stations in the area.
- **Recommendations:** near the towns of Sástago and Chiprana and, above all, Alcañiz there are numerous ornithologically very interesting *saladas* (temporary saline lagoons).
- **Further information:** maps 1:50,000, n°. 356, 385, 386 and 387 (SGE). The book *Ecología de Los Monegros*, edited by Cesar Pedrocchi (Instituto de Estudios Altoaragoneses, 1998), is highly recommendable.

mon Crossbill, Rock and Cirl Buntings.
Summer visitors: Egyptian Vulture, Short-toed and Booted Eagles, Montagu's Harrier, Lesser Kestrel, Roller, Short-toed Lark, Tawny Pipit, Northern and Black-eared Wheatear, Orphean and Spectacled Warblers.
Winter visitors: Hen Harrier, Merlin.

24. BELCHITE STEPPES

Not far from the city of Zaragoza and near the town of Belchite lies one of the best-preserved areas of steppe in the Ebro Depression. A landscape of great beauty awaits the visitor, with vast plains interrupted by *muelas* (flat-topped table mountains) and

Belchite steppes © Gabriel Sierra

ARAGON

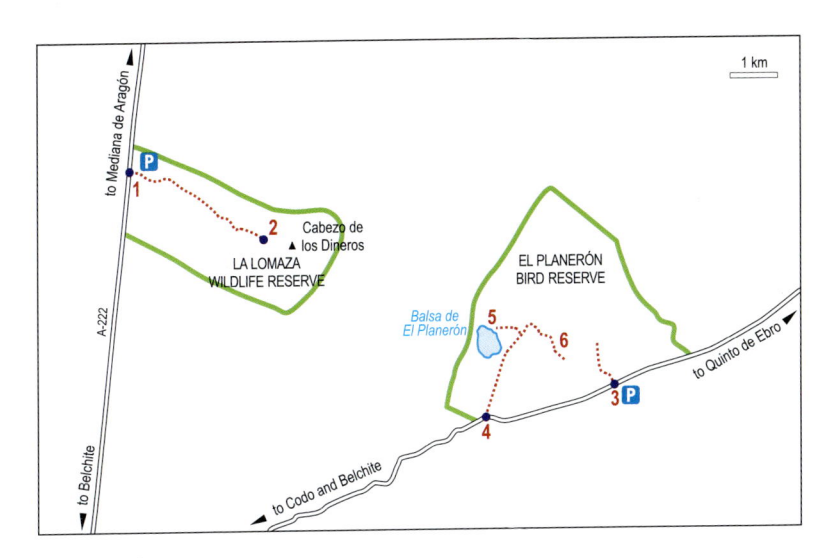

cabezos (peaks) decorated with red clayey and saline soils and topped off with contrasting white-coloured gypsum strata. Land-use varies from cereal cultivation and fallow, to scrub and steppe grasslands.

Two areas of these steppes are protected: La Lomaza (almost 1,000 ha), a *refugio de fauna silvestre* run by the Aragonese government, and El Planerón (600 ha), a *reserva ornitológica* bought by SEO/BirdLife with the help of donations from members and a number of private and public bodies. Both areas are part of an SPA.

ORNITHOLOGICAL INTEREST

This is a highly important area for steppe birds and the breeding populations of Pin-tailed Sandgrouse and Dupont's and Lesser Short-toed Larks are some of the most important in Europe. Also, Egyptian Vulture, Golden Eagle, Peregrine Falcon and Eagle Owl breed on the cliffs and there is an interesting passage of waders and groups of wintering duck when rains flood some of the inwardly draining depressions. Eurasian Dotterel regularly stop over on migration.

ITINERARIES

1. 4 km

Some 10.5 km from Belchite on the right of the A-222 as you head towards Zaragoza park in the car-park at km point 11.7 (1). From here a track heads into La Lomaza along the length of a gypsum *meseta* covered in low scrub. The route ends at a spectacular natural viewpoint (2) known as Cabezo de los Dineros. This is a magnificent site for Dupont's Lark, above all in spring when they can be heard particularly well at daybreak, Stone Curlew and both sandgrouse.

2. 4 km

El Planerón consists of a broad depression with highly saline soils covered by a mosaic of natural steppe vegetation and partially abandoned fields. This is a good place for both sandgrouse and Lesser Short-toed, Thekla and, with luck, Dupont's Larks. Golden Eagles come to hunt and Lesser Kestrels breed in the abandoned farm buildings.

> **PRACTICAL INFORMATION**
>
> - **Access:** from Zaragoza take the N-232 for 20 km; between El Burgo de Ebro and Fuentes de Ebro, turn right towards Belchite on the A-222 for access to La Lomaza. For El Planerón, take the minor road between Belchite and Quinto de Ebro.
> - **Accommodation:** in Belchite, Fuendetodos, Quinto de Ebro, Fuentes de Ebro and, naturally, Zaragoza (50 km away).
> - **Visitor facilities:** the Ebro Steppe interpretation centre run by SEO/BirdLife is in the town of Belchite. Open weekends and public holidays in spring and autumn.
> - **Visiting tips:** spring is the best period of the year for finding steppe birds. Begin early in the morning, especially if you are looking for Dupont's Lark. Do not walk or drive off the tracks and marked paths. After rains, the tracks can be very slippery.
> - **Recommendations:** a permit is needed to enter La Lomaza and can be obtained in advance from the Aragonese Department of the Environment in Zaragoza (tel. 976 714 000).
> - **Further information:** map 1:50,000, n°. 412 (SGE). The Aragoneses delegation of SEO/BirdLife (tel. 976 373 308, aragon@seo.org) will answer queries relating to El Planerón and the itineraries herein described.

The first itinerary heads off from the road between Belchite and Quinto de Ebro. Some 9.5 km after passing through the village of Codo, there is a car-park (3) for the ornithological reserve from where a marked signpost takes you through a varied landscape of cultivated fields, fallow and scrub.

On the same road, but 6 km after Codo, a gravel track heads left (north) (4) towards the ornithological reserve. After 1.5 km, park and walk along a track to the left which will take you to the lagoon of El Planerón (5). When this temporary saline lagoon – today dammed off – has water, Marsh Harrier and passage waders, as well as other aquatic birds can be found here. Return to the gravel track and 200 m further on, turn right along a marked footpath (6) that passes through another interesting area of steppe.

MOST REPRESENTATIVE SPECIES

Residents: Griffon Vulture, Marsh Harrier, Golden Eagle, Common Kestrel, Peregrine Falcon, Pin-tailed and Black-bellied Sandgrouse, Eagle and Little Owls, Dupont's, Calandra, Lesser Short-toed, Crested and Thekla Larks, Black Wheatear, Dartford Warbler.
Summer visitors: Egyptian Vulture, Montagu's Harrier, Lesser Kestrel, Stone Curlew, Short-toed Lark, Tawny Pipit, Northern and Black-eared Wheatear, Spectacled Warbler, Woodchat Shrike.
Winter visitors: Hen Harrier, Merlin.
On migration: Eurasian Dotterel, other waders if the lagoons have water.

25. LAGUNA DE SARIÑENA

Up until the 1970s, a large and ecologically singular saline lagoon, of similar characteristics to those that still survive in Los Monegros, existed next to the town of Sariñena (Huesca province). However, at the end of the 1970s irrigation systems were installed in the surrounding area to encourage rice, alfalfa and maize production and as a result the northern sector of the lagoon began to fill up with the irrigation run-off water. In all, the lagoon almost tripled in size and so in 1982 a canal had to be dug to allow excess water to drain off into the river Flumen.

Currently, the lagoon, lying between the basins of the rivers Flumen and Alcanadre, covers over 200 ha and has a diameter of over 2 km. The input of freshwater from the irrigation projects has all but eliminated the lagoon's natural salinity and its original singular characteristics have been lost. Nevertheless, it is still an interesting site for aquatic birds and is protected as a *Refugio de Fauna Silvestre* and an SPA.

ORNITHOLOGICAL INTEREST

In conjunction with the nearby paddy fields used as feeding sites by many birds, the Laguna de Sariñena (along with the Laguna de Gallocanta) is one of the most important wetlands in Aragon. Its heron communities are outstanding and include most Iberian species: there is a large Cattle Egret roost, many Purple Herons and lots of records of Great Bittern, including some from the breeding season. Red-crested Pochard, Marsh Harrier, waders, gulls and many passerines breed and this is one of the few sites in Aragon for Purple Swamp-hen.

Large numbers of Great Cormorant winter, along with many duck species and Greylag Geese. On migration, Garganey, Osprey, Common Crane and waders turn up.

ITINERARIES

1. 13 km

The lagoon can be circumnavigated by car, although the best tactic is to combine strategic stops with short walks. A good place to start is the interpretation centre (1), situated 1 km before the town of Sariñena on the A-129 from Zaragoza. A short track takes you down to the centre where more information about routes is on offer. Continuing in a clockwise direction, after passing the town pick up a perimeter track that follows the drain-

Laguna de Sariñena © Pedro Retamar

This is also one of the best places in Aragon for Marsh Harrier, Purple Heron and large numbers of wintering Great Cormorants, ducks, Eurasian Coots and gulls. The surrounding vegetation holds Reed and Great Reed Warblers and Penduline and, occasionally, Bearded Tits; Brambling and Reed Bunting also winter.

2. 30 km

It is a good idea to investigate the tracks that criss-cross the paddy fields around the villages of Capdesaso, San Lorenzo del Flumen and Alberuela de Tubo. Many of the species present on the lagoon come here to feed: look out for Little and Cattle Egrets and Black-winged Stilt, and on passage and in winter, Northern Lapwing, Ruff, Black-tailed Godwit and Spotted and Common Redshanks.

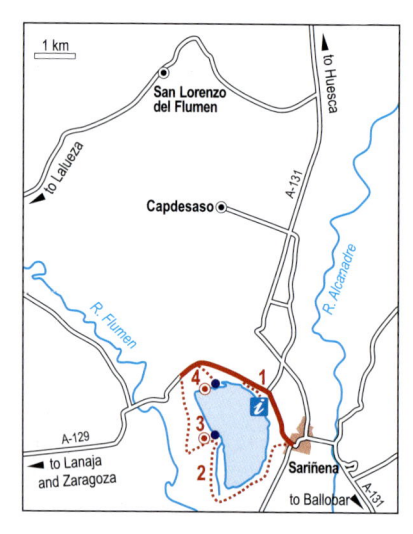

age canal (2) in the southern-most point of the lagoon and then reaches a hide (3) in the south-west corner. From here, continue along the west and north banks of the lagoon by car or on foot. Another alternative is to return to the interpretation centre and work in an anti-clockwise direction as far as a tower hide (4) in the northern sector of the lagoon.

The pride and joy of Sariñena is the Great Bittern, which can be heard more easily here between February and April than almost anywhere else in Spain. With luck you might catch it in flight over the large reed beds.

MOST REPRESENTATIVE SPECIES

Residents: Great Bittern, Cattle and Little Egrets, Red-crested Pochard, Marsh Harrier, Purple Swamp-hen, Black-headed and Yellow-legged Gulls, Bearded Tit.

PRACTICAL INFORMATION

- **Access:** the lagoon is roughly 70 km from Zaragoza along the A-120 and 50 km from Huesca along the A-131.
- **Accommodation:** Sariñena has a hotel and hostals.
- **Visitor facilities:** The interpretation centre is 1 km before Sariñena coming from Zaragoza. A track on the right leads towards the centre, open only weekends and public holidays. Otherwise, for more information or to visit at other times, call 607 849 963.
- **Visiting tips:** any time of year except July and August due to the heat and lack of bird activity. Avoid secondary dirt tracks, which can become extremely muddy after rain.
- **Further information:** maps 1:50,000, n°. 324, 325, 356 and 357 (SGE) and Sariñena Town Council (tel. 974 570 900).

ARAGON

Summer visitors: Little Bittern, Night and Purple Herons, Black-winged Stilt, Stone Curlew, Little Ringed Plover, Great Spotted Cuckoo, Reed and Great Reed Warblers, Penduline Tit.
Winter visitors: Great Cormorant, Greylag Goose, Shelduck, Common Teal, Northern Shoveler, Common Pochard, European Golden Plover, Northern Lapwing, Common Snipe, Brambling, Reed Bunting.
On migration: Squacco Heron, Great Egret, Eurasian Spoonbill, Garganey, Osprey, Common Crane, Ruff, Black-tailed Godwit, Eurasian Curlew, *Tringa* waders, Alpine Swift, hirundines.

26. GALACHOS DEL EBRO

The river Ebro begins to meander in its mid-course in the province of Zaragoza and over the years has created a number of ox-bow lakes, known locally as *galachos*. Despite being physically separate from the river, the *galachos* are still fed by subterranean filtrations from the Ebro, as well as by the aquifer and rainfall. Dense reed-beds have formed around these lakes and in parts they are lined by excellent patches of fluvial woodland.

A single *reserva natural*, downstream from Zaragoza, protects three of these *galachos*. La Alfranca, on the left-bank, is the largest and best preserved *galacho*; the *galachos* of La Cartuja and El Burgo de Ebro on the right-bank complete the protected area (also an SPA). Upstream from Zaragoza but outside the reserve, the Juslibol *galacho*, created by the great floods of 1961, hides amongst the trees on the river's left-bank.

ORNITHOLOGICAL INTEREST

The large reed-bed at La Alfranca boasts a colony of over 150 pairs of Night Heron, as well a few pairs of Purple Heron and other herons and Marsh Harrier. The birds of the fluvial woodland are also of interest and many passerines drop in on migration.

The low gypsum cliffs that line the north-bank of the river Ebro are good

Galacho de la Alfranca on the river Ebro © Fernando Barrio

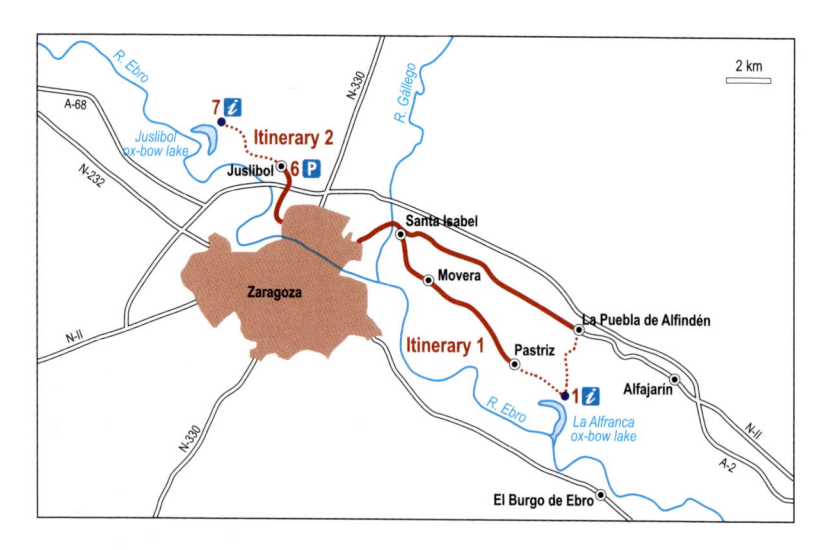

for breeding Egyptian Vulture, Common Kestrel, Peregrine Falcon and Eagle Owl, as well as rock-loving passerines.

ITINERARIES

1. 2 km 🚶 ☀ ➡

From the interpretation centre (1) a path leads past the *galacho* of La Alfranca: access to this restricted area and to the hide (2), from where there are excellent views of the Night heron colony in the reed-bed, is only possible with a monitor from the centre. Other birds visible here include Purple Heron, Little Bittern, Marsh Harrier and passerines such as Penduline Tit.

From here you return to the main path and gradually approach the banks of the river Ebro itself (3) for good views of the river's broad meanders and fluvial woodland. The route ends at a point (4) from where a path on the right takes you to the *soto* (an area of fluvial woodland) known as Rincón Falso (5). This last part of the itinerary leads you to a meander in the river graced by the shade of leafy woodland with Black Kite, Wryneck, Golden Oriole and, on migration, warblers and flycatchers.

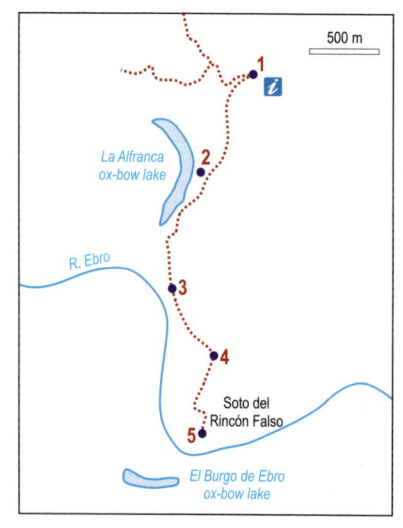

2. 8 km 🚶 ☀ ⭕

A track closed to vehicles heads towards the *galacho* of Juslibol from the car-park (6) at the far end of the village of the same name. On your right low gypsum cliffs and on your left allotments accompany you on a

ARAGON

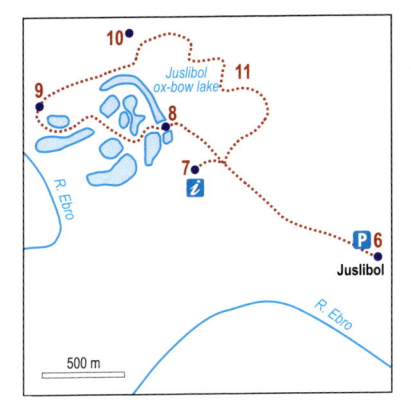

walk of less than 2 km to the visitor centre (7) of this protected area. A bridge (8) then takes you across the *galacho* and into the sector of flooded gravel pits and *sotos*, good for Purple Heron and Penduline Tit. Follow the main track as far as the end of the gravel pits (9), at which point you should turn right towards the cliffs, here topped off by a ruined castle (10). Climb up to the steppe (11) that extends beyond the line of cliffs for views of Little Owl, European Bee-eater, Thekla Lark, Crag Martin, Back Wheatear and Rock Sparrow. After walking for a little way along the edge of the line of cliffs, a path takes you back down some wooden steps to the visitor centre and then the car-park.

MOST REPRESENTATIVE SPECIES

Residents: Great Crested Grebe, Purple Heron, Cattle Egret, Marsh Harrier, Common Kestrel, Peregrine Falcon, Water Rail, Eagle and Little Owls, Crag Martin, Kingfisher, Black Wheatear, Blue Rock Thrush, Thekla Lark, Dartford and Sardinian Warblers, Rock Sparrow and Penduline Tit.

Summer visitors: Night and Purple Herons, Little Egret, Little Bittern, Black Kite, Egyptian Vulture, Little Ringed Plover, Common Sandpiper, European Bee-eater, Wryneck, Sand Martin, Golden Oriole.

Winter visitors: Great Cormorant, Common Teal and other ducks, Hen

PRACTICAL INFORMATION

- **Access:** from Zaragoza along the N-II towards Lleida as far as La Puebla de Alfindén (11 km), or through the district of Santa Isabel and the villages of Movera and Pastriz, to the unsurfaced tracks that reach the *galacho* at La Alfranca. From Juslibol, a road closed to vehicles heads to the *galacho* of the same name.
- **Accommodation:** plenty of variety in Zaragoza.
- **Visitor facilities:** the interpretation centre for the Galachos del Ebro Natural Reserve (tel. 976 070 002, 616 499 398) is near the *galacho* of La Alfranca. Open weekends and public holidays. Guiding service available for entering the restricted area. The Juslibol *galacho* also has a visitor centre, run by Zaragoza City Council (Gabinete de Educación Ambiental, tel. 976 724 230, galacho@zaragoza.es).
- **Visiting tips:** spring is best for herons. The end of summer and beginning of autumn are good for passerine migration. In winter, abundant aquatic birds.
- **Recommendations:** The 'train' *El Carrizal* links Zaragoza and the Juslibol *galacho* at weekends and on public holidays (except July and August); special trips can be arranged on weekdays for groups. Departs from Calle María Zambrano, opposite the Carrefour supermarket.
- **Further information:** maps 1:50,000, n°. 384 and 354 (SGE). Useful book: *El Galacho de Juslibol y su entorno. Un espacio singular.* (Ansar, 1996).

Harrier, Green Sandpiper, Black-headed Gull, Reed Bunting, Water Pipit, Bluethroat.
On migration: warblers, flycatchers and other passerines.

27. LAGUNA DE GALLOCANTA

Lying on the border between the provinces of Zaragoza and Teruel and high up in the Sistema Ibérico, the Laguna de Gallocanta occupies part of a large inwardly draining basin located in the midst of a plateau, mostly given over to cereal cultivation. It is the most important saline lake in western Europe and has been declared a Refugio de Fauna Silvestre, a Ramsar Site and an SPA.

It is fed principally by rain water and its water levels fluctuate so dramatically that the lake may be dry for months and even years on end. On the other hand, in wet periods, the lake can occupy 1,400 ha of land and measure over 7 km from end to end, thereby making it the largest natural lake in the Iberian Peninsula. Its shallow banks hold areas of salt-loving plants and reedbeds.

ORNITHOLOGICAL INTEREST

Gallocanta is the main staging post in western Europe for Common Cranes on their way from their breeding grounds in northern and central Europe to their winter haunts in southern Spain. Over the last 15 years censuses show a minimum of 20,000 and a maximum of 60,000 birds passing through Gallocanta; as well, around 10,000 birds winter in and around the lagoon.

The abundance and diversity of aquatic birds depends on the amount of water in the lagoon. When full in winter, the lagoon is alive with large quantities of diving duck, Eurasian Coots, waders and gulls; breeders include Shelduck, Marsh Harrier and colonies of gulls and waders. In the dry fields surrounding the lagoon small groups of steppe birds – Little and Great Bustard (the latter very scarce) and Dupont's Lark – can be found.

Laguna de Gallocanta © Fernando Barrio

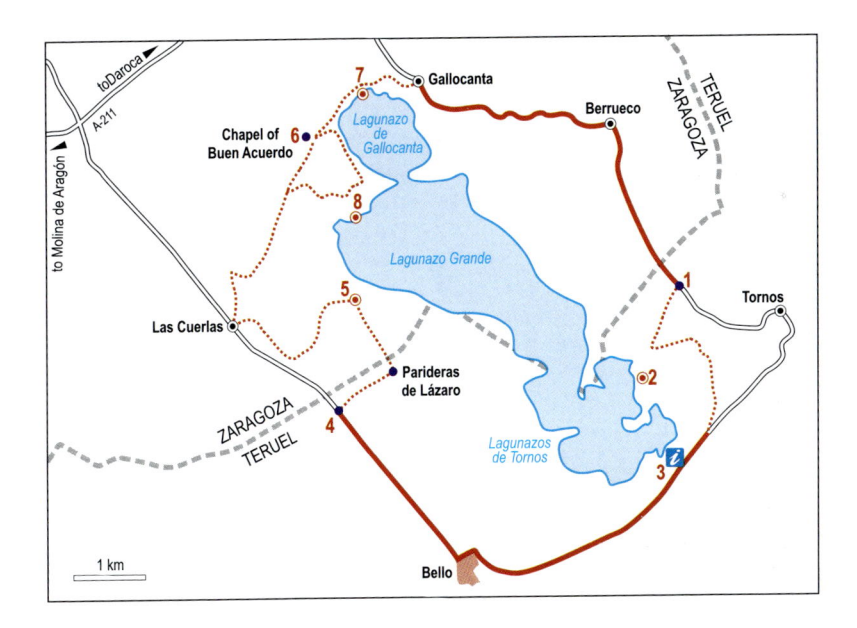

ITINERARIES

1. 30 km 🚗 ☀ 🔄

The lagoon can be circumnavigated by road and tracks. Leave Gallocanta on the road towards Berrueco, whose castle offers a wonderful panorama of the whole lagoon. Just after passing over the provincial boundary and into Teruel province, the first track on the right (1) will take you close to the south-east corner of the lagoon and the observation point of Fuente del Cañizar (2), overlooking Los Lagunazos de Tornos, a group of islands surrounded by water and marsh vegetation.

Return to the road and to the nearby interpretation centre (3). Continue on through the village of Bello and again just after crossing the provincial boundary (this time back into Zaragoza province), turn right along a track (4) that will take you in 400 m to the area known Las Parideras de Lázaro. Here, a track heading left takes you to the tower hide of La Reguera (5), overlooking the Lagunazo Grande, the central nucleus of the whole lagoon. From here, return by tracks to Gallocanta via the village of Las Cuerlas and the Ermita del Buen Acuerdo (6). The area around this church, as well as the Reguera observation point and the interpretation centre, are good places to wait for the cranes to fly in to roost at dusk.

2. 3 km 🚶 ☀ ➡

The final part of the first itinerary can be done on foot. Leave Gallocanta village and skirt the Lagunazo de Gallocanta in the extreme north-west of the site and head for the hide at Los Aguanares (7) overlooking a salt-marsh and a reed-bed. After walking up to the Ermita del Buen Acuerdo, an excellent observation point, walk down to another hide at Los Ojos (8), which overlooks the Lagunazo Grande. Return to Gallocanta along the same path, although there is no need to climb up to the

Ermita; simply skirt the base of the small hill it stands on and rejoin the main path further on.

MOST REPRESENTATIVE SPECIES

Residents: Shelduck, Gadwall, Mallard, Marsh Harrier, Little and Great Bustards, Stone Curlew, Northern Lapwing, Black-headed Gull, Black-bellied Sandgrouse, Dupont's, Calandra, Crested and Thekla Larks.

Summer visitors: Black-necked Grebe, Montagu's Harrier, Black-winged Stilt, Avocet, Kentish Plover, Gull-billed Tern, Short-toed Lark, Tawny Pipit, Yellow Wagtail, Northern Wheatear.

Winter visitors: Greylag Goose, ducks, Hen Harrier, Merlin, Common Crane, Dunlin, Little Stint, Common Snipe, Eurasian Curlew, Short-eared Owl.

On migration: Black Stork, Garganey, Honey Buzzard, Lesser Kestrel, waders.

PRACTICAL INFORMATION

- **Access:** from Madrid along the A-2 motorway towards Zaragoza. In Alcolea del Pinar, turn along the N-211 to Molina de Aragón. From here, take the road towards Daroca and after 45 km turn right to the village of Gallocanta and the north bank of the lagoon. From Zaragoza, the N-330 takes you to Daroca and from here, once over Puerto de Santed, in less than 20 km you reach the turn-off to Gallocanta mentioned above.
- **Accommodation:** most of the villages in the area have rural accommodation. Berrueco has a hotel and Tornos a hostal. In Gallocanta there is the excellent ornithological refuge 'Allucant' (tel. 976 803 137, info@allucant.com). This refuge is a meeting place for ornithologists and has a library, meeting rooms, observation point and a restaurant.
- **Visitor facilities:** the lagoon interpretation centre (tel. 978 734 031) stands next to the road between Tornos and Bello. Open all day in November and February during crane migration. The rest of the year, open only at weekends and on public holidays.
- **Visiting tips:** November and February are the best months to see large flocks of Common Cranes on migration. After heavy rain, take care on the dirt tracks, which can become very slippery. To avoid disturbing the birds, it is forbidden to approach the edges of the lagoon by vehicle or on foot.
- **Recommendations:** Grullaguía (tel. 976 803 026) offer bird guiding.
- **Further information:** maps 1:50,000, nº. 490 and 491 (SGE). The book *Guía de naturaleza de Gallocanta* (Prames, 2001) by Javier Mañas offers a complete description of the natural heritage of the lagoon and its surroundings.

WHERE TO WATCH BIRDS IN SPAIN. THE 100 BEST SITES

ASTURIAS

28. Somiedo
29. Cabo Peñas
30. Ría de Villaviciosa

Stretched out along the northern coast and running inland up to the mountains of the Cordillera Cantábrica in the south, Asturias, despite its relatively small size and much altered and densely populated central lowland strip, is famous for its wealth of natural splendours. Its upland forests and unaltered coastlines, contrasting with the relative lack of wildlife in the central zone, are the main natural attractions in this autonomous community.

The Cordillera Cantábrica tumbles steeply down to the sea in a complex succession of ridges and peaks in which snow, rain and autumn colours accompany the passing of the seasons and transform these mountains into a landscape of insuperable beauty. The steep mountainsides are covered by some of the best deciduous forests in Spain, while elsewhere the pastures, scrub and rocky outcrops provide the principal refuge in Spain for species such as the European Brown Bear and Wolf.

Rufous-tailed Rock Thrush © Juan Martín Simón

The spectacular limestone massifs of the Picos de Europa, for the most part protected by a national park, are by far the most visited part of the Cordillera and are dealt with in the chapters on Cantabria (area of Liébana) and Castilla y Leon. Asturias, however, has many other attractive upland areas: Redes, Ponga, Fuentes del Narcea and Ibias and, of courses, Muniellos, considered to be one of the best remaining sessile oak forests in Europe.

Equally exciting is the Somiedo Natural Park (28), with its singular combination of untamed nature and rural customs that in some way exemplify the way of life in the Asturian mountains. The Capercaillie, declining worryingly throughout the Cordillera, is the star of a varied avifauna that includes forest, rock and genuine alpine species of bird.

Along the almost 300 km of coastline there are numerous places of interest for the visiting birdwatcher. Some areas such as those around the industrial cities of Gijón and Avilés are heavily built up, while in other areas whole valleys have been disfigured by the large-scale planting of eucalyptus; thus, it is worth searching out the rocky promontory at Cabo Peñas (29) or at the nearby La Vaca for a more natural landscape.

These rocky salients are ideal sites for seawatching in autumn and spectacular passages of shearwaters, Northern Gannet, skuas, auks, gulls and other seabirds are frequent along this coastline. Some of the sea-cliffs and offshore islands hold small and disperse colonies of Shag and other seabirds, while in the nearby fields winter visitors as rare in Spain as Snow Bunting are annual.

The small estuaries that dot the coastline are the other attraction of the area. The Ría de Villaviciosa (30) is one of the most interesting for waders and other water birds, above all during migration periods and in winter, and is of national significance for species such as Eurasian Curlew. During storms or cold spells, divers, sea duck and gulls from more northerly latitudes take refuge in these relatively sheltered waters.

All in all, Asturias is worth a birdwatching trip at any time of the year: spring, summer and even early autumn in upland areas or winter and spring and autumn migration periods on the coast.

28. SOMIEDO

The mountains of Somiedo, deeply scored by long narrow valleys, lie in the south of Asturias and have been declared a Natural Park (30,000 ha), Biosphere Reserve and SPA. Home to a wonderful natural and scenic diversity, these mainly limestone mountains are exceedingly abrupt and exhibit in the form of corries, hanging valleys and the most important group of upland lakes in the Cordillera the traces of the region's glacial past.

Mountain slopes are covered in beech forests and, to a lesser extent, deciduous oaks and birches, while in lower areas holm and Lusitanian oaks are commoner. Up high near the summits, scrub, subalpine pastures and rock outcrops dominate. This area harbours the most important Spanish populations of European Brown Bear, and there are also Wolf, Otter, Chamois and Castroviejo's Hare. The local culture is extraordinarily rich and is based on traditional livestock-rearing on the mountainsides, where there are many groups of *brañas* (stone-built mountain huts).

ORNITHOLOGICAL INTEREST

The Capercaillie is the most important bird of the upland forests, al-

ASTURIAS

Somiedo Natural Park   © Fernando Barrio

though here as elsewhere in the Cordillera it is in serious decline. Just 20 years ago Somiedo boasted 60 males; today the species has all but disappeared. Also present but scarce are Black and Middle Spotted Woodpeckers.

In the highest areas, typical upland species such as Grey Partridge, Wallcreeper, Alpine Accentor, Alpine Chough and Snow Finch breed. A mix of raptors are present – Egyptian and Griffon Vultures (although the latter does not breed) and Short-toed and Golden Eagles – and in the village of Santa María del Puerto, one of the few pairs of White Storks in Asturias makes its nest. A good variety of passerines can be seen on migration at mountain passes such as Somiedo and La Farrapona.

ITINERARIES

1. 6 km

From the end of the village of Valle de Lago (1) a path closed to traffic climbs gently along the bottom of the glaciated valley to El Valle (2), the largest natural lake (24 ha) in the Cantabrian mountains. The abundance of limestone cliffs here make this a good site for Wallcreeper (in winter check the walls of the houses in the village!).

Other species to be seen here include Egyptian Vulture, Golden Eagle, Alpine Swift, Water Pipit, Ring Ouzel (on passage), Rufous-tailed Rock Thrush and both Alpine and Red-billed Choughs. From the lake, the keenest walkers can climb the 500 m to nearby Peña Orniz (3) (2,190 m) along a marked path and will with luck come across Snow Finch.

2. 7 km

This route links the villages of Llamardal (4) and Coto de Buenamadre via La Braña de Mumián (5). The final part of the route passes through a splendid beech forest with Goshawk and forest passerines, although the Black Woodpecker is too scarce to guarantee views. This section of forest is classified as "*de uso restringido*" (restricted use) by the Natural Park and must be crossed in silence and without leaving the path. Clearings provide the chance for observing raptors.

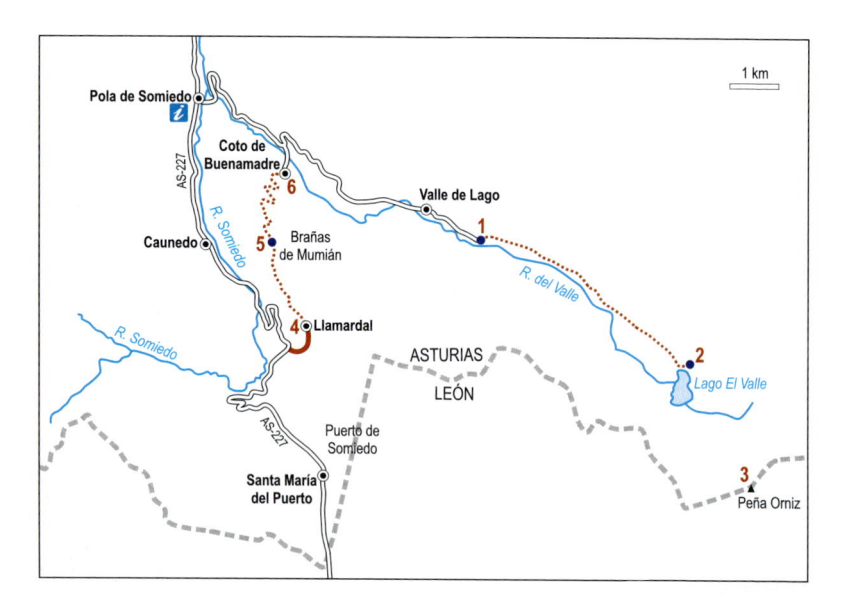

Ideally you will have a second car waiting in Coto de Buenamadre (6), from where you can return to Polo de Somiedo. Otherwise, from Braña de Mumián return to Llamardal through the beech forest.

PRACTICAL INFORMATION

- **Access:** from central Spain take the León-Oviedo motorway and turn off towards Villablino. In Piedrafita de Babia, take the road up to the Somiedo pass. From Pola de Somiedo, the local capital, a steep 8-km road heads up to the village of Valle de Lago. Alternatively, from Pola take the AS-227 back towards the Somiedo pass and after Caunedo, turn left to Llamardal. Access by car to this village is forbidden and you must park at the junction.
- **Accommodation:** rural accommodation in Pola de Somiedo, Valle de Lago and other villages in the area.
- **Visitor facilities:** the Natural Park interpretation centre is in Pola de Somiedo (tel. 985 763 758). Open all day, morning and afternoon. On weekends in July and August there is a free guiding service (reservation essential).
- **Visiting tips:** All year, but above all spring and autumn. Summer is interesting, although visitor numbers are high. The winter weather can be harsh. Access to certain areas of the park is restricted and so you should find out in advance about the 12 recommended routes in the park.
- **Recommendations:** The Somiedo Ecomuseum in Pola de Somiedo has an interesting collection of riding equipment and traditional tools; the visit includes a chance to look at the restored traditional houses in the village of Veigas. Open all year from Tuesday to Sunday.
- **Further information:** map 1:50,000, n°. 76 and 77 (SGE).

ASTURIAS

MOST REPRESENTATIVE SPECIES

Residents: Griffon Vulture, Goshawk, Hen Harrier, Golden Eagle, Peregrine Falcon, Grey Partridge, Eagle Owl, Black Woodpecker, Crag Martin, Water Pipit, Dipper, Alpine Accentor, Bluethroat, Blue Rock Thrush, Marsh and Crested Tits, Wallcreeper, Eurasian Treecreeper, Alpine and Red-billed Choughs, Snow and Citril Finches, Yellowhammer, Rock Bunting.

Summer visitors: White Stork, European Bee-eater, Black Kite, Egyptian Vulture, Short-toed and Booted Eagles, Alpine Swift, Whinchat, Northern Wheatear, Rufous-tailed Rock Thrush, Iberian Chiffchaff, Bonelli's Warbler, Red-backed Shrike, Ortolan Bunting.

29. CABO PEÑAS

The vertical cliffs (over 100 m) of Cabo Peñas, more or less equidistance between the cities of Gijón and Avilés, mark the most northerly point of the

La Gaviera at Cabo Peñas © César Álvarez

Asturian coastline. Offshore lie various islets, the largest of which is La Erbosa. Inland, pastures and scrub mix, while on the coast itself dunes and cliffs preserve valuable patches of autochthonous vegetation.

Cabo de Peñas is the heart of a glorious 20 km-long stretch of coast, protected as a *paisaje protegido* of almost 2,000 ha. To the west of the protected area in the Ría de Avilés lie a lagoon, Charca de Zeluán, and a beach, Ensenada de Llodero, both declared *monumentos naturales*.

ORNITHOLOGICAL INTEREST

This is an excellent site for observing birds such as shearwaters, Northern Gannet, skuas, auks, gulls and terns migrating east-west along the Cantabrian coast in autumn.

On the islets of La Erbosa and El Sabín there is a small colony of Shags and a few pairs of European Stormpetrel, as well as the largest Yellow-legged Gull colony in Asturias (over 500 pairs). The nearby pastures are an excellent area for winter and migrating passerines, with rare species such as Richard's Pipit and Snow Bunting regularly seen. Migrating waders frequent the temporary pools and tidal sands, and many gulls winter.

ITINERARIES

1. 3 km

From the Peñas lighthouse (1) a track leads towards the tip of the promontory where there is a car-park and bar (2). From here, walk west along the coastline as far as point (3), from where there are good views of La Sabín and its gull colony and breeding Shags.

The inland fields are frequented by Peregrine Falcon and migrating and wintering passerines such as Richard's Pipit, Whinchat, Northern Wheatear and Snow Bunting. The very tip of the cape, accessible of foot from the car-park, is a good place for seawatching.

2. 1 km

In the town of Luanco, take the avenue leading to the breakwater and beach and then the last left turn up

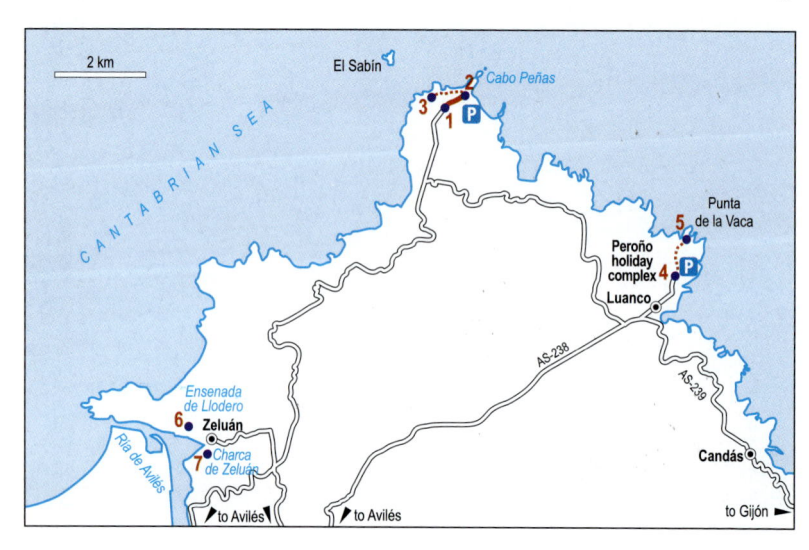

ASTURIAS

to the Peroño holiday home complex (4). At the highest point there is a car-park; from here walk along the track that takes you La Vaca (5), another coastal promontory and the best site in Asturias for autumn seawatching (August to October).

Early in the season, passage begins with Sooty and Balearic Shearwaters and Arctic, Common and Sandwich Terns, followed in mid-season by Cory's Shearwater, Common Scoter and Arctic, Great and Pomarine Skuas. The last birds to pass through are Razorbill, Guillemot, more sea-duck, Kittiwake and waders such as Oystercatcher and Whimbrel. As well, there can be spectacular numbers of Northern Gannet; Peregrine Falcons perch on the sea-cliffs.

3. 1 km

It is also worth visiting the Ría de Avilés, west of Cabo Peñas. The best site is Ensenada de Llodero (6), which has a hide. Walk from the village of Zelaún near the river mouth on the eastern bank of the estuary. In autumn and spring a great variety of waders can be seen at low tide: Oystercatchers, plovers, Ruff, godwits, Eurasian Curlew, Whimbrel, *Tringa* waders and Turnstones, amongst others. Also look for wintering gulls and, on the nearby Charca de Zeluán (7), aquatic birds.

MOST REPRESENTATIVE SPECIES

Residents: European Storm-petrel, Shag, Peregrine Falcon, Yellow-legged Gull.
Summer visitors: Tree Pipit, Yellow Wagtail, Common Whitethroat, Red-backed Shrike.
Winter visitors: Great Cormorant, European Golden Plover, Northern Lapwing, Mediterranean, Black-headed, Common, Lesser Black-backed, Great Black-backed and Herring Gulls, Richard's Pipit, Great Grey Shrike, Snow Bunting.
On migration: Cory's, Great, Sooty and Balearic Shearwaters, Northern

PRACTICAL INFORMATION

- **Access:** from Avilés, just over the Ría, at a roundabout on the AS-238 to Luanco, turn off for Cabo Peñas. Otherwise, 2 km further along the AS-238, turn left to Zeluán, from where it is also possible to reach Cabo Peñas. From Gijón, take the AS-239 through Candás to Luanco, from where a minor road takes you north to Cabo Peñas.
- **Accommodation:** in Luanco, other tourist towns in the area or, further afield, Avilés and Gijón.
- **Visitor facilities:** the environmental education centre La Noria (tel. 985 510 546) run by Avilés Town Council organises activities at the Ensenada de Llodero and the nearby Charca de Zeluán. Open Monday to Friday, morning and afternoon.
- **Visiting tips:** best between the end of summer and beginning of October for autumn seabird migration. The best conditions are windy mornings (especially if the wind is from the north-west); telescope needed. In spring wader passage is good and seabird colonies are already breeding. Keep an eye open for rarities.
- **Further information:** maps 1:50,000, n°. 13 and 14 (SGE). Telephone information from the *Servicio de Conservación del Medio Natural* of the Asturian Department of the Environment (tel. 985 105 500).

Gannet, Common Scoter, Oystercatcher and other waders, Kittiwake, Arctic, Common and Sandwich Terns, Razorbill, Guillemot, Tawny and Meadow Pipits, Skylark, Whinchat, Northern Wheatear, Pied Flycatcher.

30. RÍA DE VILLAVICIOSA

This small *ría* (estuary) is situated on the centre-west of the Asturian coast near the town of Villaviciosa. It is protected as a *reserve natural* (approx. 1,000 ha) that, for most of its 8 km, lies between the two roads running parallel to the estuary's shores. Large tidal mudflats dominate the scenery, above all on the right-bank, and there is also a dune cordon located at the river mouth. The construction of dykes has allowed dry land to be won from the estuary, and these brackish grazing marshes (known locally as *porreos*) contain patches of aquatic vegetation including sedge-, rush- and reed-beds.

ORNITHOLOGICAL INTEREST

This is an important site for wintering and migrant waders and most of the typical Spanish species can be seen here. Also present are Great Cormorants, herons, gulls, duck and auks, as well as the odd rarity in cold weather. Eurasian Curlews bred right up to the 1970s and today over 100 winter, one of the largest such groups in the Iberian Peninsula. Rails and passerines breed in the aquatic vegetation.

ITINERARIES

1. 12 km

Most of observation of the birdlife of the area is determined by the tides. At low tides, waders and other birds spread out greatly across the mudflats, canals and shallow waters; however, at high tide they group together on the few remaining patches of exposed high ground in places such as the island of El Bornizal.

Ría de Villaviciosa © Isolino Pérez Tuya

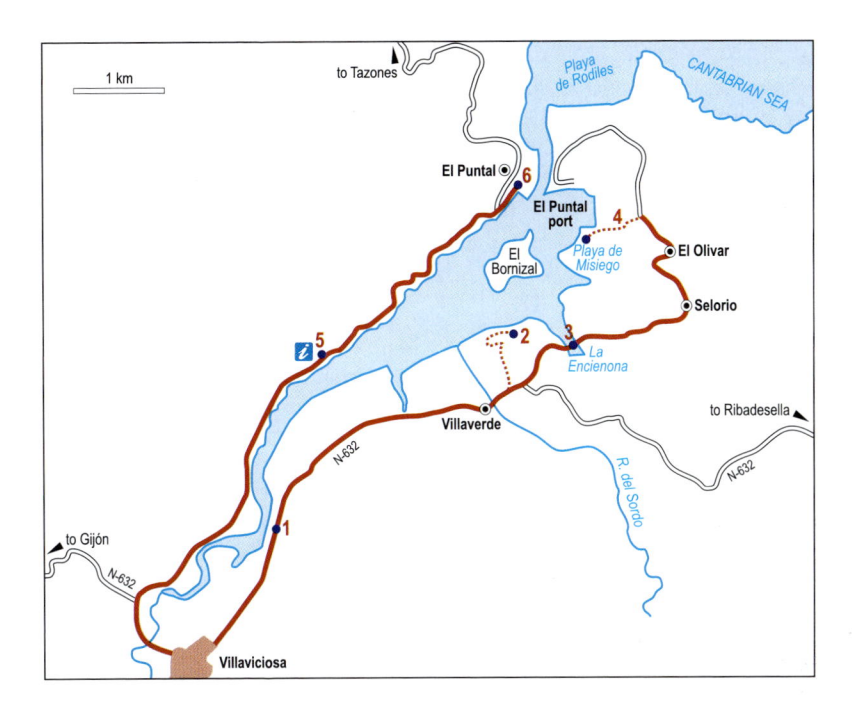

Drive the road along the right-bank of the *ría* between Villaviciosa and Playa de Rodiles, stopping when you find numbers of birds.

After taking the N-632 towards Ribadesella, make a first stop just 1 km further on next to the El Gaitero cider factory (1) to view the head of the *ría* and the first Great Cormorants, herons, waders and gulls. Once past the village of Villaverde and after crossing the bridge over the river Sebrayo, park and walk along the path (2) parallel to this tributary as far as the *ría* itself. The *porreos* here are frequented in winter by Cattle and Little Egrets, Grey Heron, Greylag Goose, Northern Lapwing and Common Snipe, and there are also patches of reeds that act as refuge for passerines and other water birds.

Almost 6 km from Villaviciosa, a left turn along a minor road takes you towards Playa de Rodiles. Almost immediately from the bridge (3) over La Encienona there are good views over the mudflats to the island of El Bornizal. You can get closer to these flats if you continue along the road through the village of El Olivar and take a path left (4) down to Playa de Misiego. At high tide whilst their main feeding areas are submerged, El Bornizal is the favourite resting place on the *ría* of Eurasian Curlew and the other waders. This is also a good site for Eurasian Spoonbill, herons, duck and gulls.

2. 8 km

The left-bank of the *ría* can be viewed from a much straighter minor road leading from Villaviciosa that stays close to the banks of the *ría*. This area is less interesting for birds, although there are a number of places where you can stop and view the *ría* and intertidal flats, as well as the mudflats on the opposite side.

The interpretation centre of the *reserva natural* (5) is 4 km from Villaviciosa. A few kilometres further on, you reach the port of El Puntal (6), whose breakwaters act as good viewing points over the *ría* and the artificial canal that joins the *ría* to the sea. Look out for Great Northern Diver, Sandwich Tern, Guillemot, Razorbill and, very occasionally, sea duck.

MOST REPRESENTATIVE SPECIES

Residents: Peregrine Falcon, Eurasian Curlew, Common Sandpiper, Black-headed, Lesser Black-backed and Yellow-legged Gulls, Kingfisher, Reed Bunting.

Summer visitors: Yellow Wagtail, Reed Warbler, Red-backed Shrike.

Winter visitors: Great Northern Diver, Black-necked Grebe, Great Cormorant, Little and Cattle Egrets, Greylag Goose and other geese and ducks, European Golden and Grey Plovers, Northern Lapwing, Dunlin, Common Snipe, Bar-tailed Godwit, Eurasian Curlew, Common Redshank, Greenshank, Mediterranean and Great Black-backed Gulls, Guillemot, Razorbill.

On migration: Eurasian Spoonbill, Garganey, Marsh Harrier, Osprey, waders, Sandwich and Common Terns, Sedge Warbler.

PRACTICAL INFORMATION

- **Access:** the N-632 between Gijón and Ribadesella passes through Villaviciosa. From Oviedo, take the N-634 and after 20 km at La Secada turn left along the AS-113 to Villaviciosa via La Campa.
- **Accommodation:** in Villaviciosa and other towns such as Selorio, El Puntal or Tazones near the *ría*. In Argüero there is the Casa del Naturalista (tel. 985 974 218). Gijón is 30 km away.
- **Visitor facilities:** the interpretation centre (tel. 687 483 378, riadevillaviciosa@sigma-sl.com) is 4 km from Villaviciosa on the left bank of the *ría* along the road to El Puntal. Open Tuesday to Sunday, morning and afternoon. There is a hide on the right-bank of the *ría*.
- **Visiting tips:** for wintering birds, visit between December and February; September and April are the best months for migrants. In terms of the position of the sun, the right-bank is best worked in the morning and the left-bank in the afternoon. Avoid walking on the treacherous mudflats and be careful not to be left stranded by high tides. Keep an eye open for rarities.
- **Further information:** maps 1:50,000, n°. 15 and 30 (SGE). The *Servicio de Conservación del Medio Natural* of the Asturian Department of the Environment (tel. 985 105 500) provides information about the reserve and the interpretation centre. Useful book: *La ría de Villaviciosa. Guía de la naturaleza*, by Luis Mario Arce (Trea, 1996).

THE BALEARIC ISLANDS

31. Serra de Tramuntana
32. S'Albufera de Mallorca
33. Menorca
34. Cabrera

The Balearic archipelago is one of the most important holiday destinations in the western Mediterranean and, indeed, in the whole world. Over 11 million visitors throng to these islands every year and tourism, which provides over half of the local GDP, is undoubtedly the corner-stone of the local economy. Unfortunately, the islands' natural heritage – which are, after all, one of the main selling points of the islands – have suffered as a result of the aggressive invasion by hotels, holiday homes, roads and other infrastructures.

The only 'positive' side to this massification is the existence of a wide range of services that speed up access to all of the larger islands of Mallorca, Menorca and Ibiza (without forgetting the smaller islands of Formentera and Cabrera). In general, the least accessible parts of the islands, for example the coastal cliffs and those coastal wetlands that have received legal protection in time, are the most worthwhile for birdwatchers.

In Mallorca, these sites are mainly distributed around the island's coasts and, above all, along the north coast. The western part of the island is occupied by the still well-conserved Sierra de Tramuntana (31), a long, rugged series of limestone mountains. This is a good area for raptors and, in particular, for Black Vultures, here in their only island breeding colony in the world, and for passage migrants moving through the Mediterranean.

These mountains form impressive sea-cliffs where they reach the sea. Both Osprey and Eleonora's Falcon breed here and, along with the Canary Islands, these are two of the only breeding populations of these raptors in Spain. They share these cliffs with Shag and a few of the very threatened Balearic Shearwater, whose largest colonies are on Formentera.

Nevertheless, the best known birdwatching site in Mallorca is S'Albufera (32), also on the north coast. When this wetland was declared a Natural Park in the 1980s, it became the first protected area in the whole of the Balears. Its heron communities, including the rare Great Bittern, are especially important and both Osprey and Eleonora's Falcon come to feed here. As well, S'Albufera holds the largest breeding population of

Eleonora's Falcon   Carlos Sánchez © nayadefilms.com

Moustached Warbler in the whole of Spain.

The whole of the island of Menorca (33) has been declared a Biosphere Reserve in an attempt to preserve its fragile natural equilibrium. It offers the birdwatcher a similar set of delights to Mallorca, with good densities of raptors such as Egyptian Vulture and Booted Eagle that here are resident (but only summer visitors to most of the Iberian Peninsula). The once abundant Red Kite is on the brink of dying out in the Balearic Islands due to high mortality rates caused by the laying of poison and electrocution on power lines.

Menorca is less populated and better conserved than its larger neighbour and most of its natural charms are to be found on its rugged north coast. Its colonies of Cory's Shearwater, the largest in the western Mediterranean, are its ornithological highlight. A number of small wetlands lie scattered along this coastline, the most important being S'Albufera des Grau near the town of Maó.

Lastly, the much smaller island of Cabrera (34) plays home to some of the most important seabird colonies in the whole of the Mediterranean. It is also a good place to find migrant raptors and passerines crossing the Mediterranean to and from Africa. Cabrera is only accessible by boat; nevertheless, protection as a National Park has enabled a public-use programme to be established that provides access to this island jewel of Mediterranean wildlife.

31. SERRA DE TRAMUNTANA

This abrupt limestone mountain chain occupies the whole of the north-west of Mallorca. Deep narrow gorges, of which some harbour the world's only populations of the *ferreret*, the endemic Balearic Midwife Toad, run down to an uninterrupted coastline of over 75 km, littered with delightfully rugged small coves.

Tramuntana combines extensive areas of Mediterranean scrub, Aleppo pine forests and patches of holm oak with endless lines of terraces buttressed by dry stone walls guarding traditional olive groves. Currently, these mountains are only protected as an SPA, although a natural park is being planned.

THE BALEARIC ISLANDS

Cap de Formentor, Serra de Tramuntana   © Francesc Jutglar

ORNITHOLOGICAL INTEREST

The northern part of the site between the town of Sóller and Cabo Formentor holds the only population of Black Vulture in the Balears (the only one in the world on an island), which is estimated to contain around 10 breeding pairs. Other breeding raptors include Booted Eagle, Osprey and Eleonora's and Peregrine Falcons, as well as the much scarcer Red Kite and Egyptian Vulture, both on the brink of disappearing from the island. Various colonies of Yellow-legged Gull, Shag and, in the extreme south-west of the mountains, Balearic Shearwater also exist. This is also an excellent area for passage raptors and migrants.

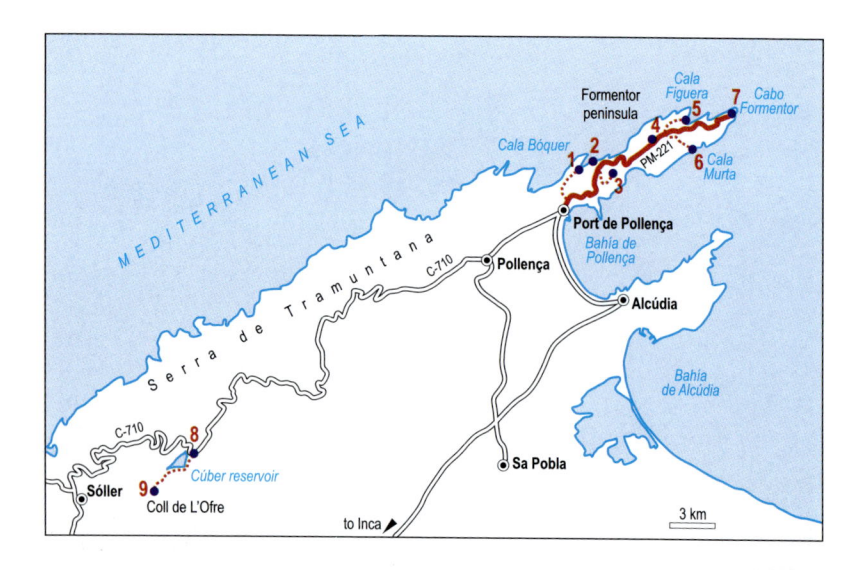

ITINERARIES

1. 20 km 🚗 🚶 ☀️ ➡️

The PM-221 road between Port de Pollença and Cabo Formentera at the far eastern end of these mountains provides ample opportunities to stop and view the coves and cliffs that appear on either side of the road as you drive the length of the Formentor Peninsula. Good stopping points from Port de Pollença onwards are: Cala Bóquer (1); a little further on at the La Creueta viewpoint (2); La Torre de Albercutx (3); after passing through a reforested area, Cases Velles (4); Cala Figuera (5); and Cala Murta (6). After the tunnel under the mountain of Es Fumat, the road reaches the Formentor lighthouse (7), one of the most impressive landscapes on the whole of the island.

Formentor and all the high observation points offer excellent views of Eleonora's Falcon during its breeding season from the end of August to October, the month in which the young birds are on the wing. Watch out too for Peregrine Falcon, Osprey (just a few breed in the area), Blue Rock Thrush on the sea-cliffs, Balearic Warbler in the scrub and Common Crossbill in the pines.

The whole route is interesting, although the area of Cases Velles is especially good for passerine migrants (including rarities such as Collared Flycatcher in spring and Yellow-browed Warbler in autumn) and raptors such as Honey Buzzard and Marsh Harrier. Shags frequent the cliffs and, according to the time of year, seawatching can produce Balearic and Cory's Shearwaters, Northern Gannet and Audouin's Gull.

2. 4 km 🚶 ☀️ ➡️

A track (at first metalled) turning off the C-710 runs along the western and southern banks of the Cúber reservoir near Sóller. Park at the barrier (8) and walk the last part

THE BALEARIC ISLANDS

PRACTICAL INFORMATION

- **Access:** Pollença or, even better, Port de Pollença are the two best starting points for the Formentor Peninsula. At around 20 km from Sóller on the C-710 towards Pollença, the track in itinerary 2 heads off to the right.
- **Accommodation:** in Port de Pollença for itinerary 1 and Sóller or Port de Sóller for itinerary 2.
- **Visitor facilities:** the botanical garden in Sóller (tel. 971 634 014) and the Museu Balear de Ciències (tel. 971 634 064) are at km 30.5 on the road from Palma to Port de Sóller. Open Tuesday to Saturday, from 10.00 to 18.00 and Sunday and public holidays from 10.00 to 14.00.
- **Visiting tips:** all year is good, although the end of summer and beginning of autumn are excellent for passage migration and for Eleonora's Falcon. The road along the Formentor Peninsula is narrow and tortuous; drive carefully, above all at weekends and in the summer months. Climate changeable; carry a waterproof at all times.
- **Recommendations:** visit the Sa Dragonera Natural Park (971 180 632), an island off the west coast reached by boat from Sant Elm that holds the most important colony of Eleonora's Falcon in Spain, as well as breeding Audouin's Gull and other seabirds.
- **Further information:** maps 1:50,000, n°. 645 and 670 (SGE). The Grup Balear d'Ornitologia (GOB, tel. 971 496 060, info@gobmallorca.com) and La Fundación para la Conservación del Buitre Negro (BVCF, tel. 971 516 620, bvcf@bvcf.org) provide information and organise activities in relation to the birds of the Serra de Tramuntana.

of the track towards Collado de L'Ofre (9).

This is one of the best sites on Mallorca for migrating and breeding raptors: Black Vultures often fly over the area, Ospreys come to fish, Booted Eagles are present and Eleonora's Falcons are occasionally seen in spring and summer. Look too for Rufous-tailed (very rare in the Balearics) and Blue Rock Thrushes on rocky outcrops and Balearic Warbler in the *garrigue* (scrub).

MOST REPRESENTATIVE SPECIES

Residents: Balearic Shearwater, Shag, Red Kite, Egyptian and Black Vultures, Booted Eagle, Osprey, Peregrine Falcon, Yellow-legged Gull, Rock Dove, Blue Rock Thrush, Balearic Warbler, Raven, Common Crossbill.

Summer visitors: Cory's Shearwater, European Storm-petrel, Eleonora's Falcon, Audouin's Gull, Alpine Swift, Rufous-tailed Rock Thrush, Subalpine Warbler.

Winter visitors: Northern Gannet, Great Cormorant, Woodcock, Robin, Black Redstart, Song Thrush, Blackcap, Common Chaffinch.

On migration: Honey Buzzard, Marsh Harrier, Sparrowhawk, Common Buzzard, Common Redstart, Whinchat, Northern Wheatear, Pied Flycatcher, Golden Oriole.

32. S'ALBUFERA DE MALLORCA

This coastal wetland on the Bahia d'Alcúdia in the north-east of Mallorca thrives on the entry of freshwater from temporary streams and the aquifer, which allows vast reed-beds and stands of great fen-sedge to de-

S'Albufera d'Alcúdia © Eduardo de Juana

velop. In lower areas nearer the coast, the penetration of saltwater enriches the flora. The remnants of the network of ditches dug in the nineteenth century to drain the wetland and convert it into agricultural land also form an integral part of the wetland.

This was the first area in the Balearic Islands to be protected (the Natural Park of around 1.700 ha was declared in 1988; today also a Ramsar Site and an SPA) and the wetland is isolated from the surrounding tourist infrastructure, industry and agriculture. Management techniques employed include grazing by cows, horses and buffaloes, designed to improve habitat quality.

ORNITHOLOGICAL INTEREST

S'Albufera is the largest wetland in the Balears and most of the herons breeding in Spain are present here, including Purple and Squacco Herons and Great Bittern, the latter in one of its best Spanish breeding sites. The small Little Bittern colony is resident and Great Egrets winter regularly. Recent successful reintroductions include Red-knobbed Coot, Purple Swamp-hen and Red-crested Pochard; less successful was the attempt at reintroducing White-headed Duck. Other breeders include Marsh Harrier, Black-winged Stilt and Little Ringed and Kentish Plovers.

Ospreys come to feed throughout the year, as do Eleonora's Falcons before they start to breed in late summer. Winter sees many duck arrive and both passage periods bring many waders, gulls, terns and hirundines.

ITINERARY

4 km

When entering the Parc Natural (1), it is easy to see the heronry with Night and Squacco Herons and Little and Cattle Egrets situated on the pines to your right. Walking alongside the Gran Canal, which divides S'Albufera into northern and southern sectors, after just over 1 km you reach the reception centre (2), surrounded by a number of hides. The first hide is the so-called CIM ob-

THE BALEARIC ISLANDS

servatory (3), overlooking the lagoon of Es Ras, where it is not difficult to see Purple Heron, Marsh Harrier and Purple Swamp-hen or, if you are lucky, hear the male Great Bitterns booming in spring. Good numbers of duck winter here. The nearby hill of Es Turó is the highest point of the park and an excellent place to watch out for Osprey and in spring and early summer groups of Eleonora's Falcon feeding on flying insects. In winter spectacular groups of Common Starling come here to roost in the reed-beds.

A track running alongside the canal of En Pujol passes between tall stands of reeds, where there are good numbers of Moustached Warbler and Reed Bunting, and takes you back to the Gran Canal. From the nearby Watkinson hide (4), Little Grebe, Red-crested Pochard and perhaps White-headed Duck can be seen.

After crossing the Gran Canal and the smaller canals that run alongside, you enter the northern sector of the Natural Park, from where a walk of over 1 km will take you to the raised hide at Punta des Vent (5) (westwards) or the Es Colobars hide (6) (northwards). On the way back to the Gran Canal it is worth taking a detour to Bishop hides I and II (7) for herons, Osprey, Black-winged Stilt and, during passage periods and in winter, ducks and waders.

If after leaving the Natural Park you take the road towards Artà, you will pass on your right a lagoon that holds many duck in winter. Just afterwards a path off to the right allows you to approach on foot Ses Salinetes (8), a group of abandoned salt-pans included within the protected area. Here you will find Black-winged Stilt and Little Ringed and Kentish Plovers and Greater Flamingo, gulls, terns and waders on migration.

MOST REPRESENTATIVE SPECIES

Residents: Great Bittern, Little Bittern, Little and Cattle Egrets, Grey Heron, Gadwall, Red-crested Pochard, White-headed Duck, Marsh Harrier, Osprey, Purple Swamp-hen, Audouin's and Yellow-legged Gulls, Moustached Warbler, Reed Bunting.

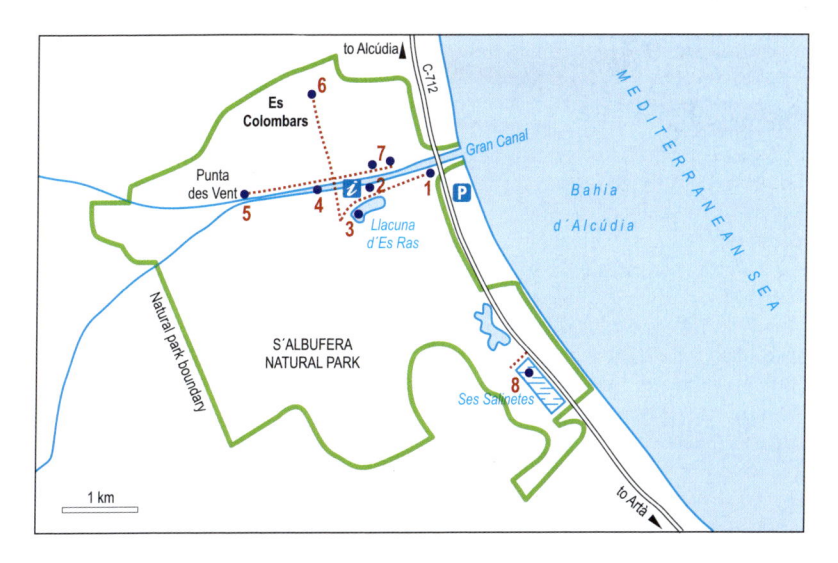

> ### PRACTICAL INFORMATION
>
> - **Access:** the entrance to the park is some 4 km from Alcúdia along the C-172 towards Artà at Pont dels Anglesos. Cars are not allowed in the park and so you should park in the car-park at the entrance or nearby.
> - **Accommodation:** in the port of Alcúdia and all of the very touristy Bahia d'Alcúdia.
> - **Visitor facilities:** the Natural Park (tel. 971 892 250, parc.albufera@wanadoo.es) has a reception and an interpretation centre: Open every day from 9.00 to 16.00.
> - **Visiting tips:** good any time of year, although summers can be very hot. Make sure you have sun cream and mosquito repellent on. The park gates open at 9.00.
> - **Recommendations:** the Natural Park organises guided visits on Saturday mornings. Occasionally the Grup Balear d'Ornitologia (GOB, tel. 971 496 060; info@gobmallorca.com) and the Associació Balear d'Amics dels Parcs (ABAP, tel. 971 465 935, associacio@amicsdelsparcs.com) organise birdwatching events.
> - **Further information:** map 1:50,000, n°. 671 (SGE). The web www.mallorcaweb.net/salbufera offers up-to-date information about the site and details of bird censuses.

Summer visitors: Night, Squacco and Purple Herons, Eleonora's Falcon, Black-winged Stilt, Little Ringed and Kentish Plovers.
Winter visitors: Great Cormorant, Great Egret, Greylag Goose and ducks, Peregrine Falcon, Common Snipe, Black-headed Gull, Spotted Redshank, Sandwich Tern, Water Pipit, Bluethroat, Common Starling.
On migration: Greater Flamingo, Glossy Ibis, Garganey, waders, Slender-billed Gull, marsh terns.

33. MENORCA

This island is relatively less densely populated than the other Balearic Islands and has been less affected by tourism. The cliffs of the rugged north coast are the most attractive area for birdwatchers, although inland the blend of pastures and Mediterranean scrub and Aleppo pine, holm oak and wild olive woodland is also of interest. The whole island has been declared a Biosphere Reserve; furthermore, a natural park and SPA (including additionally a number of offshore islands) protects S'Albufera des Grau, an interesting coastal lagoon.

ORNITHOLOGICAL INTEREST

The island is an excellent place for raptors: both Egyptian Vulture and Booted Eagle are resident and there are still 10 or so pairs of Red Kite, although this species is suffering a serious decline in numbers. A few pairs of Osprey also breed, although more birds arrive during migration periods. The north coast houses the largest colonies of Cory's Shearwater in the western Mediterranean (over 2,000 pairs), a few pairs of Balearic Shearwaters (although at least one colony has been identified as being composed of Mediterranean Shearwater, here in its only Spanish site), Shag and Audouin's Gull. In S'Albufera des Grau and other small coastal wetlands duck and other water birds winter in good numbers.

ITINERARY

60 km

From Maó, just after taking the PM-170 towards Fornells, turn right towards Es Grau. Then, 4 km after this junction and just before reach-

THE BALEARIC ISLANDS

Menorca © Xavier Ruiz

ing this small fishing village, turn left along an asphalted road towards the Shangril·la holiday complex (1). With a few exceptions, this planned residential zone was never built owing to the declaration of the S'Albufera des Grau Natural Park.

Park where the tarmac ends and walk through the 'streets' of Shangril·la to the southern edge of the estuary-like lagoon. Ospreys are present all year round, although less obviously so during the breeding season, along with Red Kite, Booted Eagle and Peregrine Falcon. Recent sightings of Purple Swamp-hen suggest that this species might be breeding.

In winter large concentrations of Great Cormorant arrive, along with herons, duck, Greylag Goose and Marsh Harrier.

Return to the PM-710 and continue north-west for 4 km to the turning on the right towards the lighthouse at Favàritx (2), also in the Natural Park. This is a good site for seawatching, above all for Cory's and Balearic Shearwaters. The cliffs are home to Peregrine Falcon and Blue Rock Thrush, while nearby pasture-land hold Thekla Lark and Tawny Pipit.

Just before reaching the lighthouse, a path off to the right (southeastwards) along the coast will take you to Bassa de Morella (3), a small wetland situated behind the beach at Capifort. Duck and other water

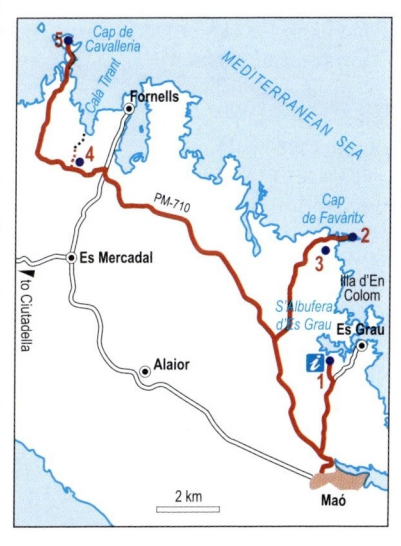

PRACTICAL INFORMATION

- **Access:** the most typical way of travelling to Menorca is by aeroplane to Maó. The island is connected Barcelona, Valencia and Palma de Mallorca by ferry.
- **Accommodation:** in Maó, Fornells and other tourist sites on the north coast.
- **Visitor facilities:** Rodríguez Femeníes reception centre (tel. 971 356 302, naturalesgrau@terra.es) in the S'Albufera Natural Park. Open Wednesday to Sunday, mornings only; also in the afternoon in the summer.
- **Visiting tips:** all year is worthwhile given the diversity of birds (seabirds, wildfowl, raptors and passerines) found on the island.
- **Recommendations:** the Grup Balear d'Ornitologia (GOB Menorca, tel. 971 350 762, info@gobmenorca.com) provides information about birds in Menorca and organises activities. Consult their book, *Aves de Menorca*.
- **Further information:** maps 1:50,000, n°. 618, 619 and 647 (SGE). The tourist information at Maó airport (tel. 971 157 115, infomenorcaeroport@cime.es) can help with your trip planning.

birds from S'Albufera visit this site and unexpected sightings cannot be ruled out.

A similar site is the nearby Bassa de Lluriac (4). Return to the PM-710 and continue towards Fornells. Just after turning right towards Fornells, turn left on a minor road that provides good views from the south of this small wetland, which normally dries out in summer. After 4 km, turn right along a track leading to Cala Tirant that once more approaches this small wetland. This is a good site for Osprey, Red Kite and Egyptian Vulture, as well as for waders, gulls and terns on passage, Purple Heron and the occasional Red-footed Falcon.

Return to the previous crossroads and continue on to Cap de Cavalleria (5) and its lighthouse, situated at the extreme northern tip of Menorca. Seawatching can be rewarding here for Shearwaters, Shag and Audouin's Gull, as well as for Montagu's and Hen Harriers and Common Buzzards on migration. Inland, the dry pastures hold Menorca's only Stone Curlews and Spectacled Warblers, while Dartford Warblers are common in the heather scrub.

MOST REPRESENTATIVE SPECIES

Residents: Little Grebe, Shag, Little Egret, Grey Heron, Red Kite, Egyptian Vulture, Booted Eagle, Osprey, Common Kestrel, Peregrine Falcon, Water Rail, Stone Curlew, Kentish Plover, Audouin's and Yellow-legged Gulls, Thekla Lark, Blue Rock Thrush, Dartford and Sardinian Warblers, Raven.

Summer visitors: Cory's and Balearic Shearwaters, Quail, Pallid Swift, European Bee-eater, Short-toed Lark, Tawny Pipit, Spectacled Warbler.

Winter visitors: Cattle Egret, Greylag Goose and other wildfowl, Marsh Harrier, Northern Lapwing.

On migration: Purple Heron, Montagu's and Hen Harriers, Common Buzzard, Red-footed Falcon, waders.

34. CABRERA

This group of limestone islands off the south coast of Mallorca boast the best conserved insular habitats in the Spanish Mediterranean. The main

THE BALEARIC ISLANDS

Cabrera © José Manuel Reyero

island of Cabrera is accompanied by the smaller island of Conejera and 15 other islets; in total they cover 1,300 ha. The highly indented 60 km of coastline includes imposing cliffs, a few small beaches and many diminutive coves. Stands of Aleppo pine are the main tree cover on the islands, and submarine habitats are very well preserved and include extensive posidonia meadows with high levels of biodiversity.

This archipelago is home to numerous Balearic endemic plants and marine invertebrates, as well 80% of the world's population of Lilford's Wall Lizard. The islands are an SPA and were declared a National Park in 1991, of which only around 10% is terrestrial.

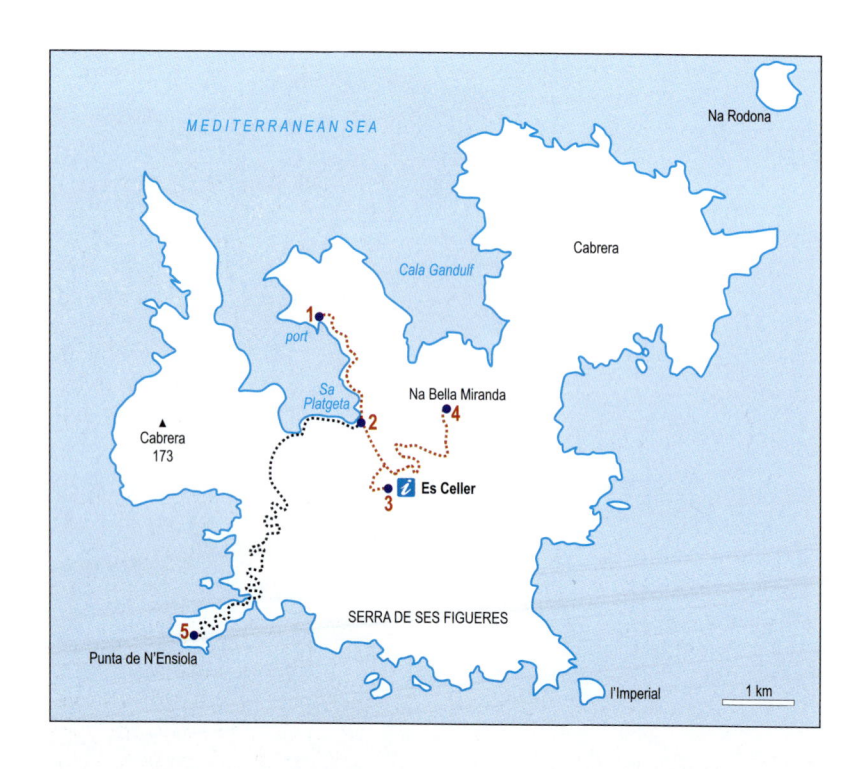

ORNITHOLOGICAL INTEREST

The islands boast important breeding colonies of Shag, Cory's and Balearic Shearwaters, European Storm-petrel and Audouin's and Yellow-legged Gulls, along with a few Ospreys and good populations of Peregrine and Eleonora's Falcons. The most interesting passerines include the endemic Balearic Warbler.

Passage sees over 100 species of bird reach the archipelago, including good numbers of Honey Buzzard, Marsh Harrier and many passerines. Song Thrush and other thrushes winter.

ITINERARIES

1. 5,5 mi

The boat crossing from Mallorca provides an opportunity to observe many of the seabirds that breed in the National Park, above all when the boat passes near to the island of Conejera and the islets to the north of the main island, where many of the main seabird colonies are. Ospreys and Eleonora's Falcons are harder to see (unless you have your own boat!), since all of the former and most of the latter breed around the southern part of Cabrera beyond the route of the tourist boats.

2. 3 km

The path from the port of Cabrera to the museum at Es Celler has to be walked with a National Park guide. The first part of the route follows the coast as far as the small beach of Sa Platgeta (2) and provides views of Shag and Audouin's and Yellow-legged Gulls; with luck an Osprey will put in an appearance.

THE BALEARIC ISLANDS

The second part of the route enters the central valley of the island and passes through a mosaic of abandoned fields, scrub and pine and juniper woodland that is excellent for warblers and other migrant passerines. Common Kestrel and Peregrine Falcon are common here, feeding on the many passage migrants that pass through. In winter, Robin, Black Redstart, Song Thrush and other fruit-eating birds are common.

For those with their own boat and who are not confined to the timetable of the tourist boat, just before Es Celler (3) a path (7-km there and back) heads off to Na Bella Miranda (4), one of the highest points of the island. This is an ideal site for Balearic Warbler and for migrant raptors. Another option is to take the guided itinerary from the port to the tip of N'Ensiola (5), a route of 4 hours that passes through some of the wildest and most rugged parts of the island.

MOST REPRESENTATIVE SPECIES

Residents: Balearic Shearwater, Shag, Osprey, Common Kestrel, Peregrine Falcon, Yellow-legged Gull, Blue Rock Thrush, Sardinian and Balearic Warblers, Greenfinch, Goldfinch, Linnet.

Summer visitors: Cory's Shearwater, European Storm-petrel, Eleonora's Falcon, Audouin's Gull, Subalpine Warbler, Spotted Flycatcher.

Winter visitors: Northern Gannet, Woodcock, Skylark, Dunnock, Robin, Black Redstart, Song Thrush, Dartford and Sardinian Warblers, Firecrest, Common Chaffinch.

On migration: Honey Buzzard, Marsh Harrier, Booted Eagle, Common Redstart, Common Whitethroat, Garden and Willow Warblers, Pied Flycatcher, Golden Oriole.

PRACTICAL INFORMATION

- **Access:** Cabrera is only accessible by sea. Between April and October tourist boats leave daily from a number of ports on Mallorca (Colònia de Sant Jordi and Porto Petro, principally). Access by private or hired boat requires permission in advance from the National Park (for both navigation and, if you intend to spend the night, anchoring).
- **Accommodation:** None. Permission is needed to anchor overnight in the port waters.
- **Visitor facilities:** the marked routes all start at the port, which is the only point on the island where landing is permitted. Other than for the beach of S'Espalmador, on all routes you must be accompanied by a guide from the National Park (reservation essential). The ethnographic museum at Es Celler has exhibitions on the history and natural and cultural heritage of the island. A visitor centre for the National Park is to be opened at Ses Salines in the south of Mallorca.
- **Visiting tips:** any time of year is good for birdwatching, although for part of autumn and all winter a private boat is needed as the tourist boats do not run. You must carry your own food and must not leave the marked paths. Camping, fishing and pet animals are all forbidden.
- **Further information:** map 1:50,000, n°. 748 (SGE). The National Park offices in Palma de Mallorca (tel. 971 725 010) issue all sailing, anchoring and diving permits. Open Monday to Friday, morning and afternoon. The book *Las aves del Parque Nacional Marítimo-Terrestre de Cabrera* (Ministerio de Medio Ambiente and GOB, 2000) is recommended. Also, www.mma.es/parques/lared/cabrera.

THE BASQUE COUNTRY

35. Álava wetlands
36. Urdaibai
37. Txingudi

The mid-altitude Montes Vascos that dominate much of the Basque Country consist of a labyrinthic series of more or less parallel, east-west running ridges sheltering numerous narrow valleys that provide a link between the Cantabrian mountains (Cordillera Cantábrica) to the west and the Pyrenees to the east.

The transitional nature of these mountains is further accentuated by the fact that the Montes Vascos are drained by rivers that flow north and south into the Cantabrian Sea and the river Ebro, respectively. The climate of the region – in general mild and rainy and thus typically Atlantic – becomes drier and more continental towards the south as you approach the river Ebro.

The population of the three densely populated Basque provinces (Vizcaya, Guipúzcoa and Álava) is divided between urban centres such as the Bilbao conurbation and the highly humanised countryside, where traditional farming techniques survive along with singular customs of great historical and cultural significance. Over the centuries the impact of farming and other forms of human intervention has greatly transformed the landscape and today very few remnants of truly natural systems

Oystercatchers   Carlos Sánchez © nayadefilms.com

actually remain. Over recent years, this destruction has been speeded up by heavy industrialisation, the building of transport infrastructures and a proliferation of pine and eucalyptus plantations.

Fortunately, the province of Álava has somewhat escaped this process of habitat loss and still boasts well preserved habitats that range from a series of limestone mountains with river gorges, deciduous forests and even in the extreme south of the province Mediterranean forests and arable land. The most interesting sites for birdwatchers in the province are Salburúa and the Ullibarri-Gamboa reservoir on the river Zadorra, a group of wetlands (35) close by the city of Vitoria that are the best area for breeding water birds (above all for duck) in the Basque Country.

The relatively long coastline is of greater interest for migrants and wintering birds. Clean wave-torn beaches alternate with sheltered fishing ports along a coastline punctuated by numerous small estuaries. The most important such river mouth is Urdaibai (36) in the north-west of Vizcaya, although the bay of Txingudi (37) on the border with France in the north-east of the Basque Country is also of interest.

Duck, waders, gulls and Eurasian Spoonbills use these coastal sites to rest during migration and during cold snaps northern species – including some extremely rare species – take shelter along this coastline.

35. ÁLAVA WETLANDS

The Salburúa wetlands consist of a number of small lagoons (*balsas*) on the outskirts of the city of Vitoria fed by the upwelling of subterranean water. Drained in the middle of the last century, the local city council has been restoring the site since 1994 and today around 60 ha of lagoon surface has been recreated. Nearby the southern part of the large Ullibarri-Gamboa reservoir on the river Zadorra has a number of shallow muddy 'arms' that have been colonised by aquatic vegetation and provide habitat for an interesting selection of birds.

Located in the heart of the plains of the Plana Alavesa, the natural value of these sites is enhanced by

Salburúa, Álava wetlands © Eduardo de Juana

its position on the imaginary line separating the Mediterranean and Atlantic biogeographical regions. The threatened European Mink has one of its main Iberian strongholds here; both sites have been declared Ramsar Sites.

ORNITHOLOGICAL INTEREST

Together these two wetlands form the most important site in the Basque Country for breeding water birds and include species that are rare elsewhere in the Iberian Peninsula – Garganey, Northern Shoveler and Tufted Duck at Salburúa and Gadwall at Ullibarri-Gamboa – or almost unknown from any other site in the Basque Country – Black-winged Stilt at Salburúa and Great Crested Grebe and Purple Heron at Ullibarri-Gamboa.

A good range of wildfowl including Greylag Geese in winter and Black Stork, Eurasian Spoonbill, waders, terns and, outstandingly, Aquatic Warbler on passage are all regular. Large groups of Red-crested Pochards carry out their post-breeding moult at Ullibarri-Gamboa and winter sees the arrival of Black-necked Grebes and unwelcome groups of Ruddy Duck, which seem to have chosen these reservoirs as their main entry port into the Iberian Peninsula.

ITINERARIES

1. 7 km

A marked itinerary (closed to vehicles) circumnavigates the two main lagoons at Salburúa. From the information centre, head anti-clockwise outside the perimeter fence around Balsa de Arkaute to the first hide of Los Fresnos (1), from where there are good views across to the artificial islands with breeding Black-winged Stilts and Little Ringed Plover. The aquatic vegetation provides shelter for breeding Northern Shoveler and Tufted Duck and habitat for hunting Marsh Harriers in winter. The path continues across flooded fields (2), where groups of Greylag Geese feed in winter, and on to the raised hide of Las Zumas (3). This hide provides good views over the lagoon's waters: look from here for wintering duck and Grey Heron (breeding) and Great Cormorants in the nearby poplar trees.

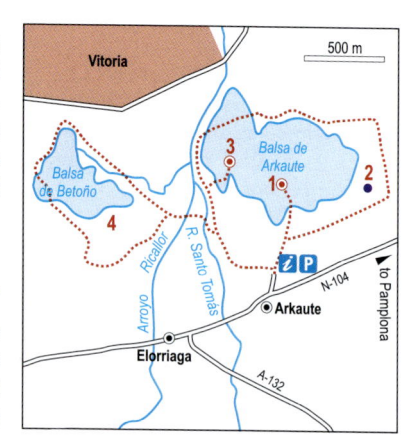

From here cross two rivers and head for the Balsa de Betoño (4), unfenced and thus closer to the water's edge. Look out for Little Bittern, easily observable in the nearby willow bushes.

2. 4 km

The Mendixur Ornithological Park in the southern sector of the Ullibarri-Gamboa reservoir has two marked itineraries. The first passes by two hides: Los Carboneros (5), lying between one of the reservoir's *sotos* (fluvial woodland) and one of its 'arms', is good for breeding ducks such as Gadwall and Purple Heron, while Los Trogloditas (6) is actually within a *soto* and is a good site for observing passerines that include Bullfinch in

THE BASQUE COUNTRY

The Garayo Provincial Park (8) lies on a spit of land between an arm of the reservoir and the river Zadorra and is criss-crossed by a network of asphalted tracks with parking spots that provide good access to the shores of the reservoir. View the open water for Black-necked Grebe, Northern Shoveler, Common Pochard and other wintering ducks. Rarities such as Great Northern Diver and Red-breasted Merganser can also turn up here.

MOST REPRESENTATIVE SPECIES

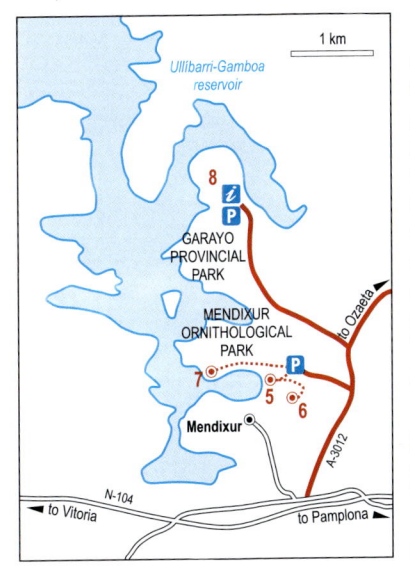

winter. The other itinerary heads for the hide of Los Buceadores (7) and is ideal for observing birds such as Great Crested Grebe, Great Cormorant and Red-crested Pochard and other duck on the open waters of the reservoir.

Residents: Little and Great Crested Grebes, Grey Heron, Gadwall, Northern Shoveler, Common Pochard, Tufted Duck, Water Rail, Zitting Cisticola.

Summer visitors: Little Bittern, Purple Heron, Garganey, Black-winged Stilt, Little Ringed Plover, Great Reed Warbler.

PRACTICAL INFORMATION

- **Access:** from Vitoria take the N-104 (former N-I) towards Pamplona. After 2 km, reach the village of Arkaute and turn left just before a petrol station to Salburúa. For Ullibarri-Gamboa, continue along the N-104 and after 10 km, take the A-3012 towards Ozaeta. Just before the village of Maturana turn off to the Garayo Provincial Park. Just before you enter this park, take a path off left to the Mendixur Ornithological Park.
- **Accommodation:** in Vitoria, rural accommodation or the Argómaniz Parador near Ullibarri-Gamboa.
- **Visitor facilities:** Salburúa has an information centre with a live video feed from the Black-winged Stilt colony. Open weekends and public holidays, mornings and afternoons; between May and September, also Monday to Friday afternoons. At Ullibarri-Gamboa, the Garayo Provincial Park information point is open every day.
- **Visiting tips:** all year. Bird numbers are higher in winter and all the typical wintering duck species found in the north of the Iberian Peninsula are present.
- **Further information:** map 1:50,000, n°. 112 (SGE). For Salburúa, get in touch with the Centro de Estudios Ambientales of Vitoria City Council (tel. 945 162 696, ceaadmin@vitoria-gasteiz.org) and its web page, www.vitoria-gasteiz.org/ceac and for Ullibarri-Gamboa, with the Provincial Council for Álava's Town Planning and Environment Department (tel. 945 181 818) or www.alava.net.

Winter visitors: Great Northern Diver, Black-necked Grebe, Great Cormorant, Greylag Goose, Red-breasted Merganser and other ducks, Marsh Harrier, Common Snipe, Kingfisher, Bullfinch, Reed Bunting.
On migration: Black Stork, Eurasian Spoonbill, Shelduck, Red-crested Pochard, Ferruginous and Ruddy Ducks, Osprey, waders, Whiskered and Black Terns, Aquatic Warbler.

36. URDAIBAI

Lying in the north-west of Vizcaya, Urdaibai is the name given to the estuary of the river Oka as it runs through the Ría de Gernika. Sheltered by Cabo Matxitxako, Urdaibai is one of the best-loved natural areas in the Basque Country and, despite its small size and relatively humanised surroundings, contains a good selection of the most typical habitats of the Basque coastline: salt-marshes, beaches and cliffs on the *ría* and on drier land, small-scale farming with pastures, arable land, market gardens, hedgerows and a few pine plantations. As well, thick stands of holm oak cover the tops of the nearby limestone mountains.

The 22,000 ha of this site is triply protected as an SPA, a Ramsar Site and a Biosphere Reserve.

ORNITHOLOGICAL INTEREST

This is the most important coastal wetland in the Basque Country for water birds and is especially significant for wintering and passage waders; along with Santoña, the Urdaibai salt-marshes are a key point for migrating Eurasian Spoonbill and in general the whole coastline in the vicinity is interesting for seabirds on migration and in winter. Breeding birds include Shag and European Storm-petrel on Cabo Ogoño and Yellow-legged Gull (1,500 pairs; the largest colony on the Cantabrian coast) and a small colony of Little Egret on the island of Ízaro in the river mouth.

The surrounding countryside consists of a mosaic of ecosystems – fields, hedgerows, copses and waterside veg-

Ría de Gernika, Urdaibai © José Manuel Reyero

THE BASQUE COUNTRY

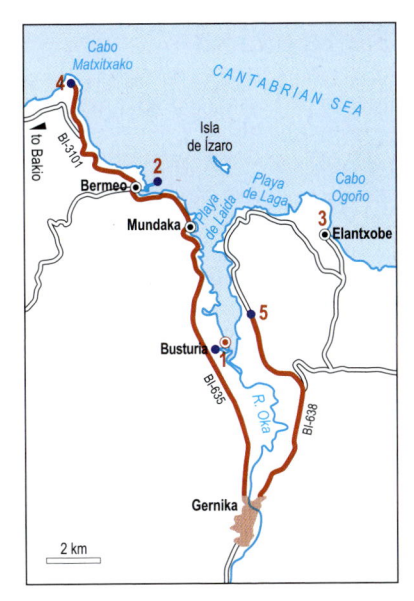

etation – that hold a great diversity of passerines. During cold spells, the estuary and nearby fields provide shelter for many wintering birds.

ITINERARY

[35 km] 🚗 🚶 ☀ ▬

From Gernika (the main town in the area) take the BI-635 to Mundaka and Bermeo along the left-hand bank of the Ría de Gernika (also possible by train) as far as San Cristóbal near Busturia and park in the railway station car-park (1). From here you can walk out onto a sandy polder fringed by reed-beds with Yellow Wagtail, Bluethroat on passage and Reed Bunting in winter. On the far side, a public hide provides good views of waders on the nearby mudflats.

Waders move with the tides: at low tide, birds are scattered all over the exposed mudflats, while at high tide they concentrate at the few points the tides do not cover. The best time to watch for waders is at the mid-point between high and low tide. This sector of the Ría is also good for Eurasian Spoonbill on passage at the end of summer.

The waters of the fishing port of Bermeo (2) are worth checking in winter for seabirds such as divers, Great Cormorants, Shag, gulls and auks. Look too on the rocks of the breakwater for Purple Sandpiper and Turnstone. The port of Elantxobe (3) on the other side of the Ría has similar species, as well as wintering Crag Martins.

North of Bermeo take the BI-3101 towards Bakio; after 3 km, turn right towards Cabo Matxitxako and its two lighthouses. This is the best sea-watching site in the Basque Country and during gales from the north-west shearwaters, Northern Gannets, scoters, Skuas, terns and auks travelling east-west are pushed close inshore. Peregrine Falcon and Blue Rock Thrush breed on the cliffs.

On the right-hand shore of the Ría take the BI-638 from Gernika northwards. After 3 km, turn left towards the beaches of Laida and Laga. Halfway you come to a hide at Kanala (5), overlooking the estuary and the damp grazing meadows and polders frequented in winter by Greylag Geese and other wildfowl and Eurasian Curlew.

MOST REPRESENTATIVE SPECIES

Residents: Shag, Peregrine Falcon, Yellow-legged Gull, Kingfisher, Crag Martin, Dipper, Blue Rock Thrush, Dartford and Sardinian Warblers, Bullfinch.
Summer visitors: Yellow Wagtail, Reed and Great Reed Warblers, Red-backed Shrike.
Winter visitors: Great Northern Diver, Great Cormorant, Greylag

PRACTICAL INFORMATION

- **Access:** Take the A8/E70 motorway from Bilbao and after 35 km turn off along the BI-635 towards Gernika and the Urdaibai Biosphere Reserve.
- **Accommodation:** in Gernika, Mundaka, Bermeo and other towns in the area. Varied rural accommodation available.
- **Visitor facilities:** Apart from the hide at San Cristóbal (left-bank of the Ría) there is no visitor centre or infrastructure for birdwatchers.
- **Visiting tips:** best September to March for passage and wintering birds. Summer is far less interesting and also there are many tourists in the area. Tide times – vital for wader watching – are given in local papers. Cabo Matxitxako is also a good site for watching for cetaceans.
- **Recommendations:** the company Aixerreku (tel. 946 870 244, aixerreku@euskalnet.net) has a team of wildlife guides and runs birdwatching outings and courses.
- **Further information:** maps 1:50,000, nº. 38 and 62 (SGE). The headquarters of the Urdaibai Biosphere Reserve is in the Palacio Udetxea in Gernika (tel. 946 257 125, urdaibai@ej.gv.es). The book *Urdaibai: Guía de aves acuáticas* by Jon Hidalgo and Joseba del Villar published by the Departamento de Medio Ambiente y Ordenación del Territorio of the Basque government (2004) provides much useful information.

Goose, Eurasian Wigeon, Red-breasted Merganser, Grey Plover, Purple Sandpiper, Dunlin, Eurasian Curlew, Greenshank, Turnstone, Black-headed and Common Gulls, Razorbill, Guillemot, Reed Bunting.
On migration: Balearic Shearwater, Northern Gannet, Eurasian Spoonbill, Osprey, waders, skuas, Mediterranean Gull, Kittiwake, Caspian Tern, Bluethroat.

37. TXINGUDI

Forming the frontier between Spain and France, the small estuary of the river Bidasoa flows into the heavily built-up bay of Txingudi, all but surrounded by the towns of Hondarribia and Irún (north-east Guipúzcoa) and Hendaya (France). However, areas of salt-marsh, artificially widened sandy beaches and sea cliffs still remain and together warrant protection as an SPA and a Ramsar Site. At the mouth of the Bidasoa (the best salmon river in the Basque country), patches of fluvial woodland, mudflats, wet meadows and reed-beds provide habitat for an interesting variety of birds.

ORNITHOLOGICAL INTEREST

Along with Urdaibai, this is the most important coastal wetland on the Basque coast. The complex of tidal lagoons protected by the Plaiaundi Ecological Park in Irún are the most significant part of the site and are home to good numbers of seabirds, wildfowl, waders, gulls and terns during migration and in winter; as well, cold snaps bring rarities to the area. Txingudi is also a good watch-point for trans-Pyrenean migrants such as raptors, pigeons and passerines and is also visited by significant numbers of Eurasian Spoonbills on passage.

ITINERARIES

1. 3 km

Before 1998, the land at the mouth of the river Bidasoa and its tribu-

THE BASQUE COUNTRY

Plaiaundi, Bahía de Txingudi © Anna Motis

tary the Jaizubía was occupied by a rubbish tip and intensive agriculture and industry. However, thanks to money from the European Union and the Basque Government (Ministry of Environment and Regional Planning), the area was transformed into the Plaiaundi Parke Ekologikoa (Plaiaundi Ecological Park) (1), a small but significant tidal wetland with two interconnected saline lagoons.

At the entrance to the park there is a car-park and an interpretation centre that overlooks a freshwater lagoon (a) with Little Grebe and Eurasian Coot, two species that are rare in tidal areas. From here an itinerary in shape of an '8' circumnavigates the twin saline lagoons of San Lorenzo (b) and Txoritegi (c), passing on its way five hides and two observation points before reaching the mudflats (Itzaberri) on the estuary: look here in winter for Great Northern Diver and Black-necked Grebe.

The whole itinerary is good for heron, wildfowl, gulls (including Mediterranean Gull), terns and, at low tide, waders. This is also an excellent place in post-breeding migration (August to October) for Eurasian Spoonbill, as well as for Osprey and Caspian Tern. The reed-beds hold Water Rail, Bluethroat and Penduline Tit on passage and Reed Bunting in winter. Little Ringed Plovers breed on the islets in the lagoons.

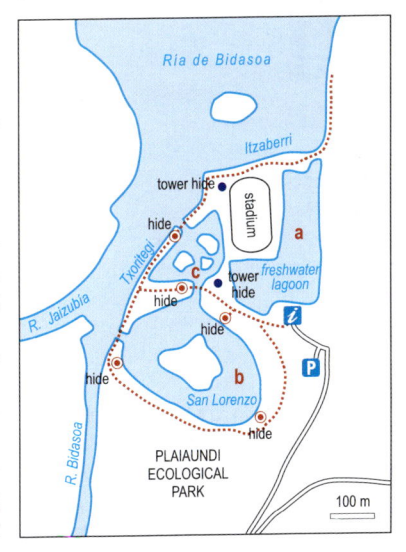

2. 4 km

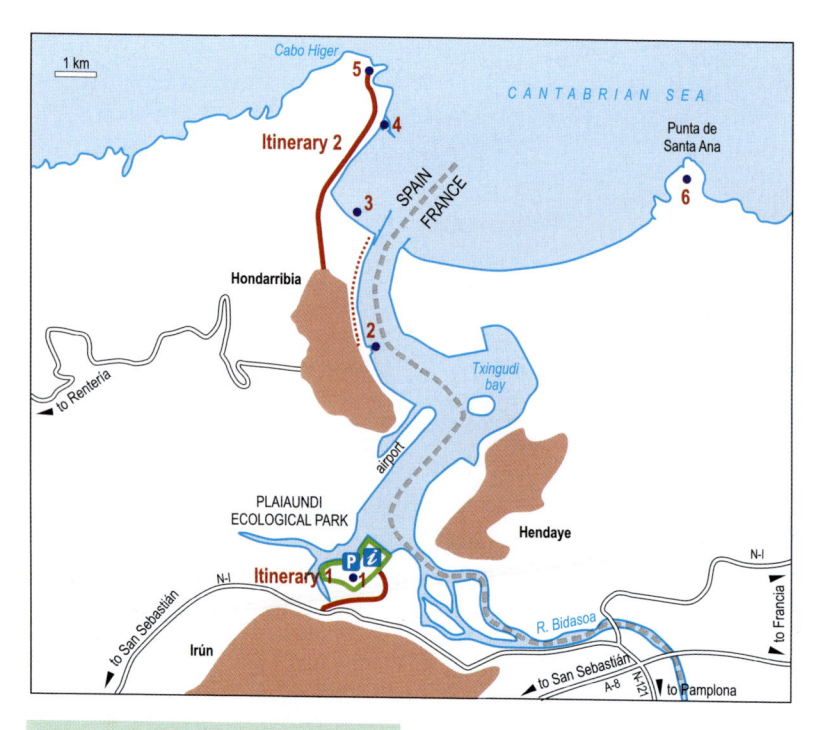

In the town of Hondarribia it is worthwhile walking along Paseo Butrón (2) from the town centre to the breakwater and beach (3). In winter this short walk gives views over the estuary and is a good site for grebes, divers and sea duck such as Red-breasted Merganser; the beach often holds groups of gulls, including Great Black-backed Gull during winter.

From here, head to the fishing port (4) that provides shelter (above all after gales) for seabirds such as Sandwich, Common and Black Terns, Razorbill and Guillemot. From the port, a road leads towards the lighthouse on Cabo Higer and ends next to a camp-site. This is an excellent seawatching site from August to November, above all when the wind is blowing strongly from the north-west: look for Balearic Shearwater, Northern Gannet, Great Skua, Little Gull, Kittiwake and other species of seabird.

Cabo Higer is the western-most point of Txingudi: a similar site – Punta de Santa Ana – lies on the eastern side of the bay (in France) and is equally good for seawatching and also has Peregrine Falcon and Shag.

MOST REPRESENTATIVE SPECIES

Residents: Little Grebe, Shag, Peregrine Falcon, Water Rail, Kingfisher.
Summer visitors: Little Ringed Plover, Grasshopper and Reed Warblers, Red-backed Shrike.
Winter visitors: Black-necked Grebe, Great Northern Diver, Great Cormorant, Shelduck, Red-breasted Merganser and other ducks, Grey Plover, Dunlin, Common Snipe, Eurasian Curlew, Razorbill, Guillemot, Reed Bunting.

THE BASQUE COUNTRY

On migration: Balearic Shearwater, Northern Gannet, Black Stork, Eurasian Spoonbill, Greylag Goose, Osprey, waders, Great Skua, Mediterranean and Little Gulls, Kittiwake, Caspian, Sandwich, Common and Black Terns, Sedge Warbler, Penduline Tit.

PRACTICAL INFORMATION

- **Access:** the N-1 passes through the northern suburbs of Irún towards the district of Behobia. At km point 478 (just before the level crossing), follow the signs to Plaiaundi. From Irún access is simple to Hondarribia and Cabo Higer.
- **Accommodation:** many places to stay in Hondarribia, Irún and Hendaya.
- **Visitor facilities:** The Plaiaundi Ecological Park has an interpretation centre (tel. 943 619 389, ekogarapen@terra.es) open every day (except Monday), morning and afternoon. Guided visits organised.
- **Visiting tips:** All year is interesting, although September to May is the best time for wintering and passage seabirds. Cars are not allowed inside Plaiaundi.
- **Recommendations:** The company Jolaski (tel. 943 616 447, jolaski@jazzfree.com) runs boat trips from Hondarribia out into the estuary.
- **Further information:** map 1:50,000, n°. 41 (SGE). Useful books: *Txingudi* by Mikel Etxaniz *et al.* and *Avifauna de Txingudi* by Josetxo Riofrío (1998 and 2000, respectively), published by the Basque Government's Departamento de Ordenación del Territorio y Medio Ambiente.

THE CANARY ISLANDS

38. Forests of Tenerife
39. Southern Tenerife
40. La Gomera
41. Fuerteventura
42. Lanzarote

The Canary Islands, as is typical in island bird communities, are relatively species-poor. However, what these seven subtropical volcanic islands lack in variety, is more than made up for by the singularity of their birdlife and these Atlantic islands have long been a favourite haunt of birdwatchers from all over the world.

Lying off the coast of West Africa and more than 1,000 km from the Iberian Peninsula, along with the Azores, Madeira and Cape Verde, this archipelago is part of the Macaronesia biogeographical zone. Most of these densely populated islands' inhabitants live on Tenerife and Gran Canaria, the two main islands, and annually millions of tourists arrive attracted by the warm and generally stable climate.

This mass tourism is the base of the local economy but also the main reason why most of the natural habitats of the archipelago – above all along the coastal strips – have been severely altered. However, many of the natural areas that have escaped development are protected by an extensive network of reserves that covers 40% of the islands and includes four national parks. One of the main functions of this network is the preservation of the natural habitats that are vital for the conservation of the islands' endemic species of birds, present here in far greater number than in continental Europe.

No less than four bird species (Bolle's and Laurel Pigeons, Fuerteventura Stonechat and Blue Chaffinch) are endemic to these islands; a fifth (the Canarian Black Oystercatcher) is now extinct and three others (Plain Swift, Berthelot's Pipit and Atlantic Canary) are endemic to Macaronesia. Furthermore, the Canary Islands can boast over 30 endemic subspecies, including two (Canary Island Chiffchaff and Canary Island Kinglet) that in the light of recent genetic studies are now generally recognised as good species.

Tenerife is the largest of the islands and the one that offers the greatest diversity of natural habitats. Its forests (38), above all composed of Canary pine, are home to interesting bird communities, of which the outstanding element is the Blue Chaffinch. Some of the picnic sites in the Teide National Park are excellent sites for this very special species.

Blue Chaffinch © Aurelio Martín

The tourist infrastructures make a visit to Tenerife very easy. In the highly touristy southern part of the island (39), some of the artificial pools and a number of coastal sites are of interest for waterfowl, including migrant and wintering waders and regular rarities from, above all, North America (over 50 recorded).

Visitors to Tenerife can still enjoy some of the island's few remaining patches of laurisilva forest, a relict from the Tertiary Era that survived the Quaternary glacial periods. Nevertheless, the best forests of this type are to be found on La Gomera (40) in the Garajonay National Park, created to protect this habitat and its Bolle's and Laurel Pigeons.

Fuerteventura (41) and Lanzarote (42) are the eastern-most islands of the archipelago and, as such, the nearest to Africa. Their stark volcanic landscapes lack any forest cover and are extremely arid; however, these islands' sandy and stony plains (known as *jables*) are breeding sites for Houbara Bustard, Cream-coloured Courser and other steppe species, whose cryptic colouration serves them well in the desert-like habitats they frequent.

The rugged landscapes of Fuerteventura, scored by myriads of gullies, are frequented by the endemic Fuerteventura Stonechat and the *guirre*, the local name for the Egyptian Vulture. Lanzarote and the offshore islands of the Chinijo peninsula are home to seabirds such as Bulwer's Petrel, Little Shearwater and White-faced and Madeira Storm-petrels, species that are unknown in the Mediterranean. Osprey and Eleonora's Falcon also breed here, along with the Barbary Falcon, found throughout the Canary Islands, North Africa and Asia.

38. FORESTS OF TENERIFE

Densely populated and eminently touristy in nature, Tenerife is the largest (2,000 km^2) and also the most diverse of the Canary Islands. The

View of El Teide, Forests of Tenerife © José Manuel Reyero

central part of the island is dominated by El Teide, a volcano of over 3,700 m (the highest point in Spain), and the vast caldera, known as Las Cañadas, that surrounds it. The whole of this spectacular landscape is protected by a National Park.

The bare volcanic landscapes are surrounded by extensive forests, mostly included in the Corona Forestal Natural Park (almost 50,000 ha), which has also been designated as an SPA. Canary pines dominate, although on the humid slopes of the north-facing La Orotava valley there are dense patches of sub-humid forest (*monteverde*) and relict stands of laurisilva.

ORNITHOLOGICAL INTEREST

The forests of Tenerife harbour most of the species and subspecies that are endemic to these islands, some in abundance and all generally well-distributed. Three important species – Blue Chaffinch and Bolle's and Laurel Pigeons – are the avian highlights of the island; the first is found in the pine forests and the two pigeons in the laurisilva forests. Other forest birds include island forms of Sparrowhawk, Common Buzzard, Great Spotted Woodpecker, Robin and Blue Tit, as well as Canary Islands Kinglet and the omnipotent Atlantic Canary.

ITINERARY

[70 km]

The TF-21 crosses the island and links the towns of La Orotava in the north and Granadilla de Abona in the south, passing through Las Cañadas del Teide and the island's main forest habitats on the way.

Leave Orotava and after a first climb of 8 km, in the village of El Camino de Chasna (1) turn right to Los Realejos and Benijos (2). After a further 4 km, look for another left turn, this time signposted to the picnic spot of Chanajiga. Park and con-

THE CANARY ISLANDS

the frequent low clouds may hamper observations. Other forest species include Sparrowhawk, Common Buzzard, Canary Islands Chiffchaff and the Canary Islands subspecies of Robin, Blackbird and Common Chaffinch.

Back on the TF-21 continue to climb towards Las Cañadas. At approximately 9 and 24 km from La Orotava it is worth stopping at two more picnic spots, La Caldera (4) and Ramón el Caminero (5), the latter fully in the pine forests and the haunt of Great Spotted Woodpecker and Blue Chaffinch.

Just over 30 km from La Orotava and just after the junction with the TF-24 (road from La Laguna), you reach the northern entrance to the National Park at the interpretation centre of El Portillo (6). By this time, the pine forests have given way to bare montane slopes and in the gardens around the centre and the car-park it is easy to find Berthelot's Pipit. Common Kestrel and the local

tinue along the track, stopping at each bend for views over the forest.

This walk takes you through the laurisilva forest of Tigaiga, the best such forest remaining in the valley of La Orotava. Here, both Bolle's and Laurel Pigeons are present, although

PRACTICAL INFORMATION

- **Access:** reaching La Orotava by road is easy from all the airports and main towns in Tenerife. A car is essential for getting around.
- **Accommodation:** great variety, above all on the coast, but also rural accommodation in the interior. In Las Cañadas del Teide there is a Parador (tel. 922 386 415).
- **Visitor facilities:** the National Park has two visitor centres, open every day from 9.00 to 16.00: El Portillo (tel. 922 356 000), the main centre, and Cañada Blanca. Pre-booking is essential for guided trips in the park. The central offices (tel. 922 290 129) of the park are in Santa Cruz de Tenerife, although they are to be moved to La Orotava and a new interpretation centre opened. A chair-lift runs up to the top of El Teide.
- **Visiting tips:** the most interesting birds are present all year, although spring and summer are the best periods. Take care on the minor roads, which are narrow and very tortuous and are often covered by the '*mar de nubes*' (low-lying cloud). The climate is changeable; take precautions against the heat and the cold. In winter, some of the roads in the National Park are closed by snow.
- **Recommendations:** the Canary Islands branch of SEO/BirdLife (tel. 922 252 129, canarias@seo.org) provides information and organises bird-related activities.
- **Further information:** maps 1:50,000, n°. 1089, 1093, 1094 and 1098 (SGE). Also, the web page www.mma.es/parques/lared/teide.

subspecies of Southern Grey Shrike occur in the area, while drinking pools especially designed for birds are frequented by Turtle Dove and Atlantic Canary. In summer Plain Swifts overfly the area.

The TF-24 continues on into the grandiose scenery of Las Cañadas and passes by the Parador (7) and a second interpretation centre (Cañada Blanca), before reaching the junction with the TF-38 from Chío at Boca de Tauce (8). Keep on the TF-24 towards Vilaflor and soon you will leave the high mountain landscape and re-enter the pine forests, here much more south-facing and thus drier.

The picnic spot of Las Lajas (9) is on the right of the road and there is room to park and walk. Many birds haunt the forest here, including Great Spotted Woodpecker, Canary Island Kinglet, Blue Tit and Atlantic Canary. It is also, however, *the* spot for birdwatchers wanting to see the Blue Chaffinch. Avoid midday and, above all weekends and public holidays, when visitor noise and numbers makes birdwatching difficult.

MOST REPRESENTATIVE SPECIES

Residents: Sparrowhawk, Common Buzzard, Common Kestrel, Barbary Partridge, Rock Dove, Bolle's and Laurel Pigeons, Great Spotted Woodpecker, Berthelot's Pipit, Grey Wagtail, Robin, Blackbird, Blackcap, Sardinian Warbler, Canary Islands Chiffchaff, Canary Islands Kinglet, Blue Tit, Southern Grey Shrike, Raven, Spanish Sparrow, Blue and Common Chaffinches, Atlantic Canary, Linnet.

Summer visitors: Turtle Dove, Plain Swift.

39. SOUTHERN TENERIFE

The extreme south of Tenerife is characterised by its barren, semi-

Southern Tenerife © Eduardo de Juana

desert landscapes of dry gullies, *malpaíses* (lava flows) and sandy and stony plains covered by a highly specialised vegetation. The coastal strip is much lower and less rugged than in the north of the island and has long sandy beaches, sand-banks and a few natural lagoons. However, much of this area has been altered by human activity, ranging from traditional forms of cultivation employed in the past and more modern greenhouse-based cultivation, to the sprawling tourist and urban development of today. Two *reservas naturales* – Malpaís de Rasca and Montaña Roja – protect part of the area.

ORNITHOLOGICAL INTEREST

Many interesting aquatic birds, above all waders, stop over on migration or spend time in winter in the artificial lakes, sand-banks and coastal lagoons in the south of the island. Tenerife's only Kentish Plovers breed here, and vagrant American waders turn up regularly. Otherwise, this section of the coastline has breeding Cory's Shearwater and other seabirds. A good selection of steppe birds – mostly endemic subspecies – frequents the inland plains and rough ground.

ITINERARIES

1. 8 km

Turn off the Autopista del Sur (TF-1) at km 56 (junction 22) (1) towards the tourist development of El Médano. Continue along the main street as far as the old wharf (2), where you should park. Check out the sand-banks on both sides of the breakwater for Whimbrel, Turnstone and other waders at low tide.

From El Médano, take the road towards the village of Los Abrigos. After 1 km, you pass a hotel and then after almost another kilometre, park on the left in a car-park. Continue on foot towards a small bunker, with Montaña Roja, the eroded volcanic cone that dominates the whole area, on your right. Halfway along this stretch a small lagoon (5), La Mareta, will appear on your left. This is a good place for waders, gulls and terns, as is the coast at the end of the path. This is the only site in Tenerife for Kentish Plover and there are often good numbers of other waders to be seen here.

2. 12 km

From the Autopista del Sur (TF-1), turn off at km 62 (junction 24) (6) to-

PRACTICAL INFORMATION

- **Access:** The Reina Sofía Airport is in the south of Tenerife. Access to the Autopista del Sur (TF-1) is easy from anywhere on the island and all junctions for the itineraries are well signposted. A car is essential for getting around.
- **Accommodation:** good choice, above all in the tourist resorts of Los Cristianos and Playa de las Américas.
- **Visitor facilities:** no visitor centre or infrastructure for birdwatchers.
- **Visiting tips:** All year is of interest. Temperatures are relatively high all year around and so take the appropriate precautions.
- **Further information:** maps 1:50,000, nº. 1106 and 1107 (SGE). *Las aves de El Médano* by Juan Antonio Lorenzo and Julio González (Atan, 1993) is of interest.

wards Las Galletas. Once there, continue on to Guaza, although after a little while, turn left into the district known as El Fraile. Pick up the road nearest to the coast; at the end of this road, just before a football pitch, turn left along a wide track (7) that will take you towards an area of greenhouses and a barrier, where you can park. Here you will see the artificial lake known as El Fraile or Los Bonny (8), surrounded by a wall with plenty of gaps that allow for good viewing. As this private property, please be discreet and respectful.

Eurasian Coot and Little Ringed Plover breed (this is one of the few sites in the Canary Islands for the former), although the real fame of this pool is the surprising and regular presence of vagrant birds from the Americas, above all ducks (Ring-necked Duck and Lesser Scaup) and waders (American Golden Plover, Buff-breasted Sandpiper and White-rumped Sandpiper). The fact that these rarities share habitat with common European species allow birdwatchers to make good comparisons between species.

At the opposite end of the pool to the access point there is another small pool; pass by this second pool and cross an area of abandoned fields and *malpaís* (9), where Berthelot's Pipit is common and other species of open areas such as Trumpeter Finch and Barbary Partridge can be found. You end up on an asphalted track; turn left and in less than 1 km you will reach the Rasca lighthouse (10). From here, look out for Cory's Shearwater, Yellow-legged Gull and other birds passing offshore; Bulwer's Petrel and Little Shearwater are seen on occasions.

MOST REPRESENTATIVE SPECIES

Residents: Common Kestrel, Barbary Falcon, Barbary Partridge, Eurasian Coot, Stone Curlew, Kentish Plover, Yellow-legged Gull, Rock and Collared Doves, Feral Pigeon, Hoopoe, Berthelot's Pipit, Grey Wagtail, Spectacled Warbler, Southern Grey Shrike, Spanish Sparrow, Trumpeter Finch.

Summer visitors: Cory's Shearwater, White Stork, Little Ringed Plover, Turtle Dove, Plain Swift.

Winter visitors and on migration: Little Egret, Grey Heron, gulls, terns and waders, as well as passerines on migration.

40. LA GOMERA

Much of the subtropical laurisilva in the central mountains of this small abrupt circular island (370 km^2) re-

THE CANARY ISLANDS

Garajonay National Park © Pedro Retamar

mains intact and most is today protected by the Garajonay National Park (almost 4,000 ha), the cornerstone of a network of almost 20 protected areas covering nearly a third of the island. The National Park is also an SPA, as are the coastal cliffs of Alajeró in the south of the island and Los Órganos in the north.

Garajonay is surrounded by a belt of pastures, hillsides and deep gullies, which is also where most of the island's population live. Traditional agriculture and stock-raising, well adapted to the island's complex relief, are still important and are typified by the serried ranks of terraces that climb the steep hillsides.

ORNITHOLOGICAL INTEREST

The laurisilva forests are home to Bolle's and Laurel Pigeons, two true Canary Islands endemics, as well as endemic forms shared with the other archipelagos of species such as Common Buzzard, Common Kestrel and Sparrowhawk, Blue Tit, Raven and Common Chaffinch. On the coastal cliffs, all but inaccessible, a few pairs of Osprey and Barbary Falcon breed, along with important colonies of shearwaters, petrels, storm-petrels and gulls.

ITINERARY

[75 km]

From the island capital of San Sebastián, port for the ferries to and from Tenerife, take the TF-711 (Carretera General del Norte) towards Hermigua. On this first stretch of road, stop in places with good visibility (1, 2, 3) for Common Buzzard, Common Kestrel, Barbary Partridge, Plain Swift, Berthelot's Pipit, Raven and Atlantic Canary.

After passing through the tunnel of Cumbre, on your right you will find the viewpoint of Las Caboneras (4), a classic site for Bolle's and Laurel Pigeons. Both species, as well as others from the *monteverde* (sub-humid forest), can be seen here or if you turn left a little further on along a road to the viewpoints of El Rejo (5) and El Bailadero (6), both in the National Park.

Continuing along this road you will come to the TF-713 (Carretera Central or Dorsal); turn right and, ignoring all side roads, carry on as far as the picnic site of La Laguna Grande (7). This is a good place for searching the forests for the local forms of Blackbird, Blue Tit and Common Chaffinch, as well as Canary Islands Chiffchaff, Canary Islands Kinglet.

A few metres before the entrance to the picnic site, a road off to the right passes through excellent laurisilva formations, and it is worth stopping at the viewpoints of Vallehermoso (8) on your left and Lomo del Dinero (9) on your right. From here the road continues directly to Juego de Bolas, the National Park visitor centre (10), and then down again to the TF-711. Turn right and return through the villages of Agulo and Hermigua to San Sebastián.

Once back at the port and if you have time before the ferry leaves, pay a visit to La Torre del Conde and its gardens to look for migrants, or the mouth of the Barranco de La Villa, where small pools attract waders and the Canary Island subspecies of the Grey Wagtail. On the nearby beach, gulls and terns rest and ospreys occasionally pass by.

Behind the breakwater and just opposite the ferry terminal you can gain access to Playa de la Cueva, where you can spend you last minutes on La Gomera seawatching. Cory's Shearwater and Yellow-legged Gull are common, but with luck you may spot Bulwer's Petrel or Little Shearwater. And if you are out of luck here, the return boat trip offers even better opportunities.

MOST REPRESENTATIVE SPECIES

Residents: Sparrowhawk, Common Buzzard, Osprey, Common Kestrel, Barbary Falcon, Yellow-legged Gull, Rock and Collared Doves, Bolle's, Laurel and Feral Pigeons, Berthelot's Pipit, Grey Wagtail, Robin, Blackbird,

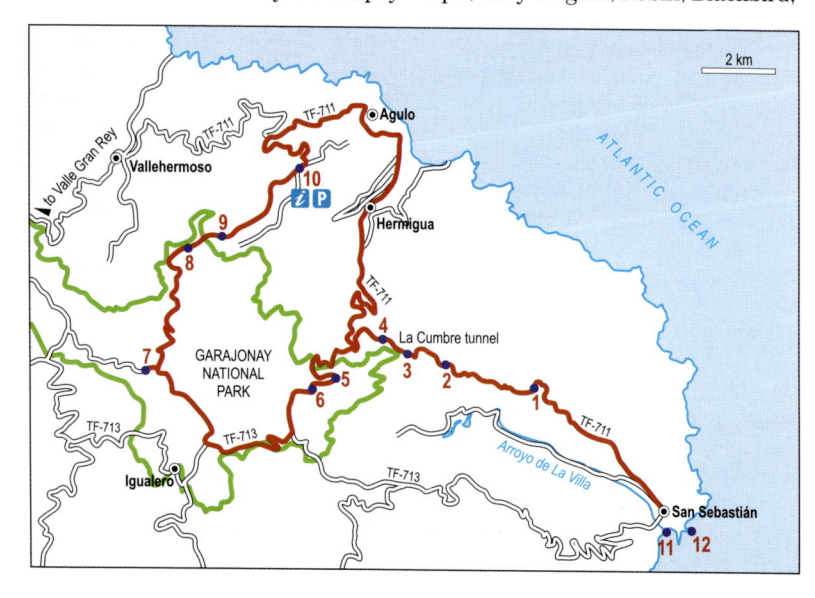

THE CANARY ISLANDS

> **PRACTICAL INFORMATION**
>
> - **Access:** La Gomera (San Sebastián) can be reached by ferry from Los Cristianos in the south of Tenerife. For short stays, a car is essential, and can be hired on the island.
> - **Accommodation:** in San Sebastián, Playa Santiago and Valle Gran Rey. There is also a lot of rural accommodation.
> - **Visitor facilities:** The National Park has a visitor centre – Juego de Bolas (tel. 922 800 993) – located in the north of the island. Open, Monday to Sunday, 9.30 to 16.30. Guided walks are organised on Saturdays.
> - **Visiting tips:** Good all year, but above all in spring and summer. The roads, narrow and with many tight bends, head up into the central part of the island where low clouds are frequent and care must be taken. Changeable climate; carry waterproof and warm clothing at all times.
> - **Recommendations:** Garajonay is a must. Reach it via the main roads of the island in order to get to know the laurisilva, a singular habitat with many endemic species of flora and fauna.
> - **Further information:** maps 1:50,000, n°. 1097 and 1105 (SGE).

Blackcap, Sardinian Warbler, Canary Islands Chiffchaff, Canary Islands Kinglet, Blue Tit, Raven, Spanish Sparrow, Common Chaffinch, Atlantic Canary, Linnet.
Summer visitors: Bulwer's Petrel, Cory's and Little Shearwaters, European Storm-petrel, Turtle Dove, Plain Swift.
Winter visitors: Grey Heron, Little Egret, waders, Black-headed and Lesser Black-backed Gulls, Sandwich Tern.

41. FUERTEVENTURA

For most of its 100-km length, the landscapes of this long, desert-like island are dominated by sandy (*jables*) or stony plains, only interrupted in a few places by volcanic cones, mountains and dry gullies. The ingenious windmills and the ranks of terraces (*gavias*) in the agricultural estates (*rosas*) reflect an agricultural system that has adapted to the harsh natural conditions by exploiting the scarce rains and frequent winds to the full. Nevertheless, today the countryside is all but totally abandoned and the main economic activity on the island is tourism. There are still many goats, however, whose milk is used for cheese production, although the fact that they graze freely only aggravates the inherent aridity of the island.

ORNITHOLOGICAL INTEREST

Highly interesting avifauna that includes Houbara Bustard, Cream-coloured Courser and Black-bellied Sandgrouse, the latter only found here in the Canary Islands. Island forms of Stone Curlew, Lesser Short-toed Lark and Trumpeter Finch occur. The mountain areas are home to Barbary Falcon and the endemic subspecies of Egyptian Vulture, Common Buzzard, Common Kestrel and the scarce Barn Owl. In a few places, the eastern Canary race of the Blue Tit and the Fuerteventura Stonechat – endemic to the island – are found and there are a few sites for breeding Ruddy Shelduck, the only sites in Spain, and very occasionally, Marbled Duck.

Betancuria, Fuerteventura © Pedro Retamar

ITINERARIES

1. ⏐60 km⏐ 🚗 🚶 ☀ ➡

Just over 20 km from Puerto del Rosario along the FV-10 you will find the village of La Oliva. From here, continue on the FV-10 towards Lajares. After 2 km on a bend, park just where a road branches off left next to a small estate called Rosa de los Negrines (1), where Houbara Bustards come to feed in the morning and evening. Wait by the road as the bustards come down from the slopes on the left. Also here, look for migrant passerines in the fields.

Another option consists of taking the left turn southwards (2) off the FV-10, 15 km from Puerto del Rosario. A little after passing through Tefía, turn right towards Puertito de Los Molinos. After the small hamlet of Las Parcelas (Colonia García Escámez), take a track off to the left (3) just before a sharp right-hand bend. Continue along this track alongside Barranco de Los Molinos as far as a small dam (4), where you should park and continue on foot.

This dam is a good place for Ruddy Shelduck, the occasional Marbled Duck, Eurasian Coot (rare in the Canary Islands) and waders on migration and in winter (above all at the end of the reservoir). The surrounding hills hold Fuerteventura Stonechat and the plains are a good place for steppe birds.

2. ⏐25 km⏐ 🚗 🚶 ☀ ➡

From Puerto del Rosario – or the nearby airport – head south along the FV-2 through the tourist town of Caleta de Fustes. Just 3 km further on you reach a small group of saltpans (5), where you should look for waders, gulls and terns. As well, walk northwards along the beach for birds feeding at low tide on the sandbanks.

THE CANARY ISLANDS

From the salt-pans, continue along the FV-2 for 3 km more until you find a dirt track on your left (6). Park with care and walk parallel to the road for 700 m until you reach the point at which the road crosses the semi-dry gully of La Torre. Drop down into the gully and walk downstream (seawards) for a couple of kilometres through a zone with pools of water and abundant patches of tamarisk. Fuerteventura Stonechats breed here, and this is also a good site for Ruddy Shelduck, Little Ringed Plover, Sardinian and Spectacled Warblers and Trumpeter Finch.

3. 40 km 🚗 🚶 ☀️ ➡️

From the town of Antigua, some 20 km from Puerto del Rosario, head for Betancuria, the former capital of the island, situated up in one of the most important mountain ranges on the island. A few miles after Betancuria, turn right and continue through the village of Vega de Río Palmas and on to the small reservoir of Las Peñitas.

Both the tamarisk bushes at the end of the reservoir and the surrounding fields and hills are rewarding places for birdwatching. The endemic form of the Blue Tit is easy to find and in the rocky outcrops you should look for Egyptian Vulture, Common Buzzard and Common Kestrel. Park near the point the road crosses the gully of Las Peñitas and walk as far as the dam.

MOST REPRESENTATIVE SPECIES

Residents: Egyptian Vulture, Common Buzzard, Common Kestrel, Barbary Falcon, Barbary Partridge, Eurasian Coot, Houbara Bustard, Cream-coloured Courser, Stone Curlew, Little Ringed and Kentish Plovers, Black-bellied Sandgrouse, Lesser Short-toed Lark, Berthelot's Pipit, Fuerteventura Stonechat, Sardinian and Spectacled Warblers, Blue Tit, Southern Grey Shrike, Raven, Spanish Sparrow, Atlantic Canary, Greenfinch, Linnet, Trumpeter Finch.

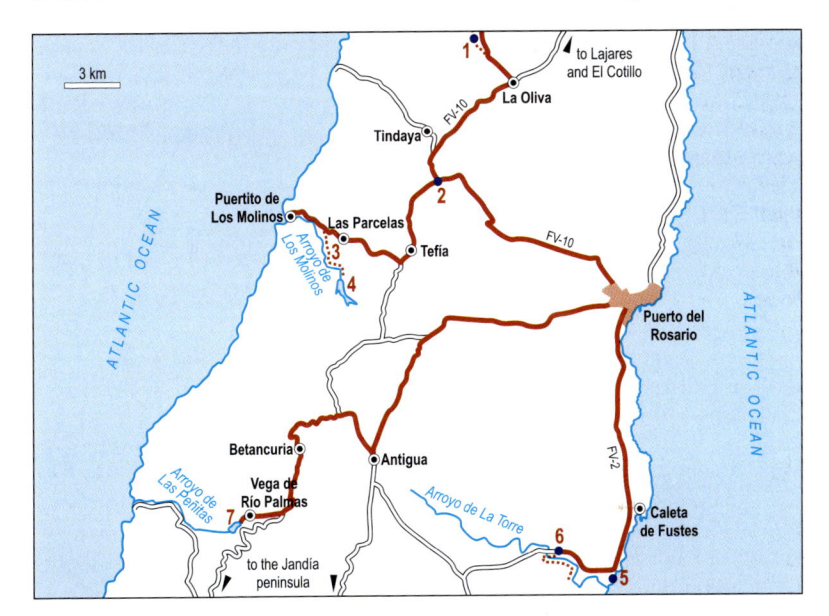

PRACTICAL INFORMATION

- **Access:** from Puerto del Rosario, the island capital, good roads lead to all the itinerary starting points.
- **Accommodation:** in the tourist towns on the coast and in rural accommodation inland.
- **Visitor facilities:** no visitor centre or infrastructure for birdwatchers.
- **Visiting tips:** best in winter and spring, especially in steppe areas. The great diurnal temperature range means that you must protect yourselves against heat and cold. All the main roads are good, but take care on minor roads.
- **Recommendations:** In the south of Fuerteventura the sandy plains and the mountains of Jandía, the vast beach of Barlovento are part of an SPA. From Corralejo in the north of the island, boats leave for trips around the nearby islet of Lobos.
- **Further information:** maps 1:50,000, n°. 23/24-19/20 (SGE). The local tourist board run by the local administration (Cabildo de Fuerteventura) has a number of tourist information offices that can help you plan your visit: for example, in Puerto del Rosario (tel. 928 530 844) and the airport (tel. 928 860 604).

Summer visitors: Cory's Shearwater, Ruddy Shelduck, Marbled Duck, Quail, Black-winged Stilt, Turtle Dove, Plain and Pallid Swifts.

Winter visitors and on migration: herons, waders, gulls and terns along the coast; passerines in gardens and stands of trees (tamarisks).

42. LANZAROTE

The spectacular combination of landscapes on Lanzarote – for example, the endless sandy plains of El Jable de Famara and the mountains and cliffs of Famara in the north of the island – has stimulated the imagination of many visitors. A large part of the island is protected as an SPA and forms part of the Chinijo Archipelago Natural Park, along with the nearby islands of La Graciosa, Montaña Clara and Alegranza. Lanzarote has also been declared a Biosphere Reserve.

ORNITHOLOGICAL INTEREST

The island's sandy plains (*jables*) and cultivated fields are home to a number of steppe birds, including the Houbara Bustard, the scarce Cream-coloured Courser and endemic forms of Stone Curlew, Lesser Short-toed Lark and Trumpeter Finch. Sea-cliffs hold a number of seabird colonies – shearwaters and storm-petrels, for example – and a few pairs of Barbary Falcon and Osprey. A few pairs of the very scarce Canary Island subspecies of the Egyptian Vulture also breed.

The Chinijo archipelago is the best site for breeding seabirds and raptors and boasts the only colony of Eleonora's Falcon in the Canary Islands and the only Spanish White-faced Storm-petrels.

ITINERARIES

1. 25 km 🚗 ☀️ ➡️

The Jable de Famara is one of the best places in the Canary Islands to observe steppe birds. From the village of Teguise (12 km north of Arrecife), take the LZ-30 towards Yaiza. After 3 km, turn right (1) towards Caleta de Famara, at which point the fields and the sandy plains of the *jable* will be on your left (2).

THE CANARY ISLANDS

View of Isla Graciosa from the north coast of Lanzarote   Carlos Sánchez © nayadefilms.com

Although there are a number of tracks that cross the plains towards the village of Sóo, for finding the majority of special species of the area, it is probably best to simply stop here and there on the road and scan the plains with binoculars or, even better, with a telescope. The best time is morning and evening. Do not leave the road or the tracks so as to avoid disturbing the area's breeding birds; one of the greatest potential threats they face is human disturbance.

2. 90 km

To enter into the mountains of Famara, return to Teguise and take the KZ-10 towards Haría. As you climb you leave the *jable* behind and, after passing through Los Valles and just before Haría, stop at a viewpoint (3) where there are areas of cultivation with Atlantic Canary and other passerines. It is worth stopping awhile in and around this picturesque village to search for the endemic race of the Blue Tit and other resident and migratory species.

Continue on towards Máguez and, once through this small settlement, turn off towards Guinate, where the road ends at another viewpoint (4).

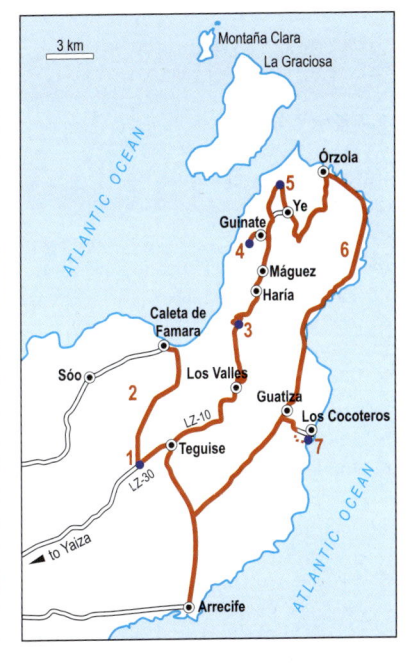

The views from here of the crags and cliffs falling away from your feet and the islets of Chinijo further away are superlative: look out for gulls, Raven and, at any moment, one of the numerous Barbary Falcons that breed here.

Return to the road and continue towards Ye. Just before this village, turn left to the viewpoint of El Río (5), from where the views are even more spectacular. This is another good site for Barbary Falcon, as well as for Eleonora's Falcon in summer and, with luck, Egyptian Vulture.

Instead of returning via Haría, continue to Órzola, where you can park. From here you can walk down to the coast along a breakwater to check out the sand-banks at low tide for migrant waders and other aquatic birds. Ospreys occasionally drop in to fish.

A coastal road through a stunning volcanic landscape (6) takes you to Guatiza, where you should head for the small holiday complex of Los Cocoteros. Just before you reach there, turn right to an area of salt-pans (7), with breeding Kentish Plover and herons, waders, gulls and terns on passage.

MOST REPRESENTATIVE SPECIES

Residents: Egyptian Vulture, Osprey, Common Kestrel, Barbary Falcon, Barbary Partridge, Houbara Bustard, Cream-coloured Courser, Stone Curlew, Kentish Plover, Rock and Collared Doves, Feral Pigeon, Barn Owl, Lesser Short-toed Lark, Berthelot's Pipit, Sardinian and Spectacled Warblers, Blue Tit, Southern Grey Shrike, Raven, Spanish Sparrow, Atlantic Canary, Linnet, Trumpeter Finch, Corn Bunting.

Summer visitors: Cory's Shearwater, Eleonora's Falcon, Quail, Turtle Dove, Plain and Pallid Swifts.

Winter visitors: White Wagtail, Black Redstart, Song Thrush, Robin, Chiffchaff.

On migration: herons, gulls, terns and waders, passerines.

PRACTICAL INFORMATION

- **Access:** Teguise, the starting point for the itineraries, is easily accessible from the capital Arrecife.
- **Accommodation:** ample variety in Arrecife, Costa Teguise, Puerto del Carmen and Playa Blanca.
- **Visitor facilities:** no visitor centre or infrastructure for birdwatchers.
- **Visiting tips:** best in winter and spring, especially in steppe areas. The great diurnal temperature range means that you must protect yourselves against heat and cold. All the main roads are good, but take care on minor roads.
- **Recommendations:** The Timanfaya National Park, famous for its volcanic landscapes, and the Janubio salt-pans, one of the most important sites for waders in the Canary Islands, are in the south of the island. The latter are accessible from the road between Yaiza and Playa Blanca. Another option is to visit La Graciosa on the daily boat from Órzola for its seabirds and raptors.
- **Further information:** maps 1:50,000, n°. 24-17/18 and 24-18/18 (SGE). The local tourist board run by the local administration (Cabildo de Lanzarote) has a number of tourist information offices that can help you plan your visit: for example, in Arrecife (tel. 928 811 762) and the airport (tel. 928 820 704).

CANTABRIA

43. La Liébana

44. Marismas de Santoña

Despite the fact that its mountain ranges are somewhat gentler and its natural habitats more intensely modified by human activity, Cantabria is otherwise fairly similar to neighbouring Asturias – sea to the north, mountains to the south and blessed with a mild, humid climate. Its birdlife is similar as well, with the most interesting areas for birdwatchers found in the inland montane forests and in the cliffs, islets, dunes, beaches and estuaries of the coast.

The Cantabrian Mountains are transitional between the high Asturian peaks and the gentler mountains in the Basque country to the east. The most rugged of all Cantabrian Mountains lie in the south-east of the region in the *comarca* of La Liébana (43), where Peña Vieja (2,613 m) tops off the eastern-most sector of the Picos de Europa. Lower down, its extensive montane forests are home to good populations of Black and Middle Spotted Woodpeckers.

In the most mature and undisturbed forests, a few Capercaillie hang on, although as elsewhere in the Cantabrian Mountains, this species is declining both in the number of birds and the size of its area of distribution. Around 20 male birds were counted in the 1980; today a handful survives, distant from the rest of the Cantabrian population.

Above the tree line there are a number of interesting alpine species. The cable car at Fuente Dé gives quick and easy access to a high rocky wastelands frequented by Alpine Accentor, Alpine Chough, Snow Finch and the elusive Wallcreeper. Nevertheless, it must be said that this type of infrastructure, orientated towards tourists wanting their slice of wilderness, also inevitably lead to a damaging massification of the very biodiversity that is brandished as a tourist attraction.

Most of the Cantabrian population lives in a 200-km long coastal strip that includes the main towns – Santander and Torrelavega – and all of the region's industry. Despite this coastal development, there are still a number of coastal sites such as the estuary of San Vicente de la Barquera and the beach at Oyambre, the Liencres dunes and the limestone cliff of Monte Candina (home to a singular Griffon Vulture colony) that have conserved their natural interest.

The human pressure on the Cantabrian coastline was somewhat halted by the European Court of Justice in Luxemburg with its sentence condemning Spain for having breached

Eurasian Spoonbill Carlos Sánchez © nayadefilms.com

the Birds Directive by allowing a road to be built through the magnificent Santoña salt-marshes (44), the most important wetland on the northern coast of Spain and known as the 'Doñana of the north'.

Despite being surrounded by built-up areas, industry and roads, Santoña is still a vitally important passage and wintering site for many of the water birds that breed in northern and central Europe, amongst which the Eurasian Spoonbill takes pride of place. In the environs there are important relict holm oak woods of great ecological interest that are home to species such as Sardinian Warbler, infrequent at these latitudes.

43. LA LIÉBANA

La Liébana lies in the far west of Cantabria and consists of a group of abrupt valleys lying in the shadow of the Central and Eastern massifs of the mountains of the Picos de Europa, the most imposing of all the Cantabrian Mountains. Up high, bare rugged limestone mountains are omnipresent; lower down, meadows, scrub and beech, Pyrenean, holm and even cork oak forests describe a cultural landscape that cannot fail to fascinate the visitor. In all, 20% of the Picos de Europa National Park (65,000 ha) – also declared an SPA – lies within Cantabria; the rest is shared between Asturias and Castilla-León.

CANTABRIA

La Liébana © Fernando Cámara, Foto-Ardeidas

ORNITHOLOGICAL INTEREST

As part of the Picos de Europa, La Liébana is also the best area in Cantabria for alpine birds such as Alpine Accentor, Alpine Chough, Snow Finch and Wallcreeper. Its forests hold the last few Cantabrian Capercaillies, as well as an important population of Black and Middle Spotted Woodpeckers. There are also good populations of cliff-breeding raptors such as Egyptian and Griffon Vultures and Golden Eagle and forest raptors such as Honey Buzzard and Short-toed Eagle. As well, birdwatchers should keep an eye out for Lammergeier.

ITINERARIES

1. 7 km

From the town of Panes, take the N-621 through the gorge of La Hermida in the north-east of the National Park. The imposing cliffs on either side of the road here hold Griffon Vulture and Golden Eagle and were also where Lammergeier and Bonelli's Eagle last bred in the Picos de Europa.

Just before the village of La Hermida, a footpath (1) heads off to the right. The first stretch of the path after the small HEP station follows the gorge of the river Urdón (look out for Dippers) and is flanked by woods with Black Woodpecker. Eventually, the track leaves the river and begins to climb steeply towards the remote village of Tresviso.

On the way there are various natural viewpoints such as El Balcón de Pilatos (2), which gives good views over the upper part of the gorge where there are a few Griffon Vulture nests. Also look out for Egyptian Vulture, Peregrine Falcon, Alpine Swift, Crag Martin, Rufous-tailed and Blue Rock Thrushes and Red-billed and Alpine Choughs.

Another worthwhile option is to continue on to Potes and take the road up to Fuente Dé. After 2 km and just before the village of Turieno, turn right (3) to Argüébanes. Park in the village and follow the path into a magnificent oak forest, where Middle Spotted Woodpecker, Pied Flycatcher, Common Redstart and Marsh Tit all breed.

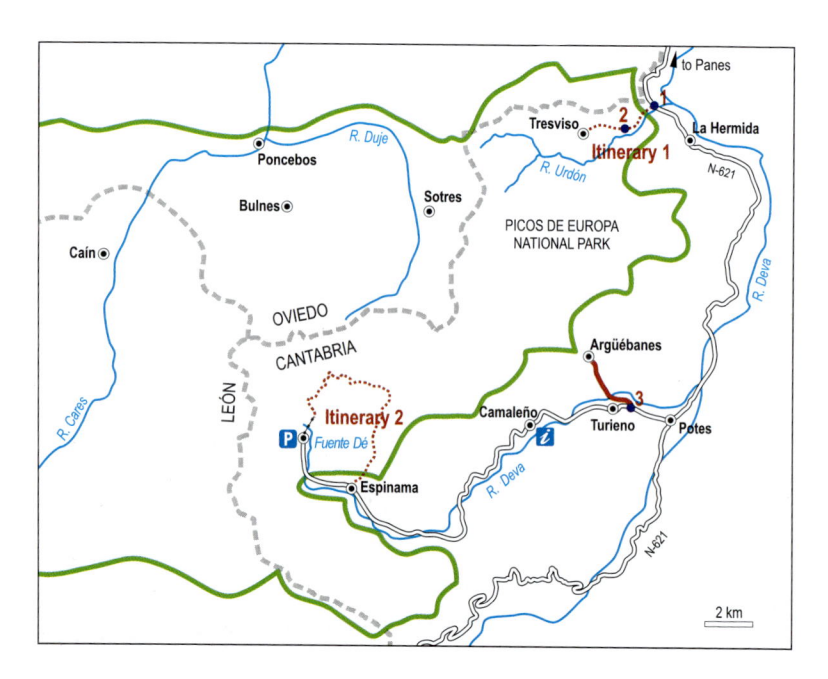

2. 12 km 🚶 ☀ ↻

From Fuente Dé in the south-east of the National Park, a cable car (known locally as 'El Cable') departs from next to the Parador and rises over 700 m without any apparent effort and deposits you in alpine wastelands with birds to match. From the upper station (4) at over 1,800 m, the main track rises to a pass, Horcadina de Covarrobres (5), before dropping down to the hotel-mountain hut of Áliva (6), surrounded by alpine pastures with Water Pipit and Northern Wheatear. Golden Eagle and other raptors fly over from time to time.

The route continues southwards, passing by a group of barns for cattle (Invernales de Igüedri, 7) and into an area of meadows and copses frequented by Red-backed Shrike and other passerines. The itinerary ends in the village of Espinama, where you should aim to leave a second vehicle (or take a taxi) to avoid the 3-km walk along the road back to Fuente Dé.

Once up at the top of the cable car, look for Alpine Accentor, Alpine and Red-billed Choughs and Snow Finch. Nevertheless, the best option is to head towards Horcadina de Covarrobres and then to turn left just before this pass along a track towards the cliffs of La Vueltona (8) where Wall-

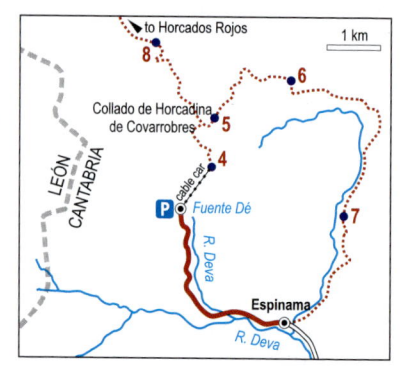

CANTABRIA

PRACTICAL INFORMATION

- **Access:** the N-621 between Panes in the north and the pass of San Glorio in the south provides access to all these itineraries. Turn off in Potes for the 20-km drive to the cable car at Fuente Dé.
- **Accommodation:** in Fuente Dé, Potes and other towns and villages in the area. The hotel-mountain hut at Áliva (tel. 942 730 999) is open from June to mid-October.
- **Visitor facilities:** Casa Dago (tel. 985 848 614) in Cangas de Onís (in Asturias and thus some distance from theses itineraries) is the main National Park interpretation centre. Open every day, morning and afternoon. There are also nearer information offices in the village of Tama and in Camaleño (tel. 942 730 555), both in La Liébana.
- **Visiting tips:** late spring and summer (crowded) are the best periods. In this abrupt area of mountains, it is best not to leave the paths, above all if the fog – frequent in this changeable and often wet mountain climate – comes down.
- **Recommendations:** in the summer the National Park organises guided walks throughout the Park.
- **Further information:** maps 1:50,000, n°. 56 and 81(SGE). Many books and maps on sale. Data on the natural heritage and public use of the Park at www.mma.es/parques/lared/picos.

creepers are often seen. If energy permits, continue on to the pass of Horcados Rojos, although in one day this option and the walk down to Espinama is too long and you will have to stay the night in the hotel-mountain hut of Áliva.

MOST REPRESENTATIVE SPECIES

Residents: Griffon Vulture, Golden Eagle, Common Kestrel, Peregrine Falcon, Black and Middle Spotted Woodpeckers, Crag Martin, Water Pipit, Dipper, Dunnock, Alpine Accentor, Bluethroat, Blue Rock Thrush, Marsh and Crested Tits, Wallcreeper, Eurasian Treecreeper, Red-billed and Alpine Choughs, Snow Finch.

Summer visitors: Honey Buzzard, Egyptian Vulture, Short-toed Eagle, Alpine Swift, Wryneck, Tawny and Tree Pipits, Northern Wheatear, Rufous-tailed Rock Thrush, Bonelli's Warbler, Red-backed Shrike.

44. MARISMAS DE SANTOÑA

Despite being surrounded by a dozen towns and villages, industry and roads, this wetland lying in the east of Cantabria is of exceptional natural interest and consists of the confluence of a number of estuaries, of which the most important is that of the river Asón (or Treto). Intertidal flats dominate the landscape and blend in with brackish grazing pastures, sand-banks, beaches and cliffs. The nearby hills are covered by some of the best coastal holm oak forests in Cantabria. The whole intertidal area is protected as a *reserva natural* (3,500 ha) and is also a Ramsar Site and part of a larger SPA.

ORNITHOLOGICAL INTEREST

This site is the most important refuge for migrating and wintering seabirds and wildfowl on the whole of the north coast of Spain. Sizeable groups of Eurasian Spoonbill appear

Santoña salt-marshes   © Eduardo de Juana

in autumn (a few winter), and there are also good numbers of herons, ducks, waders, gulls and auks, as well as birds from northern Europe such as divers, Eider, scoter and mergansers that are rare in Spain. Small numbers of Shag, Purple Heron and, recently, Shelduck and Cattle and Little Egrets breed.

ITINERARIES

1. 12 km 🚗 ☀ ➡

In the port of Santoña (1), make your way to the main breakwater, a good place in winter to observe divers, gulls and auks at close range. Nearby lies the Santoña industrial estate, from where you can explore the Bengoa salt-marshes along an earth bank (2) that leads across the marshes to a hide. This sector is best at high tide, when waders and other birds congregate here on the patches of land not covered by the incoming tide.

The same can be said of La Arenilla (3), an island that is used by birds such as Eurasian Spoonbill at high tide. Leave Santoña on the S-401 and after crossing the river Boo on the first bridge, there is a hide on the left which offers good views of this sector. Continue for a couple of kilometres to the turn off to the convent of Montehano (4), where you should park and check the Canal de Ano. This part of the estuary is deeper and is frequented by Red-necked Grebe and other diving species such as Great Northern, Arctic and Black-necked Divers, Common Eider, Common and Velvet Scoters and Red-breasted Merganser.

A little way further on, turn right towards Escalante and at a quarry (5) at the base of the limestone mountain of Montehano, stop to view the Escalante salt-marshes. The nearby brackish grazing marshes are visited by Greylag Goose, European Golden Plover and Eurasian Curlew and hunted over by the Peregrine Falcons that breed nearby.

Return to the S-401, which takes you on to the N-634. After just a few metres along the road towards Bil-

CANTABRIA

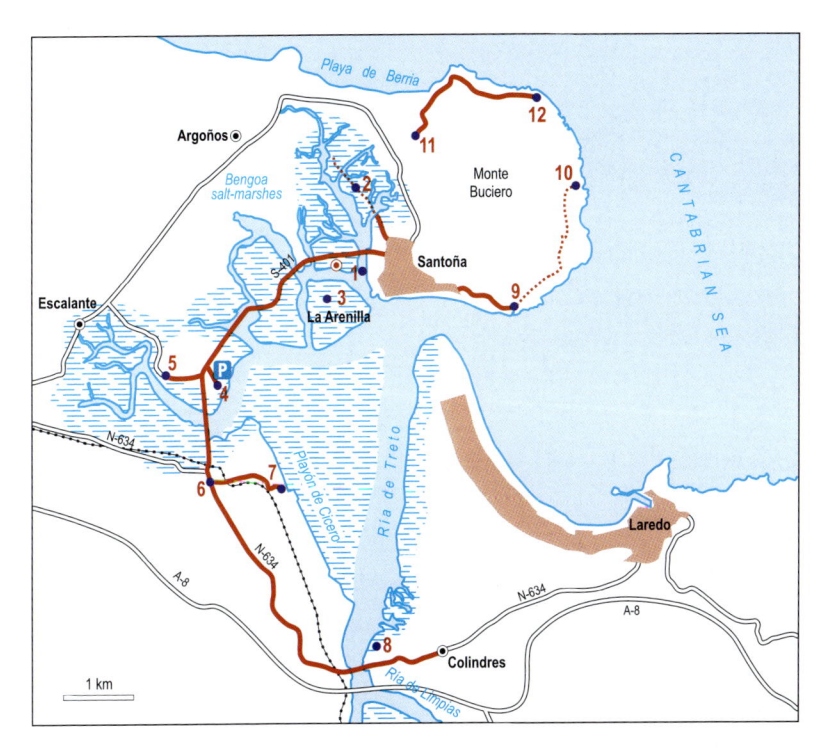

bao, turn left (6) and cross over and continue alongside a railway line. Approach the sand-banks of Playón de Cicero, the heart of this large wetland. The anchovy processing factory (7) is a good reference point: park and view the large concentrations of duck, above all of Eurasian Wigeon. At low tide, Dunlins and other waders feed here.

A final sea walk from the town of Colindres (8) will give good views of the Treto estuary and an island consisting of a breakwater that was once joined to dry land with an artificial islet at its far end. This structure is a good place to look for Avocet, Mediterranean Gull, terns and waders.

Monte Buciero near the town of Santoña is covered by a splendid holm oak forest whose edges are home to one of the few Sardinian Warbler populations in Cantabria. Park at Fuente de San Carlos (a natural spring) (9) and walk up to the lighthouse of Caballo (10) along the cliff-edge frequented by Shags. Another route begins in the district of El Dueso (11) or the nearby beach of Berria, from where a road will take you to the lighthouse of El Pescador (12), a good site for seawatching in autumn for Northern Gannet, shearwaters, skuas and other seabirds.

MOST REPRESENTATIVE SPECIES

Residents: Shag, Eurasian Spoonbill, Shelduck, Gadwall, Peregrine Falcon, Blue Rock Thrush, Sardinian Warbler.

Winter visitors: Red-throated, Black-throated and Great Northern Divers,

Red-necked and Black-necked Grebes, Great Cormorant, Greylag and Barnacle Geese, Eurasian Wigeon, Common Eider, Common and Velvet Scoters, Red-breasted Merganser, Marsh Harrier, Oystercatcher, European Golden and Grey Plovers, Dunlin, Black-tailed and Bar-tailed Godwits, Whimbrel, Eurasian Curlew, Common Redshank, Mediterranean Gull, Sandwich Tern, Razorbill, Guillemot.

On migration: Northern Gannet, herons, Osprey, Avocet and other waders, Caspian, Little and Black Terns.

PRACTICAL INFORMATION

- **Access:** leave the A-8 motorway from Bilbao or Santander at the turn-offs to Gama, Cicero or Colindres. The N-634 running parallel to the motorway is another possible means of access.
- **Accommodation:** in Laredo, Santoña and other nearby towns. The hostal run by the Santoña town council (tel. 942 662 008, alberguesantona@iespana.es) organises wildlife activities for groups.
- **Visitor facilities:** La Arenilla hide on the S-401 on the way out of Santoña.
- **Visiting tips:** any time of year is of interest. Autumn passage and winter are especially attractive times to visit; September is the best month for Eurasian Spoonbill.
- **Recommendations:** the Cantabrian delegation of SEO/BirdLife (tel. 942 223 351, cantabria@seo.org) organises wildlife walks, summer camps and other ornithological activities. More information on www.telefonica.net/web/seocantabria.
- **Further information:** maps 1:50,000, n°. 35 and 36 (SGE); the offices of the Marismas de Santoña y Noja Natural Reserve (tel. 942 233 503, santona@oapm.es). Useful book: *Aves marinas y acuáticas de las marismas de Santoña, Victoria, Joyel y otros humedales de Cantabria*, by Juan Manuel Pérez de Ana (Fundación Marcelino Botín, 2000).

CASTILLA-LA MANCHA

- 45. Tablas de Daimiel
- 46. La Mancha wetlands (I)
- 47. La Mancha wetlands (II)
- 48. Laguna de Pétrola
- 49. Castrejón and Azután
- 50. Valle del Tiétar
- 51. Alto Tajo
- 52. Cabañeros

The enormous sedimentary plain of La Mancha, a land of broad and all but perfect horizons, occupies the southern part of the central Spanish Meseta. As the largest area of plains in Spain, La Mancha is without doubt the natural and scenic element that best defines the autonomous community of Castilla-La Mancha, one of largest and least densely populated areas in Spain. La Mancha is also the economic and cultural centre of the region, the scene of the fictional adventures of Don Quijote and home to a thriving agricultural economy.

The natural landscapes of the region have been severely transformed (despite little industrial development) and today land-use is dominated by cereal cultivation, vineyards, olive groves and, increasingly, irrigated crops. Nevertheless, few other areas in central Spain – and so close to Madrid – can offer birdwatchers as much as Castilla-La Mancha. To begin with, La Mancha itself, along with other areas of plains such as La Sagra, Llanos de Oropesa, Campo de Montiel and Camp de Calatrava, boasts excellent communities of steppe birds that include Great Bustard and the Iberian Peninsula's best populations of Little Bustard, Stone Curlew and Pin-tailed Sandgrouse.

As well, the area is dotted with wetlands, of which the best known is Las Tablas de Daimiel (45). This National Park, lying at the confluence of the floodplains of the rivers Cigüela and Guadiana, attracts many visi-

Pin-tailed Sandgrouse © Gabriel Sierra

tors, despite the fact that the surrounding intensive agriculture so deprives Las Tablas of water (altering its natural hydrological balance in the process) that supplies must be pumped from the distant river Tajo to ensure its survival.

Las Tablas are also a Biosphere Reserve, Ramsar Site and an SPA, the same legal protection as that enjoyed by the groups of shallow lakes known as La Mancha Húmeda (46 and 47). The majority of these small wetlands lie in two areas: one near the town of Alcázar de San Juan and the other further east where the provinces of Toledo, Ciudad Real and Cuenca meet. Outside area but still essentially similar is La Laguna de Pétrola (48) in Albacete province. Despite the ecological importance of these temporary brackish lagoons, many have been drained to create agricultural land or are increasingly adulterated by the input of run-off water from towns and/or irrigation projects.

Nevertheless, their ornithological singularity is still immense, as they act as oasis for migrant birds travelling to and from Africa and for breeding birds in an otherwise highly arid landscape. Their breeding populations of Shelduck, White-headed Duck, Marsh Harrier, Black-headed Gull, Gull-billed Tern and Bearded Tit, as well as the wintering duck and passage waders that pass through, are nationally important.

This discussion of the wetlands of Castilla-La Mancha would not be complete without mention of some of the region's reservoirs – for example, Azután and Castrejón (49) on the river Tajo and Rosarito and Navalcán on the river Tiétar (50) in the piedmont of the Sierra de Gredos – whose water surfaces and surrounding pastures and forests are attractive to tens of thousands of birds.

Castilla-La Mancha is bordered by a series of imposing mountain ranges. The Sistema Central to the north and the Sierra Morena to the south mark the edge of the region's plains, while to the east its boundaries are formed by the high rounded plateaux of the Sistema Ibérico, a mountain area slashed open by deep limestone gorges along the upper reaches of rivers such as the Tajo (51) that are a paradise for cliff-breeding raptors and forest avifauna.

CASTILLA-LA MANCHA

Los Montes de Toledo, a much more modest range of mountains, spread out east-west through the centre of the region and separate the region's two major rivers, the Tajo and Guadiana. Abrupt quartzite ridges interrupt the gentler mountainsides that hold some of the Peninsula's best-preserved Mediterranean forests, often fenced-off as hunting estates and home to the Spanish Imperial Eagle and many other raptors.

The best-known area of Los Montes de Toledo is Cabañeros (52), earmarked as a bombing range in the 1980s, but saved at the last movement thanks to a popular protest that led to the declaration of a national park in 1995. Today, the Black Vultures and other creatures of Cabañeros are safe to enjoy the bounties of La Raña, a vast savannah-like grassland that has become the symbol of the park.

45. TABLAS DE DAIMIEL

In 1973 Las Tablas de Daimiel (Ciudad Real province; 1,900 ha), the smallest of all Spain's national parks, was created to protect the remaining acres of the flood plain lying at the confluence of the rivers Cigüela – brackish and seasonal – and Guadiana – fresh and permanent – from being drained and converted into agricultural land. Despite this legal protection, the over-exploitation of the area's water table has continued and today these once splendid wetlands have to be maintained in dry years by supplies of water from the Tajo-Segura canal.

Even so, this intricate site, full of channels, lagoons and islands, is the most important wetland in central Spain. The splendid aquatic vegetation of Las Tablas includes the largest extensions of great fen-sedge in Western Europe, in retreat as the reed-beds advance, and stands of tamarisk, all but the only tree to be found in the National Park. The hinterland is given over to irrigated agricultural land. The area is also a Biosphere Reserve, a Ramsar Site and an SPA.

Tablas de Daimiel National Park © Fernando Barrio

ORNITHOLOGICAL INTEREST

Las Tablas are the best site in Castilla-La Mancha for water birds, although species numbers depend to a large extent on water levels. Almost all the heron species found in Spain breed, including the scarce Great Bittern in some years and over 100 pairs of Purple Heron, and a mixed colony of around 1000 nests (including around 10 pairs of Squacco Heron) exists in a stand of tamarisks.

Duck numbers are important in winter and breeding duck include Red-crested Pochard (300 breeding pairs) and both Marbled and White-headed Ducks also often manage to raise young. Other breeders include Black-necked Grebe, Marsh Harrier, Purple Swamp-hen, colonial gulls, terns and waders, and numerous small reed-bed passerines.

Large build-ups of White Stork occur in late summer and over 3,000 Common Cranes winter. Wader passage is of interest and in the surrounding fields species such as Montagu's Harrier, Lesser Kestrel, Little Bustard and Black-bellied and Pin-tailed Sandgrouse can be found.

ITINERARIES

1. 2 km

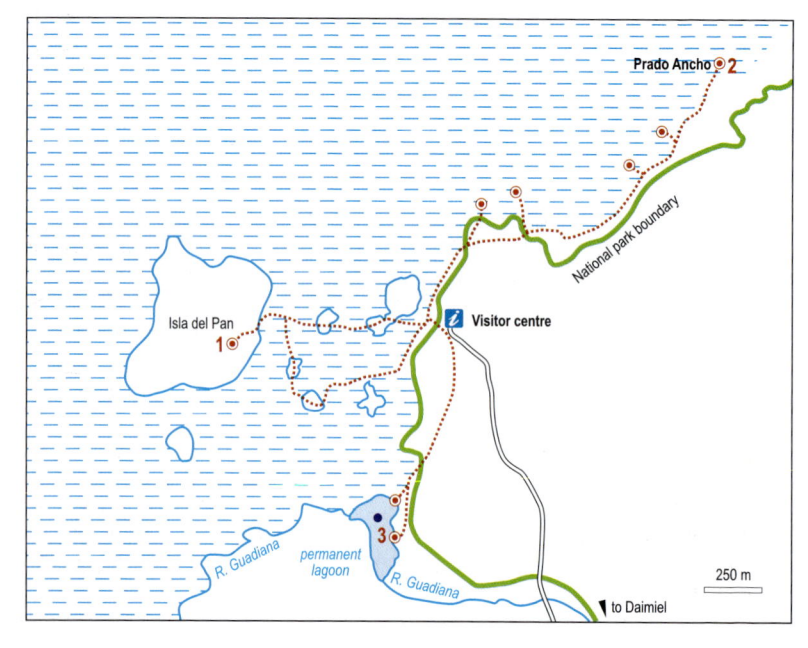

A network of raised boardwalks criss-crossing the National Park and linking some of the islands permit visitors to cross the flooded area of the park with ease. This is the best way to observe much of the birdlife of the reed-beds and lagoons: herons in summer, duck in winter, waders in autumn and spring and reed-bed passerines all year round. One option takes you to the hide (1) located on the Isla del Pan, an exceptional viewpoint for observing Marsh Harrier and, in winter, Common Crane returning to their roosting sites.

2. 1,5 km

A path follows the edge of Las Tablas eastwards between the reeds and fields. After passing a series of hides, you reach a tower hide (2) known as Prado Ancho, from where you can fully appreciate the immensity of the park's reed-beds: look out for water birds, raptors, Common Cranes and in the surrounding fields, Little Bustard in winter and Collared Pratincole in spring.

3. 1 km

A path leading south-west from the visitor centre takes you to La Laguna Permanente (3), a meander in the river Guadiana with water the whole year. Two hides give good views of grebes, herons, duck and waders and, in winter, the coming and goings of a large Great Cormorant colony located in a nearby poplar grove.

MOST REPRESENTATIVE SPECIES

Residents: Black-necked Grebe, Red-crested Pochard, White-headed Duck, Marsh Harrier, Purple Swamphen, Little Bustard, Northern Lapwing, Black-headed Gull, Black-bellied and Pin-tailed Sandgrouse, Moustached Warbler, Bearded and Penduline Tits, Reed Bunting.

PRACTICAL INFORMATION

- **Access:** coming from the north at Puerto Lápice or at Manzanares from the south (both on the Autovía de Andalusia), turn off along the N-420 and N-430, respectively, to the town of Daimiel. From here a metalled road leads in 11 km to the National Park.
- **Accommodation:** in Daimiel and Villarrubia de los Ojos. Further afield, Almagro and Ciudad Real.
- **Visitor infrastructure:** the National Park has a visitor centre (tel. 926 693 118), the starting point for these itineraries. Open every day, morning and afternoon. On the outskirts of the town of Daimiel, there is an interpretation centre run by the local town council (open Monday to Friday in the morning) centred around the subject of water and the wetlands of La Mancha (tel. 926 260 633, cidahm@aytodaimiel.es).
- **Visiting tips:** any time of year is good with the exception of summer (lack of water and heat). The National Park can only be visited along the itineraries herein described. Avoid midday, weekends and public holidays (excess of visitors). Phone for information on water levels before visiting.
- **Recommended:** the National Park employs wildlife guides for organised groups: reservations made via the visitor centre.
- **Further information:** maps 1:50,000, nº. 737 and 760 (SGE). Useful book: *Las Aves del Parque Nacional de las Tablas de Daimiel y otros humedales manchegos*, by José Jiménez *et al*. (Lynx, 1991). More background information at www.mma.es/parques/lared/tablas.

Summer visitors: Little Bittern, Night, Squacco and Purple Herons, White Stork, Garganey, Montagu's Harrier, Lesser Kestrel, Black-winged Stilt, Avocet, Collared Pratincole, Little and Whiskered Terns, Savi's Warbler.
Winter visitors: Great Cormorant, duck, Hen Harrier, Common Crane, Common Snipe.
On migration: Black Stork, Glossy Ibis, Eurasian Spoonbill, Greater Flamingo, Greylag Goose, Osprey, waders, Black Tern, Bluethroat.

46. LA MANCHA WETLANDS (I)

The town of Alcázar de San Juan (Ciudad Real province), in the heart of the area known as La Mancha Húmeda, lies at the centre of a broad expanse of plains containing a number of shallow lagoons. Most are endorheic (inwardly draining), seasonal and brackish, and are home to numerous specialist plant species, well-suited to living in these harsh conditions. However, these ecologically singular plant communities have been severely disturbed by the input of waste water from nearby towns.

The lagoons of Alcázar de San Juan and the nearby lagoon of Salicor are Ramsar sites are *reservas naturales*. All of La Mancha Húmeda has been declared a Biosphere Reserve and an SPA.

ORNITHOLOGICAL INTEREST

This area (along with Las Tablas de Daimiel) is the most important area for water birds in central Spain. Whenever water levels are high, good numbers of Black-necked Grebe, Shelduck, Collared Pratincole, Avocet, Black-headed Gull and Gull-billed and Whiskered Terns breed. Greater Flamingo are present all year round and even bred in La Laguna de Manjavacas in 2004.

Laguna de La Veguilla, La Mancha wetlands © Rubén Moreno-Opo

CASTILLA-LA MANCHA

The greatest numbers of birds and species occurs in winter, when duck and hundreds of Common Crane flock to the area. Migration periods are important for waders, while the surrounding plains are home to Lesser Kestrel, Great Bustard, Pin-tailed Sandgrouse and Dupont's Lark.

ITINERARY

50 km

1. *Laguna de El Longar.* Saline and seasonal, this lagoon on the outskirts of Lillo (Toledo), nevertheless, receives water from the nearby village.

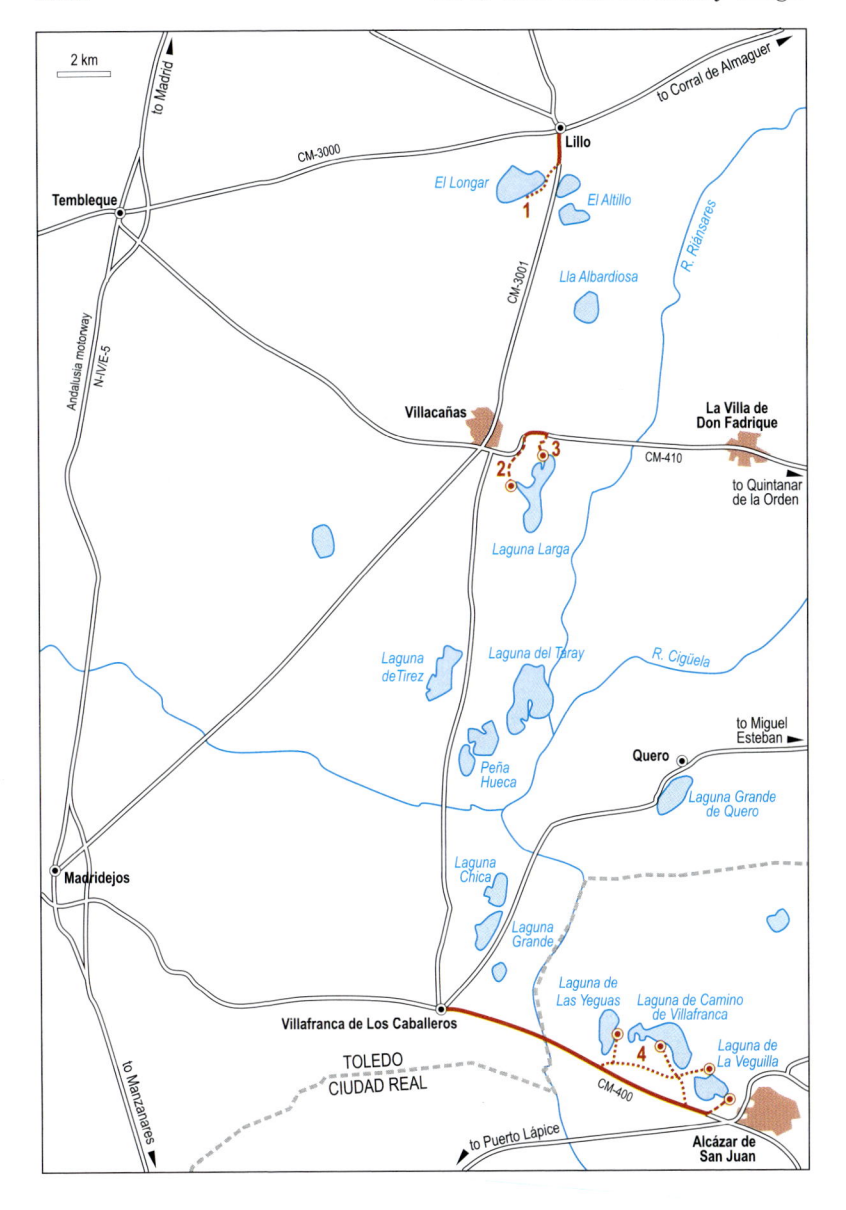

Access is from the CM-3001 towards Villacañas: 1 km south of Lillo, just after a petrol station, a dirt track (1) on the right approaches the lagoon. Surrounded by a broad belt of halophytic vegetation, this lagoon holds one of the best colonies of Gull-billed Terns in La Mancha and is a habitual drinking site for Pin-tailed and, with luck, Black-bellied Sandgrouse. Shelduck are present all year and waders pass through on passage. The surrounding steppe is alive with larks, including – if you are lucky – Dupont's Lark.

2. *Laguna Larga.* Next to the town of Villacañas (Toledo), lies Laguna Larga, today permanent as a result of the input of waste water and recently restored with UE money. Two hides have been built: La Canastera on the west and El Flamenco on the north, and are reached along marked paths. Leave Villacañas on the CM-410 towards Villa de Don Fadrique: the path to the first hide (2) heads off from next to the sewage farm and the bridge over the railway line, and to the second (3), from just after a petrol station and a door factory.

This is the main site in the area for breeding Shelduck (20+ pairs); numbers build up even higher in winter. Greater Flamingos are present all year, Collared Pratincole breed, Common Crane pass through in winter and passage periods bring many waders.

3. *Lagunas de Alcázar de San Juan.* On the outskirts of this town (Ciudad Real province) lie a succession of saline seasonal lagoons – Las Yeguas, Camino de Villafranca and La Veguilla – that are some of the best for wader observation during migration periods. Spring passage of Ruff can be spectacular with males in full breeding plumage, while in winter, Greylag Goose and other species of wildfowl, lots of Black-headed Gulls and a regular group of Eurasian Curlew on Las Yeguas are present. Good water levels encourage waders, gulls and terns to breed: La Veguilla holds Whiskered Tern and Purple Swamphen in its vegetated margins.

The best way to visit these three lagoons is to leave Villafranca de Los Caballeros along the CM-400 and head for Alcázar de San Juan. Once over the river Cigüela (between km

PRACTICAL INFORMATION

- **Access:** from Tembleque (Toledo) on the Autovía de Andalucía, head for Lillo on the CM-3000 (24 km). From here, take the CM-3001 to Villacañas (13 km from Lillo). From Villacañas head south for Villafranca de Los Caballeros, where you turn off along the CM-400 for Alcázar de San Juan.
- **Accommodation:** in Villacañas, Villafranca de Los Caballeros, Alcázar de San Juan and Campo de Criptana.
- **Visitor infrastructure:** hides overlooking the lagoons near Villacañas and Alcázar de San Juan.
- **Visiting tips:** any time of year is good with the exception of summer; winter and passage periods are especially recommended. To avoid disturbance to birds, do not leave the tracks or approach the shores of the lagoons too closely. After heavy rain, tracks may be come impassable.
- **Further information:** maps 1:50,000, nº. 659, 687 and 713 (SGE). The town councils of Villacañas (tel. 925 160 428) and Alcázar de San Juan (tel. 926 550 005) will provide visiting naturalists with information. Contact also the naturalist group Esparvel (925 232 153, esparvel@jazzfree.com) with offices in Toledo.

CASTILLA-LA MANCHA

points 89 and 90), a track heads off to the left (4). Along the 5 km of this driveable track view the southern shore of Las Yeguas and Camino de Villafranca and the western and northern shores of La Veguilla from the hides built on the banks of each lagoon.

MOST REPRESENTATIVE SPECIES

Residents: Black-necked Grebe, Greater Flamingo, Shelduck, Marsh Harrier, Purple Swamp-hen, Great Bustard, Kentish Plover, Northern Lapwing, Black-headed Gull, Black-bellied and Pin-tailed Sandgrouse, Dupont's, Lesser Short-toed and Calandra Larks.
Summer visitors: Lesser Kestrel, Black-winged Stilt, Avocet, Collared Pratincole, Gull-billed and Whiskered Terns, Short-toed Lark.
Winter visitors: Greylag Goose and other wildfowl, Peregrine Falcon, Common Crane, Eurasian Curlew, Reed Bunting.

On migration: Glossy Ibis, Common Teal, White-headed Duck, Osprey, Ringed and Grey Plovers, Little Stint, Dunlin, Curlew Sandpiper, Ruff, Black-tailed Godwit, Spotted and Common Redshanks, Greenshank, Green, Wood and Common Sandpipers, Black Tern.

47. LA MANCHA WETLANDS (II)

The landscape of vineyards, cereal fields and irrigation projects straddling the borders of the provinces of Toledo, Ciudad Real and Cuenca is enriched by the presence of a group of variable-sized lagoons. As in the previous site, the lagoons in this eastern sector of La Mancha Húmeda receive waste water from nearby towns and thus few maintain their original levels of salinity. Even so, some lagoons are still surrounded by belts of halophytic vegetation, while in others reed-beds have developed.

La Laguna de Manjavacas – along with three other smaller lagoons

Laguna de Pedro Muñoz, La Mancha wetlands © Fernando Barrio

nearby – and La Laguna de La Vega del Pueblo (Pedro Muñoz) are *reservas naturales* and Ramsar sites. All of the Mancha Húmeda is part of a Biosphere Reserve and an SPA; as well, SEO-BirdLife runs a reserve at Los Charcones de Miguel Esteban.

ORNITHOLOGICAL INTEREST

The lagoons of this eastern sector are as important as those described in the previous itinerary. Black-necked Grebe, Red-crested Pochard, White-headed Duck, Marsh Harrier, Black-headed Gull and Gull-billed and Whiskered Terns breed and waders are abundant either as breeders – Black-winged Stilt, Avocet, Collared Pratincole, Kentish Plover and Northern Lapwing – or on passage.

Greater Flamingos do not breed regularly, although they can be seen throughout the year, and in winter, Greylag Goose and other wildfowl and Common Crane are present. The surrounding dry fields hold Lesser Kestrel, Little Bustard and Pin-tailed Sandgrouse, amongst others.

ITINERARIES

1. *Los Charcones*. Near Miguel Esteban, this once temporary wetland has been transformed by the building of dykes to retain waste water from the nearby town. The vegetation that has sprung up as a result now provides habitat for Black-necked Grebe and White-headed Duck. When water levels are high here and in a nearby low-lying area, this site is excellent for duck, coots, waders and gulls.

Two kilometres from Miguel Esteban along the road to El Toboso, take a track (1) opposite a small church

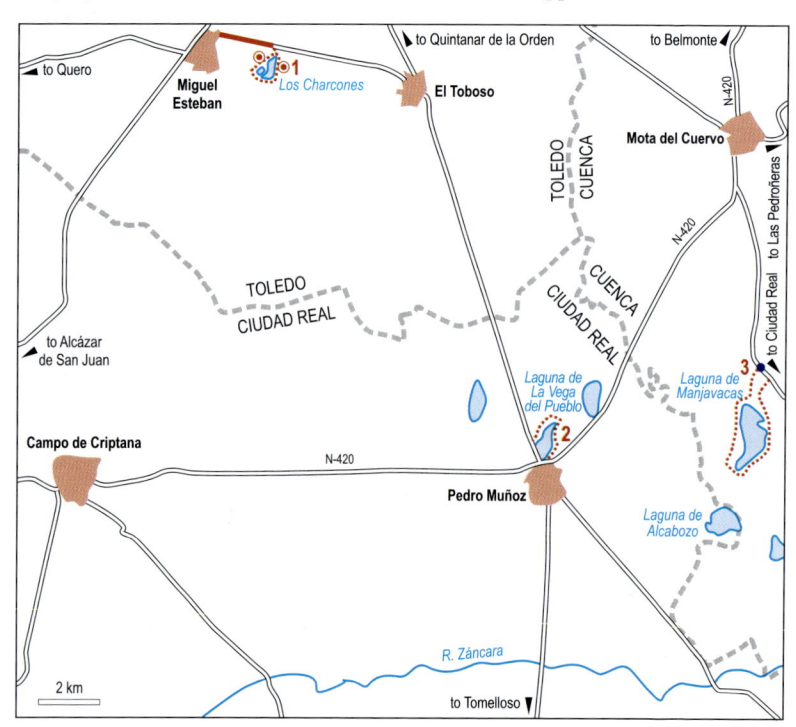

CASTILLA-LA MANCHA

towards the lagoon and park. A broad track circumnavigates the lagoon and passes by a number of hides.

2. *La Laguna de la Vega del Pueblo.* Located next to the town of Pedro Muñoz (Ciudad Real), this lagoon also receives fresh water from the nearby village and is no longer temporary or particularly saline. However, the 'freshening up' of this relatively small lagoon now means that breeding birds include Black-necked Grebe, Shelduck, Red-crested Pochard, White-headed Duck, Black-headed Gull, Whiskered Tern and Bearded Tit. Greater Flamingos are seen all year, duck arrive in winter and passage periods are enlivened by the presence of Black Stork, Glossy Ibis, Osprey and waders.

A track from the town itself heads towards the lagoon and provides good viewing from all sides. Alternatively, right at the edge of the town on the N-420 towards Mota del Cuervo, a track off to the left allows you to drive to the top of a nearby hill for excellent views over the whole lagoon.

3. *Laguna de Manjavacas.* Set in an area of steppe, this is one of the best of all the lagoons in La Mancha and covers over 200 ha when full. Although waste water from nearby Mota del Cuervo (Cuenca) has led to the growth of a reed-bed in its northern part, this lagoon is still slightly saline and subject to great seasonal changes in water levels. Broad muddy banks, shallow water and a number of islets make this ideal habitat for breeding and passage waders. Other breeders include Marsh Harrier and Gull-billed Tern. Groups of Greater Flamingo, Greylag Geese, Shelduck and other duck and Common Crane appear in winter.

At 3 km from Mota del Cuervo along the minor road to Las Mesas, take the driveable track off right (3) from next to a church and circumnavigate the lagoon (best on foot).

MOST REPRESENTATIVE SPECIES

Residents: Black-necked Grebe, Greater Flamingo, Shelduck, Red-crested Pochard, White-headed Duck, Marsh Harrier, Little Bustard, Kentish Plover, Northern Lapwing, Black-headed Gull, Pin-tailed Sandgrouse, Lesser Short-toed and Calandra Larks, Bearded Tit.

Summer visitors: Lesser Kestrel, Black-winged Stilt, Avocet, Collared Pratincole, Gull-billed and Whiskered Terns, Short-toed Lark.

PRACTICAL INFORMATION

- **Access:** Tembleque, Madridejos and Puerto Lápice on the Autovía de Andalucía are access points for La Mancha Húmeda. The N-420 crosses the area and passes through Pedro Muñoz (itinerary 1) and Mota del Cuervo (itinerary 2). From Pedro Muñoz to Miguel Esteban (itinerary 3) take the minor road through El Toboso (20 km).
- **Accommodation:** in Campo de Criptana, Pedro Muñoz and Mota del Cuervo.
- **Visitor infrastructure:** hides on some of the lagoons described.
- **Visiting tips:** any time of year is good with the exception of summer, although winter and passage periods are especially recommended. Most of the visitable lagoons are fenced off: remember, to avoid disturbance to birds, do not leave the tracks or approach the shores of the lagoons too closely. After heavy rain, tracks may be come impassable.
- **Further information:** maps 1:50,000, n°. 688 and 714 (SGE).

Winter visitors: Greylag Goose and other wildfowl, Hen Harrier, Peregrine Falcon, Common Crane, Reed Bunting.
On migration: Glossy Ibis, Common Teal, Osprey, Grey, Ringed and Little Ringed Plovers, Little Stint, Dunlin, Ruff, Black-tailed Godwit, Spotted and Common Redshanks, Greenshank, Green, Wood and Common Sandpipers, Black Tern.

48. LAGUNA DE PÉTROLA

Scattered around the plains of eastern Albacete province lie a group of small, endorheic lagoons, of which most are saline and seasonal. The largest is that of Pétrola (1.5 km long), whose hyper-saline waters were until recently used for salt production. Today it is a *Refugio de Fauna*, while part of the nearby cereal steppe, dotted with patches of holm oak and vineyards, have been declared an SPA for their steppe birdlife. Nevertheless, irrigation schemes are slowly encroaching on the area.

ORNITHOLOGICAL INTEREST

This is an excellent site for breeding waterfowl: White-headed Duck breed occasionally and gather here in numbers in the autumn. Other breeders include Black-necked Grebe, Shelduck, Red-crested Pochard, Marsh Harrier and colonial breeders such as Black-winged Stilt, Avocet, Northern Lapwing, Black-headed Gull and Gull-billed Tern. Greater Flamingos are present all year and bred in 1999 and 2000. The surrounding cereal steppe holds Great and Little Bustards, Stone Curlew and Black-bellied and Pin-tailed Sandgrouse.

ITINERARIES

1. 3 km 🚗 ☀️ ➡️

Two driveable tracks head northwards from the road that skirts the village of Pétrola. One (1) provides views over the eastern part of the lagoon and takes you to a hide (2) located next to the old salt-works. From

Laguna de Pétrola   © Fernando Barrio

CASTILLA-LA MANCHA

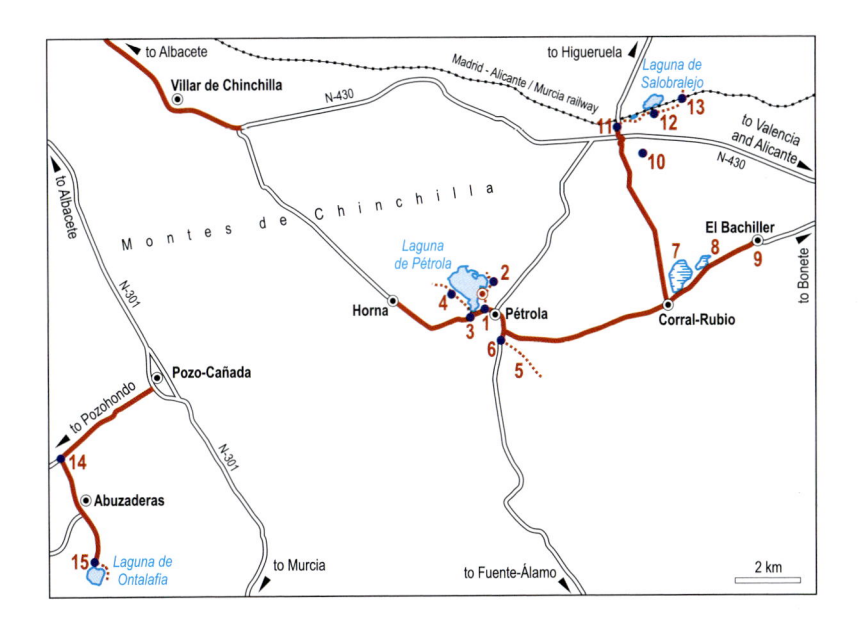

here there are good views along a dyke crossing the lagoon that is frequented by resting Greater Flamingos and other aquatic birds. Beyond the salt-works the track approaches the lagoon again and provides good unobstructed views.

The other track (3) heads for the western part of the lagoon and an area of reed-beds favoured by Bluethroats and Reed Buntings in winter. A little further on a small hill (4) provides a good viewing point for groups of ducks (including Redcrested Pochard) and hunting Marsh Harrier. When water levels are high enough, an island appears in the lagoon and is used by breeding waders and Gull-billed Terns.

2. 35 km

Leaving Pétrola south towards Fuente-Álamo, carry straight on at the junction with the road to Corral-Rubio. Turn left almost immediately on a track (5) heading towards a wind farm and continue into what is one of the best places in the province for Great Bustard.

Alternatively, head for Corral Rubio (6); just past this village there are a couple of depressions (7, 8) that attract good numbers of birds when full of water. Just 5 km further on, the hamlet of El Bachiller (9) sits on a small hill that is a good vantage point for searching for bustards and other steppe birds.

From Corral-Rubio it is worth heading for the lagoon of Salobralejo, similar to that of Pétrola. Take the road north towards Higueruela, passing Los Cerros de Monpichel (10) on your right before crossing the dual-carriageway (N-430). Just before reaching the Madrid-Alicante-Valencia railway line, turn right along a track (11) that will take you to Las Casas de Salobralejo (12), a good site for viewing the lagoon. A kilometre further on, the bridge (13) over the railway line provides further views of the lagoon. As the lagoon is private, do not leave the track or attempt to approach the lagoon.

PRACTICAL INFORMATION

- **Access:** from Albacete take the Autovía de Levante (referred to as N-340 Mérida-Valencia) towards Almansa. Just after Chinchilla (13 km), turn off towards Horna and then Pétrola.
- **Accommodation:** in Albacete.
- **Visitor infrastructure:** there is a hide at Laguna de Pétrola.
- **Visiting tips:** winter brings the greatest diversity of birds, although wind and cold, as well as hunters (only at Pétrola is hunting banned), make birdwatching less easy. After heavy rain, the tracks can become impassable.
- **Recommended:** La Sociedad Albacetense de Ornitología for information and activities concerning Laguna de Pétrola and surrounding areas (tel. 608 163 591, buzon@sao-albacete.org).
- **Further information:** maps 1:50,000, n°. 791, 792 and 817(SGE).

3. 1 km

Despite being a little distant, it is worth investigating the Laguna de Ontalafia, which receives run-off water from nearby irrigation projects and so almost always has water. Black-necked Grebes breed in good numbers and recently a few White-headed Duck have started breeding.

Take the N-301 from Albacete to Murcia and in Pozo Cañada, pick up the local road towards Pozohondo. After almost 6 km, turn left (14) towards Abuzaderas and then, just beyond this hamlet, keep left at a fork. This road will take you to a small settlement next to the lagoon, from where you can take a track around its eastern edge.

MOST REPRESENTATIVE SPECIES

Residents: Greater Flamingo, Shelduck, Red-crested Pochard, White-headed Duck, Hen and Marsh Harriers, Water Rail, Great and Little Bustards, Stone Curlew, Northern Lapwing, Pin-tailed and Black-bellied Sandgrouse, Great Spotted Cuckoo, Lesser Short-toed, Thekla and Calandra Larks.

Summer visitors: Black-necked Grebe, Montagu's Harrier, Black-winged Stilt, Avocet, Little Ringed and Kentish Plovers, Black-headed Gull, Gull-billed Tern, Short-toed Lark, Yellow Wagtail, Northern and Black-eared Wheatears, Reed, Great Reed and Spectacled Warblers.

Winter visitors: Wildfowl, Bluethroat and Reed Bunting.

On migration: Little Egret, Garganey, Curlew Sandpiper, Little Stint, Dunlin, Black Tern.

49. CASTREJÓN AND AZUTÁN

These two reservoirs lie along the middle reaches of the river Tajo in the province of Toledo near the foothills of the Montes de Toledo. Castrejón, 25 km downstream from the city of Toledo, consists of a couple of large flooded meanders, while Azután is much narrower and snakes its way between the towns of Talavera de la Reina and El Puente de Arzobispo. Both are framed by mudstone and sandstone cliffs, particularly spectacular at Castrejón, and contain wooded islets. Sedimentation is taking its toll in both reservoirs and permits the growth of good-sized reed-

CASTILLA-LA MANCHA

Azután reservoir   © Miguel A. Sánchez, Foto-Ardeidas

beds along the banks; both also enjoy the shade of well-constituted *sotos* (areas of fluvial woodland), replete with poplars, willows and tamarisks. An SPA has been declared to protect these wetlands, although currently this is the only legal protection they enjoy.

ORNITHOLOGICAL INTEREST

Most of the Iberian species of heron frequent these reservoirs, including the occasional Great Bittern. Night, Grey and Purple Heron breed (a dozen or so pairs each), and there are also good densities of Marsh Harrier, Purple Swamp-hen and Red Avadavat, an escaped bird of Asian origin, in the reed-beds. Wintering birds include Great Cormorants, groups of Greylag Geese and other wildfowl and gulls.

Juvenile raptors, attracted by the abundance of rabbits, visit the area from their nearby breeding grounds – look out for Spanish Imperial and Golden Eagles, as well as for Black Vulture. The surrounding dry plains hold breeding colonies of Lesser Kestrel, large flocks of Great Bustard and other steppe birds.

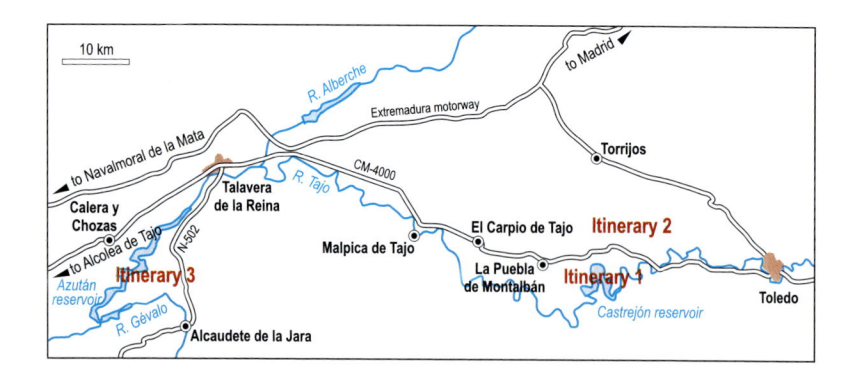

ITINERARIES

1. [4 km] 🚶 ☀ ○

Just past the km point 24 on the CM-4000 (travelling from Toledo towards La Puebla de Montalbán), turn left on a track (1) that takes you to a car-park in 300 m. From here pick up the wildlife footpath opened up by Acmaden-Ecologistas en Acción, a local wildlife conservation group. After a short climb, you come to the top of Las Barrancas (2), the spectacular cliffs that form the northern bank of the reservoir. In winter it is easy to hear Eagle Owl calling at dusk. Two viewing points – El Cambrón and Los Enebros – have been set up and provide excellent views over the water and reed-beds. The footpath continues down to the shore of the reservoir and to a hide. From here, the path returns to the car-park along the small stream Arroyo de Alcubillete.

2. [30 km] 🚗 ☀ →

North of the river Tajo, the road between Albarreal de Tajo and Gerindote passes through dry cereal cultivation where there are breeding Great and Little Bustards, Montagu's Harrier, Lesser Kestrel and Pin-tailed and the here scarce Black-bellied Sandgrouse. Rabbits are abundant in the granite plains between La Puebla de Montalbán and San Martín de Montalbán and so this is a good area for Mediterranean raptors such as Spanish Imperial Eagle.

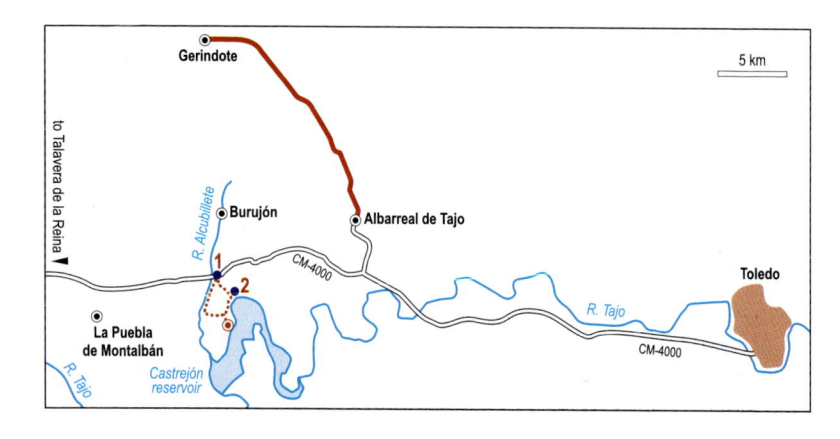

CASTILLA-LA MANCHA

3. 70 km

This itinerary starts on the river Tajo in Talavera de la Reina, where you can walk along the right-bank of the river from Puente Viejo (3). In winter the *sotos* on the opposite bank (4) are home to a large Great Cormorant and Cattle Egret roost.

From Talavera de la Reina take the N-502 towards Alcaudete de la Jara and after 10 km turn right (5) towards Las Herencias. From this village you can walk downstream along the banks of the river Tajo – here already part of the reservoir (6) – and search for Great Cormorants, herons, wildfowl and gulls. It is worth climbing up to the top of the cliffs (7) to south of the village, where Common Kestrel and Eagle Owl breed.

Return to the N-502 and continue as far as Alcaudete de la Jara and take the CM-4160 towards Calera y Chozas. Once over the river Gévalo, the road continues alongside this flooded river almost to its confluence with the Tajo. The ash woodland is frequented by Golden Orioles and other forest species, while the reed-beds are home to passerines and Marsh Harrier. Four kilometres after crossing the Puente de los Silos (8) over the Tajo, turn right (9) opposite a church along a road that borders the reservoir in places and takes you to Alberche del Caudillo, and from here Talavera de la Reina. All along this sector of the reservoir there are many White Stork nests.

PRACTICAL INFORMATION

- **Access:** from the city of Toledo, along the CM-4000 and 6 km before La Puebla de Montalbán for itinerary 1. Continue along this road to Talavera de la Reina (also accessible from the Autovía de Madrid, 110 km from Madrid), starting point for itinerary 2.
- **Accommodation:** in Toledo, 30 km from Castrejón, or in Talavera de la Reina, near Azután.
- **Visitor infrastructure:** Castrejón has a wildlife footpath, Azután has no infrastructure for visiting birdwatchers.
- **Visiting tips:** in winter there is a greater diversity of wildfowl and many herons, raptors and steppe species are also present. In summer, the heat can be prohibitive. Most of the land beyond the itineraries is private and thus not accessible.
- **Recommended:** the Puebla de Montalbán delegation of Ecologistas en Acción (tel. 619 948 631, toledo@ecologistasenaccion.org) provides information about species and routes, as well as a guiding service. This group also organises bird counts and other activities.
- **Further information:** maps 1:50,000, n° 626, 627, 628, 654 and 656 (SGE).

MOST REPRESENTATIVE SPECIES

Residents: Great Bittern, Cattle and Little Egrets, Grey Heron, Black Vulture, Marsh Harrier, Spanish Imperial, Golden and Bonelli's Eagles, Purple Swamp-hen, Little and Great Bustards, Stone Curlew, Black-bellied and Pin-tailed Sandgrouse, Eagle Owl, Calandra Lark, Penduline Tit, Red Avadavat.
Summer visitors: Little Bittern, Night and Purple Herons, Montagu's Harrier, Lesser Kestrel, Black-winged Stilt, Short-toed Lark, Sand Martin, Red-rumped Swallow, Golden Oriole.
Winter visitors: Great Cormorant, Greylag Goose and other wildfowl, European Golden Plover, Black-headed and Lesser Black-backed Gulls, Reed Bunting.
On migration: Black Stork, Eurasian Spoonbill, Osprey, Common Crane.

50. VALLE DEL TIÉTAR

The north-west of the province of Toledo is cut through by the rivers Tiétar and Guadyerbas, both of which have been damned to create, respectively, the reservoirs of Rosarito and Navalcán. The valley of the river Tiétar and its hinterland extend to the foothills of the Sierra de Gredos and into the provinces of Ávila and Cáceres, and boasts some of the best lowland wood pastures (*dehesas*) in Spain. As you approach the Tajo valley around Oropesa, the *dehesas* give way to pasture lands and cereal cultivation, much of which has been transformed by irrigation schemes. Two SPA have been declared in the area: in the north, the Tiétar valley and in the south, the Oropesa plains.

ORNITHOLOGICAL INTEREST

This is an excellent sit for Mediterranean raptors: the Spanish Imperial Eagle seems to have recovered after having been on the brink of extinction in the 1990s due to the illegal use of poisoned bait, two pairs of Bonelli's Eagle breed and Black and Griffon Vultures frequent the area. There are well-established breeding populations of Black and White Storks and the two reservoirs also

Rosarito reservoir, Tiétar valley © Fernando Barrio

CASTILLA-LA MANCHA

hold important post-nuptial concentrations of storks, some of which remain all winter.

In winter Common Crane feed in the *dehesas* and holm oak forests and roost on the two reservoirs, along with Great Cormorant (which also breeds on Rosarito in one of its few Spanish breeding colonies), Greylag Geese and other wildfowl, Osprey and gulls. To the south, birds of the plains such as the relatively abundant Black-winged Kite and Lesser Kestrel can be found.

ITINERARY

80 km

At Oropesa (km point 148 of the Autovía de Extremadura), turn right at the roundabout on the CM-5150 towards Candeleda (Ávila). The first part of this road heads north towards the impressive Gredos Mountains though pastures and cereal fields frequented by Black-winged Kite, Montagu's and Hen Harriers and Roller. Once in the village of La Corchuela, wander through the nearby holm oak and fluvial woodland.

From this village, pick up an asphalted track that heads east to Dehesón del Encinar (2), an agricultural research centre belonging to the government of Castilla-La Mancha. This sector is frequented by numerous raptors, including Spanish Imperial Eagle, and Common Crane in winter.

Return to La Corchuela and continue northwards along the CM-5150. Turn right towards the village of Navalcán and stop at the dam on the Guadyerbes river that forms the Navalcán reservoir (4): this is a good spot for grebes, Great Cormorants, duck, coots, gulls and, with luck, Osprey.

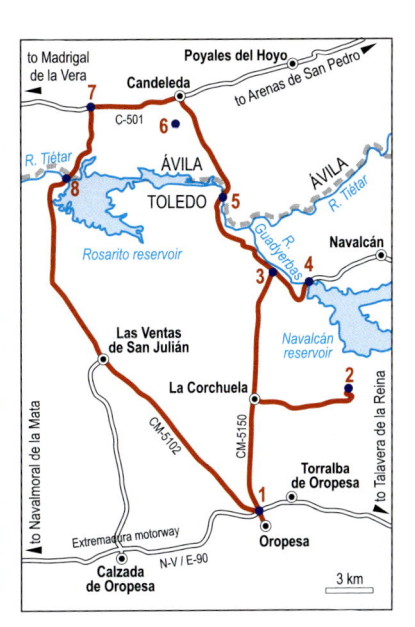

Return to the road and continue towards Candeleda alongside the river Guadyerbas and then river Tiétar through magnificent holm oak forests that are home to Black Stork and raptors such as Spanish Imperial Eagle. Common Crane overfly in winter and the *sotos* are home to many forest birds. After crossing the Tiétar (5), the landscape changes and small pastures, hedgerows and oak copses take over from the *dehesas*. Just before the town of Candeleda, on the left of the road there is a loose White Stork colony (6) on trees and buildings.

From Candeleda, pick up the C-501 towards Madrigal de la Vera (Cáceres) and after 9 km, turn left (7) along a road that will take you down to a good vantage point over Rosarito (8). A little further ahead, join the CM-5102 heading for Oropesa and pass through rolling countryside and *dehesas* that eventually give way to open pastures after the village of Las Ventas de San Julián. The itinerary ends in Oropesa, where White Stork and Lesser Kestrel (one

PRACTICAL INFORMATION

- **Access:** Oropesa is at km point 148 on the Autovía de Extremadura (N-V) between Talavera de la Reina and Navalmoral de la Mata.
- **Accommodation:** in Oropesa (Parador). Also in Talavera de la Reina, Candeleda, Arenas de San Pedro, Madrigal de la Vera and Navalmoral de la Mata. The apartments El Mirador de la Vera (tel. 927 570 730, info@elmiradordelavera.com) in Robledillo de la Vera (Cáceres) provide information and activities for birdwatchers.
- **Visitor infrastructure:** no visitor centre or infrastructure for birdwatchers.
- **Visiting tips:** in spring the *dehesas* are at their best, and in winter there are large numbers of wintering birds including Common Crane. Heavy spring and autumn rain can make tracks impassable.
- **Recommended:** The wildlife centre El Vado de los Fresnos (tel. 920 377 223) provides excellent information: on the C-801 heading towards Madrigal de la Vera, turn left 2 km after Candeleda where signposted. Follow the asphalted track for 6 km to the centre, where native species of Iberian fauna are kept in captivity. Open all day in summer (except Monday); the rest of the year, only open at weekends and on public holidays.
- **Further information:** maps 1:50,000, n°. 600, 601, 625 and 626 (SGE).

of the most important colonies in Castilla-La Mancha) breed in the town itself.

MOST REPRESENTATIVE SPECIES

Residents: Great Cormorant, Black-winged and Red Kites, Griffon and Black Vultures, Spanish Imperial and Bonelli's Eagle, Stone Curlew, Dartford Warbler, Azure-winged Magpie, Spanish Sparrow, Cirl Bunting.
Summer visitors: Black and White Storks, Short-toed and Booted Eagles, Montagu's Harrier, Lesser Kestrel, Little Tern, Roller, Black-eared Wheatear, Spectacled and Orphean Warblers.
Winter visitors: Cattle Egret, Grey Heron, Greylag Goose and other wildfowl, Hen Harrier, Osprey, Common Crane, European Golden Plover, Northern Lapwing, Common Snipe, Black-headed and Lesser Black-backed Gulls.
On migration: Eurasian Spoonbill, Shelduck, waders.

51. ALTO TAJO

This site, the largest protected area in Castilla-La Mancha (Alto Tajo Natural Park, over 100,000 ha), consists of the best collection of river gorges in the region. Grouped around the river Tajo and its confluents, these spectacular limestone gorges (sandstone in the case of the rivers Gallo and Arandilla) boast a number of karstic phenomena, including the largest tufa structure in Europe (Puente de San Pedro).

Extensive black pine forests and distinctive cleared Spanish juniper formations on the *parameras* (high plateaux) cover much of the area; Lusitanian oaks thrive in the gorges on shady lower slopes and stunted holm oaks cling to the walls of the most inaccessible parts of the canyons. The gorges are too narrow, however, for much fluvial woodland to be present, but by way of compensation, their clean rapid rivers hold Otter, Trout and the White-clawed Crayfish, and the surrounding mountains deer and Spanish Ibex.

CASTILLA-LA MANCHA

Laguna de la Parra in the Alto Tajo Natural Park © Pedro Retamar

ORNITHOLOGICAL INTEREST

The Alto Tajo is home to important numbers of cliff-breeding raptors, with an estimated 500 pairs of Griffon vulture complemented by good populations of Egyptian Vulture, Golden Eagle, Peregrine Falcon and, to a lesser extent, Bonelli's Eagle and Eagle Owl. Red-billed Choughs are abundant, as is the avifauna in general in all the area's three principal habitats – rivers, pine forests and juniper woodland.

ITINERARIES

1. 20 km

This itinerary takes you through the central sector of the Natural Park between the sheer cliffs and pine forests of the river Tajo gorge. Coming from Molina de Aragon, just before the village of Poveda de la Sierra and once across the bridge over the Tajo, turn right along a good (but busy in summer) dirt track (1). The track follows the river downstream and the best stopping points for raptors and forests birds are as follows: Puente de Peñalén (2), where you can cross the river and walk up the valley of the river Cabrillas (3), Las Fuentes de Las Tobas (4), La Teja (5), La Parra (6) and La Falaguera (7).

The track reaches another road (8) from Molina de Aragón: turn left towards Zaorejas and after 1 km, turn right along a semi-asphalted road as far as a small dam, where you should park. Cross the dam and walk upstream along the right-bank of the Tajo as far as the waterfall of El Campillo (9), a good place for Dipper and other water birds. Return to the road and a little way further on towards Zaorejas, turn left along a driveable track (10) that in 2 km will take you to an excellent position overlooking the cliffs of the Tajo gorge (nearby colony of Griffon Vultures and other raptors).

2. 5 km

In the north-east of the Natural Park, just to the north of the village of Riba de Saelices, take a track to near the picnic area at Cueva de Los Casares

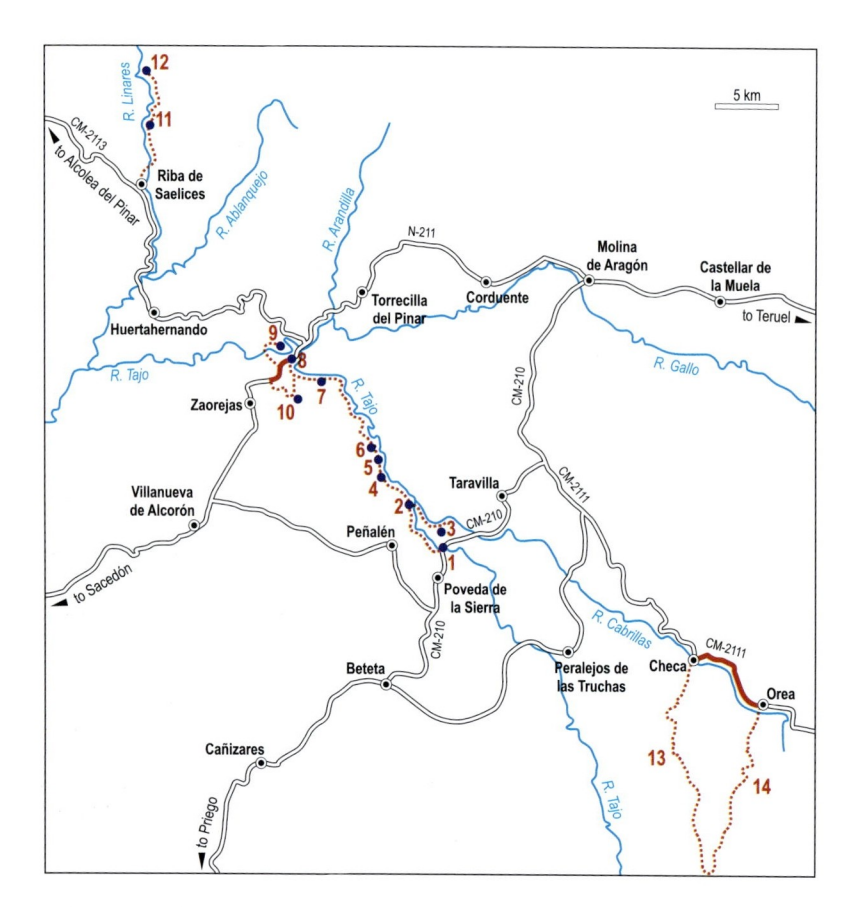

(11). To see all the area's raptors except Bonelli's Eagle and many forest birds, continue up the ever-narrowing gorge of Arroyo de Linares on foot along a flat but winding path that takes you to another cave, Cueva de la Hoz (12). The walk can be extended to Santa María del Espino if you have a second car for the return journey.

3. 56 km 🚗 ☀ ◯

In the south-east of the Natural Park, a track (13) from the village of Checa crosses the river Hoz Seca and drops down to the river Tajo as it flows between the provinces of Teruel, Guadalajara and Cuenca. Return along another track (14), almost parallel to the first, that will take you out to the village of Orea, from where you can return by road (8 km) along the river Cabrillas to Checa.

This damp and cold *paramera* (high plateau) over 1,500 m is covered in Scots pine interspersed with stunted junipers and is an excellent site for Goshawk, Citril Finch, Siskin, Common Crossbill and other forest birds. Ring Ouzel and Song and Mistle Thrushes, Redwing and Fieldfare all winter.

MOST REPRESENTATIVE SPECIES

Residents: Griffon Vulture, Goshawk, Golden and Bonelli's Eagle, Peregrine Falcon, Eagle and Long-

CASTILLA-LA MANCHA

PRACTICAL INFORMATION

- **Access:** from Alcolea del Pinar on the Autovía de Zaragoza, the N-221 heads to Molina de Aragón (60 km). Take the CM-201 towards Poveda de la Sierra for itinerary 1. From Alcolea del Pinar take the CM-2113 towards the Alto Tajo and Riba de Saelices (20 km) for itinerary 2. Twenty kilometres from Molina de Aragón along the CM-210 take the CM-2111 towards Albarracín (Teruel) for Checa and itinerary 3.
- **Accommodation:** best in Molina de Aragón, but also some in Peralejos de las Truchas, Poveda de la Sierra and other villages in the area.
- **Visitor infrastructure:** the first of a network of Natural Park interpretation centres will open in Corduente.
- **Visiting tips:** best in late spring and early summer. After heavy rain, take care on tracks and walking in river gorges.
- **Further information:** maps 1:50,000, n°. 488, 513, 514, 539 and 540 (SGE). More information from the tourist office in Molina de Aragón (tel. 949 832 098).

eared Owls, Kingfisher, Crag Martin, Dipper, Blue Rock Thrush, Dartford Warbler, Crested and Coal Tits, Red-billed Chough, Rock Sparrow, Citril Finch, Siskin, Common Crossbill, Hawfinch and Cirl and Rock Buntings.

Summer visitors: Black Kite, Egyptian Vulture, Short-toed and Booted Eagles, Northern Wheatear, Rufous-tailed Rock Thrush, Subalpine Warbler, Ortolan Bunting.

Winter visitors: Ring Ouzel, Redwing, Fieldfare.

52. CABAÑEROS

Cabañeros is *the* great sanctuary of Mediterranean wildlife in Castilla-La

Cabañeros National Park © Pedro Retamar

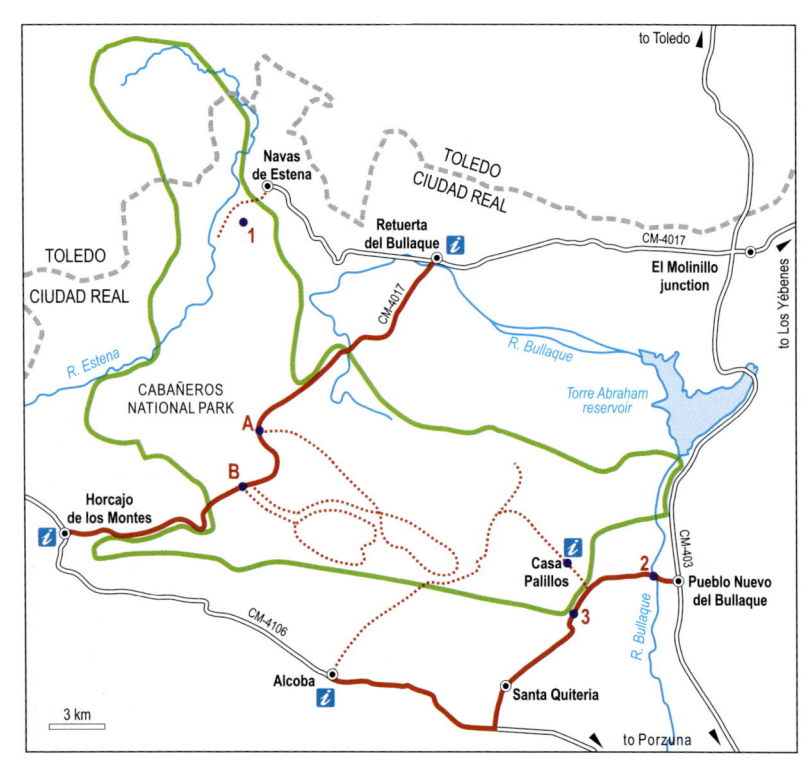

Mancha. Hidden away in the Montes de Toledo (mostly Ciudad Real province; part in Toledo province), Cabañeros spreads as far as the river Bullaque in the east and the river Estena in the west and consists essentially of a series of quartzite ridges covered by dense forests and Mediterranean scrub surrounding a large grassy plain known as La Raña (8,000 ha) dotted with a few oaks. After having escaped the indignity of use as a bombing range, Cabañeros was declared a Natural Park in 1988 and then upgraded to National Park in 1995 (40,000 ha). It lies within a much larger SPA (12,000 ha) that covers most of Los Montes de Toledo.

ORNITHOLOGICAL INTEREST

Above all Cabañeros is important for its Mediterranean raptors: the 140 pairs of Black Vulture form the largest known colony in the world, and there is still room for 2 pairs of Spanish Imperial Eagle (one in, one nearby the National Park), Egyptian Vulture, Golden Eagle and abundant forest raptors such as Short-toed Eagle. There are also good numbers of Black and White Storks (also post-breeding concentrations), while La Raña holds Black-winged Kite, Montagu's Harrier, Little Bustard and Common Crane (only in winter).

ITINERARIES

1. 60-70 km

To enter the National Park, you have to join one of the three 4-hour routes in 4-wheeled drive vans guided by official National Park drivers/guides (morning or afternoon). The first route (A) departs from Retuerta del Bullaque

CASTILLA-LA MANCHA

and passes through magnificent scrub formations, enters a valley with an interesting Pyrenean Oak forest and ends up in La Raña: good for Golden Eagle. The second route (B) starts in Horcajo de los Montes and takes in superb Mediterranean forests, habitat of the Spanish Imperial Eagle and other forest raptors, and the Laguna de los Cuatro Morros, frequented by Black Storks. The third route (C) departs from Alcoba de los Montes and visits La Raña, and provides the opportunity to see Black-winged Kite (all year), Merlin (winter), Little Bustard (spring) and Roller (at the beginning of summer).

During all three itineraries it is easy to see Black Vulture in flight or on La Raña, often feeding in company with Griffon Vultures. The display flights of the Spanish Imperial Eagle in February and March alone would warrant a visit, while in summer hundreds of White Storks (until mid-July) and Montagu's Harrier (August onwards) flock to La Raña to feed on grasshoppers.

2. [15 km]

In the north-west of the park you can walk with or without guide through the gorge of Boquerón del Estena (1), a stretch of river with Eagle Owl, Kingfisher, Crag Martin and Dipper.

In the eastern part of the park, it is worth driving the road from Pueblo Nuevo del Bullaque to Santa Quiteria (13 km). Examine the fluvial woodland as you cross the river Bullaque (2) and keep an eye on the roadside bushes for groups of Spanish Sparrows. Halfway along the road you reach Casa Palillos, the interpretation centre, while 2km further on a hide provides good views of a White

PRACTICAL INFORMATION

- **Access:** from Madrid and Toledo the Montes de Toledo can be reached via Las Ventas con Peña Aguilera. From here, continue along the CM-403 for 18 km to the junction at El Molinillo: continue south for Pueblo Nuevo del Bullaque, Alcoba de los Montes and Horcajo de los Montes, or west (CM-4017) for Retuerta del Bullaque and Navas de Estena in the north of the park.
- **Accommodation:** hostals and rural accommodation in many of the villages in the area.
- **Visitor facilities:** the Casa Palillos interpretation centre lies 5 km from Pueblo Nuevo del Bullaque towards Santa Quiteria. Open every day, morning and afternoon. There are also information points in Retuerta del Bullaque, Alcoba de los Montes and Horcajo de los Montes.
- **Visiting tips:** all year is interesting (but avoid the summer heat). Outside the National Park, the rest of Los Montes de Toledo is also of interest to ornithologists.
- **Recommendations:** the guided 4-wheel drive visits must be pre-booked (preferably well in advance) via the centre in Horcajo de los Montes (tel. 926 775 384). For the guided walk, make your reservation in the park's central office in Pueblo Nuevo del Bullaque (also information point: tel. 926 783 297). Open morning and afternoon from Monday to Friday, but only mornings at weekends and on public holidays.
- **Further information:** maps 1:50,000, n°. 683, 684, 709, 710 and 711 (SGE). Excellent ornithological information in *Aves de Cabañeros y su entorno* by José Jiménez (Ecohabitat, 1995) or in the chapter on fauna in *Parque Nacional de Cabañeros* (Esfagnos, 2000). Basic data on the park at www.mma.es/parques/cabaneros.

Stork colony situated on Lusitanian and holm oaks and, in winter, Common Cranes coming to roost.

MOST REPRESENTATIVE SPECIES

Residents: Black-winged and Red Kites, Griffon and Black Vultures, Spanish Imperial and Golden Eagles, Little Bustard, Stone Curlew, Eagle Owl, Kingfisher, Calandra and Thekla Larks, Crag Martin, Dipper, Dartford and Sardinian Warblers, Southern Grey Shrike, Azure-winged Magpie, Spanish Sparrow, Corn Bunting.

Summer visitors: Black and White Storks, Black Kite, Egyptian Vulture, Short-toed and Booted Eagles, Montagu's Harrier, Great Spotted Cuckoo, Roller, Northern and Black-eared Wheatears, Spectacled, Subalpine and Orphean Warblers, Woodchat Shrike.

Winter visitors: Hen Harrier, Merlin, Peregrine Falcon, Common Crane, European Golden Plover.

WHERE TO WATCH BIRDS IN SPAIN. THE 100 BEST SITES

CASTILLA Y LEÓN

53. La Moraña
54. Lagunas de Villafáfila
55. Laguna de La Nava
56. Azud de Riolobos
57. Hoces del Duratón
58. Montejo de la Vega
59. Picos de Europa
60. Hoces del Alto Ebro
61. Valle del Arlanza
62. Sierra de Gredos
63. La Granja and Segovia
64. Arribes del Duero

Over 90,000 km² shared between nine provinces and occupying almost 20% of the land surface of peninsular Spain are the bare statistics lying behind the largest autonomous community in Spain. Within this immensity, the northern half of the great Meseta, stretching north and south of the mighty river Duero, occupies the northern two-thirds of this autonomous community and is its most outstanding natural and geographical feature.

The characteristic landscape of this autonomous community is that of interminable cultivated plains whose monotony is only broken by isolated patches of pines and holm oaks and a number of important *sotos* (fluvial woodland) running along some of the main rivers. Traditionally, dry cereal farming and sheep-raising has dominated the region's agriculture, although in recent years the implantation of irrigation projects has been going ahead apace and has negatively affected one of the region's main ornithological attractions – the steppe birds.

Nevertheless, Castilla y León still harbours almost half of all Spanish Great Bustards (the most important population in the world). One of the best sites for these birds is the area of La Moraña (53) in the province of Ávila, where you can also find Montagu's Harrier, Lesser Kestrel, Little Bustard, Black-bellied Sandgrouse and Calandra Lark, amongst other steppe species.

Great Bustards © Gabriel Sierra

Add to this a series of lagoons scattered throughout the cereal steppe and you have a wonderful area for ornithologists. The most important group of lagoons are those at Villafáfila (54) in the Tierra de Campos (Zamora province), made even more attractive by the world's greatest density of Great Bustards. This wetland, the largest in Castilla y León, is frequented during migration periods and in winter by large contingents of duck and other water birds, including many thousands of Greylag Geese.

Further east in Tierra de Campos lies the Laguna de La Nava (55), home to similar birds as Villafáfila and now happily restored to its former glory after having been drained in the last century. In the north of Salamanca province the recently created reservoir of Azud de Riolobos (56) has quickly gained renown in birdwatching circles and provided more than one pleasant surprise for birdwatchers.

Other smaller but singular steppe areas exist in this region, for example, the high limestone plains on the south-east of the Meseta that have been carved open by steep river gorges. A prime example in Segovia province are the gorges of Duratón (57), near Sepúlveda, and Riaza, near the village of Montejo de la Vega (58). Some of the largest Griffon Vulture colonies in Europe exist here, along with other cliff-loving birds and Dupont's Larks on the surrounding plains.

An almost continuous belt of sizeable mountains borders the region to the north, south and east, grading into intermediate zones at lower altitudes where Atlantic and Mediterranean influences mix and provide a greater mixture of habitats.

Examples of mountain sites include the Picos de Europa (59), with impressively rugged mountains and, in the transition zone between the Meseta and the main mountain ridges, the gorges of the Alto Ebro (60) and the valley of Arlanza in the foothills of the Sistema Ibérico (61). In the first of these three sites, forest and high-level birds such as the ever-scarcer Capercaillie and Grey Partridge, here in one of its Spanish strongholds, are characteristic, while in the latter two you can expect to find a greater variety of species given

CASTILLA Y LEÓN

the diversity associated with such transitional zones and the mosaic of habitats present, including many limestone cliffs.

This autonomous community is closed off to the south by the Sistema Central, a long mountain range consisting of the Sierra de Gredos (62) and Sierra de Guadarrama, whose heights hold healthy numbers of breeding Bluethroats. The high peaks tumble down onto a broad piedmont zone and then merge gently into the Meseta, where Pyrenean oak and pine forests spread northwards and are replaced in lower areas by scrub formations, wood pastures and a mosaic of fields and hedgerows. Of interest for the ornithologist are the Valle del Tormes and the area around La Granja near Segovia (63), with the world's most northerly populations of the threatened Black Vulture and Spanish Imperial Eagle.

The only flank of the region not lined by mountains is the western, where Spain meets Portugal. The deep granite river gorges such as the grandiose and markedly Mediterranean Arribes de Duero (64) are popular amongst birdwatchers, who are attracted by Black Stork, Black and Egyptian Vultures and Bonelli's Eagle.

53. LA MORAÑA

La Moraña is the name given to a large area of dry cereal farming (although with ever-increasing amounts of irrigated land) in the north of the province of Ávila. Small and medium-sized saline lagoons called *lavajos*, stands of maritime and stone pines, a few good-sized extensions of holm oak woodland and a number of *sotos* (for example, along the river Adaja) add variety to the landscape.

This site forms part of the plains of Tierra de Campiñas, stretching into the neighbouring provinces of Valladolid and Salamanca, which have been declared an SPA.

ORNITHOLOGICAL INTEREST

This is one of the most important sites for steppe birds in the region, above all for the over 500 Great Bustards that inhabit these plains. Other breed-

La Moraña © Gabriel Sierra

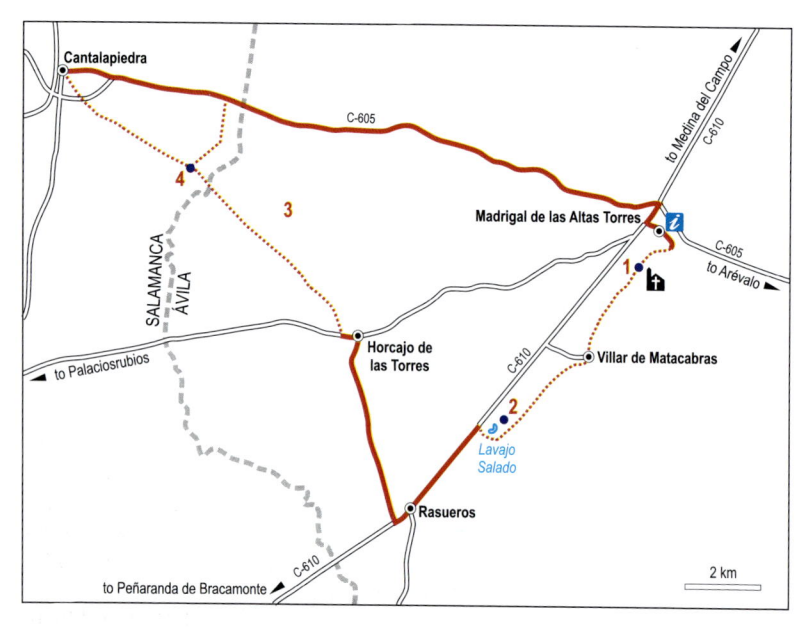

ing birds include Montagu's Harrier, Little Bustard (declining in numbers with the loss of fallow land) and Black-bellied and the here scarce Pin-tailed Sandgrouse. There are also numerous Lesser Kestrel colonies and winter Red Kite roosts.

The reserve run by the SEO/BirdLife at the lagoon at El Oso is one of the most important wintering sites for Common Crane (500 + birds) in the region and also acts as a stop-over site for Greylag Geese in winter moving between Doñana and the Villafáfila lagoons.

reached from Extramuros by following a track running parallel to the C-610. Continuing along the same track and just before reaching the road near Rasueros, you will pass by Lavajo Salado (2), a saline seasonal lagoon of a certain interest: Black-winged Stilt breed occasionally and both sandgrouse come here to drink.

Black and Red Kites, Common Buzzard and Common Kestrels perch on the telegraph poles next to the road from Rasueros to Horcajo de las Torres; once in the latter village, take the track that leads to

ITINERARIES

1. 25 km

Leave the village of Madrigal de las Altas Torres towards the ruin of the nearby monastery of Extramuros (1), which is home to a colony of at least 20 pairs of Lesser Kestrel. Another colony (almost 40 pairs) exists in the church of the abandoned village of Villar de Matacabras, which can be

CASTILLA Y LEÓN

Cantalpiedra (Salamanca province). Great Bustard frequent the plains in this area (3) and it is not difficult to see groups of dozens of males. Also here, look out for Montagu's Harrier, Little Bustard, Stone Curlew, Black-bellied Sandgrouse and Calandra and other Larks.

Turn right at any junction of tracks (4), for example on the Ávila-Salamanca border, or pick up the C-605 in Cantalpiedra to return to Madrigal de las Altas Torres.

2. 2 km

From the village of El Oso two tracks lead to two hides overlooking the lagoon of the same name. One option is to take the track (5) towards San Pascual, although it is best to park next to the information point at the beginning of the itinerary. Walk the track as far as the tower hide, which is normally locked – ask for the key in the bar El Sol in the village of El Oso. The other track (6) heads for Hernansancho, asphalted at first and driveable to the car park next to the second hide (always open).

El Oso is best in winter: the most spectacular sight is that of the Common Cranes coming in to roost. Greylag Geese and other wildfowl are abundant, as are wintering raptors: there is a Red Kite roost near the village of San Pascual.

MOST REPRESENTATIVE SPECIES

Residents: Red Kite, Common Buzzard, Common Kestrel, Peregrine Falcon, Great and Little Bustards, Black-bellied and Pin-tailed Sandgrouse, Little and Long-eared Owls, Calandra Lark, Skylark.

Summer visitors: Black Kite, Montagu's Harrier, Booted Eagle, Lesser Kestrel, Hobby, Black-winged Stilt, Stone Curlew, Roller, Short-toed Lark, Tawny Pipit, Yellow Wagtail, Northern Wheatear, Subalpine Warbler.

Winter visitors: Greylag Geese and other wildfowl, Hen Harrier, Merlin, Common Crane, European Golden

PRACTICAL INFORMATION

- **Access:** take the Autovía de A Coruña (A-6) to Arévalo, from where you pick up the C-605 to Madrigal de las Altas Torres, or to Sanchidrián for the CL-803 to El Oso.
- **Accommodation:** Arévalo, Madrigal de las Altas Torres and Riocabado near El Oso all have hostals.
- **Visitor infrastructure:** The town councils of Arévalo (tel. 920 300 001) and Madrigal de las Altas Torres (tel. 920 320 001) have both opened up nature interpretation centres. The former is devoted to the White Stork and the latter to the Lesser Kestrel, and both have live video feeds from nearby nests.
- **Visiting tips:** the best time to see Great Bustards is early spring between March and April, when the crops are not too high and the males are displaying. The wildfowl at the lagoon of El Oso is best in winter. After heavy rains, some of the tracks can become impracticable.
- **Recommended:** the agency Enclave Rural organises occasional ornithological outings to La Moraña (tel. 915 061 707, info@enclaverural.com).
- **Further information:** maps 1:50,000, n°. 454 and 481 (SGE). The site's species and a number of itineraries are described in *Guía de las aves de La Moraña y Tierra de Arévalo* by Luis José Martín and Gabriel Sierra (Asodema, 1999).

Plover, Northern Lapwing, Common Snipe, Short-eared Owl.
On migration: Ringed Plover, Dunlin, Ruff, Common Redshank, Green and Common Sandpipers.

54. LAGUNAS DE VILLAFÁFILA

Near the village of Villafáfila in the heart of Tierra de Campos (north-east Zamora province) lie a group of shallow saline lagoons that dry up in the summer. The three most important are Salina Grande (180 ha max.), Barillos (110 ha max.) and Las Salinas (70 ha max.) and are all protected as a *reserva regional de caza* (almost 33,000 ha) and internationally as a Ramsar Site and SPA.

Very little fringing vegetation grows around the lagoons, although there are interesting halophytic communities. When full, the lagoons contrast starkly with the surrounding treeless cereal plains – interrupted only by numerous picturesque adobe dovecotes – that extend as far as the eye can see.

ORNITHOLOGICAL INTEREST

This is the most important wetland in Castilla y León and is renown for the 30,000 Greylag Geese, a total only surpassed by Doñana in the Iberian Peninsula, that winter here. Other, rarer geese mix in with the Greylags, which are joined by large numbers of duck. The recipe for an outstanding birdwatching site is completed by the Eurasian Spoonbill, Garganey, Common Crane and waders that pass through on passage, the post-breeding groups of White Storks that sometimes form, and the breeding Marsh Harrier, Black-winged Stilt (100+), Northern Lapwing and Gull-billed Tern.

The surrounding arable land is of vital importance for steppe birds. The over 2,000 Great Bustard represent almost 10% of the whole world's population and the greatest density anywhere on Earth. Furthermore, Villafáfila boasts the largest population of Lesser Kestrels in Castilla y León (300 pairs), as well as Montagu's Harrier, Little Bustard, Black-bellied Sandgrouse and winter roosts of Red Kite.

Dovecot at Lagunas de Villafáfila © Anna Motis

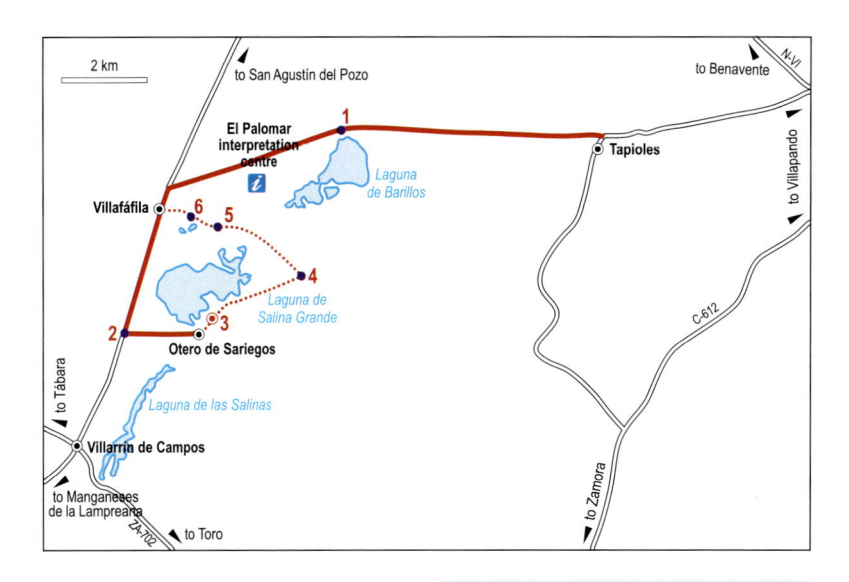

ITINERARIES

1. 〔25 km〕 🚗 ☀️ ➡️

After Villalpando and then from the village of Tapioles onwards towards Villafáfila, the road passes through an area much frequented by Great Bustards. Stop at the Laguna de Barillos (1) and visit its two hides for good concentrations of Greylag Geese, Eurasian Wigeon, Northern Shoveler and other duck. The interpretation centre at El Palomar lies a little way ahead on the same road, 1.5 km before the village of Villafáfila itself. Here turn left on the road to Villarrín de Campos and after 3 km, turn left (2) on a road to the all but deserted village of Otero de Sariegos, home to Villafáfila's largest colony of Lesser Kestrels (60 pairs).

Just 300 m from Otero de Sariegos a tower hide (3) offers the best views of La Salina Grande, the largest of the lagoons. This is a perfect point for watching the movement of Greylag Geese from their feeding areas in the surrounding fields and pastures to their roosting sites.

2. 〔5 km〕 🚗 🚶 ☀️ ➡️

From this tower hide, you can continue along a track, by car or on foot, around the eastern and northern side of La Salina Grande, where European Golden Plover, Northern Lapwing and Eurasian Curlew are found and the fields are frequented by Great Bustard.

Take the second track on the left (4) and head for Villafáfila village, passing through an area of pastures that in wet years attracts good numbers of waders: view also from the restored Roman bridge (5). Just before Villafáfila you cross a dyke – Regato de Tío Hachero – and a number of small lagoons (6), home to many waders on passage, including Little Stint, Dunlin and Curlew Sandpiper.

MOST REPRESENTATIVE SPECIES

Residents: Gadwall, Marsh Harrier, Great and Little Bustards, Northern Lapwing, Common Redshank, Black-headed Gull, Black-bellied Sand-

PRACTICAL INFORMATION

- **Access:** Leave the Autovía de A Coruña (A-6) between Mota del Marqués and Benavente (km point 236) and head for Villalpando. Cross the village and then, 2 km after a petrol station, turn right towards the village of Villafáfila, 18 km away.
- **Accommodation:** in Villalpando there is a hotel and hostals. Nearer the lagoons, Villafáfila has a hostal and rural accommodation. Other possibilities include the towns of Benavente and Zamora.
- **Visitor infrastructure:** the interpretation centre of El Palomar (tel. 980 586 046), architecturally inspired by the local dovecotes, is situated 16 km from Villalpando and 1.5 km before Villafáfila. Three artificial lagoons have been dug and there is a trail and hides. Open Wednesday to Sunday, morning and afternoon; however, between November and May, on weekday mornings the centre is only open for groups.
- **Visiting tips:** best from October to March. For steppe birds, spring is better. Do not leave the described routes to avoid disturbing the steppe birds. After heavy rain, vehicles may get stuck if you leave the main tracks.
- **Further information:** maps 1:50,000, n°. 308, 309 and 340 (SGE). The book *Guía de las aves de las lagunas de Villafáfila y su entorno* by Joaquín Sanz-Zuasti and Tomás Velasco (Adri-Palomares, 1997) is an excellent description of the birds of the area. Also of interest is *Guía de fauna de la Reserva de las lagunas de Villafáfila* by Jesús Palacios and Mariano Rodríguez (Junta de Castilla y León, 1999). For more descriptions and itineraries, go to www.villafafila.com.

grouse, Barn and Short-eared Owls, Calandra Lark.

Summer visitors: Black Kite, Montagu's Harrier, Lesser Kestrel, Black-winged Stilt, Avocet, Little Ringed Plover, Gull-billed Tern, Short-toed Lark, Yellow Wagtail, Northern Wheatear.

Winter visitors: Pink-footed, White-fronted, Barnacle and Greylag Geese, Shelduck and other wildfowl, Red Kite, Hen Harrier, Peregrine Falcon, European Golden Plover and other waders.

On migration: White Stork, Eurasian Spoonbill, Garganey, Common Crane, waders, Whiskered Tern.

55. LAGUNA DE LA NAVA

In 1990, a group of naturalists, now part of the Fundación Global Nature, put forward the idea of restoring the Mar de Campos, a vast steppe lagoon near the village of Fuentes de Nava (Palencia) that had been drained for agricultural use during the previous century. The restoration of this lagoon – generally considered to be the most important inland wetland in the north of the Iberian Peninsula – was carried out by the Castilla y León government with support from the UE and has so far transformed 300 ha of agricultural land into an artificial lagoon covering 10% of the area once occupied by the Mar de Campos. In 1998, the Fundación Global Nature began the restoration of the Laguna de Boada, a smaller wetland that was also drained a number of decades ago. Both lagoons are part of a vast SPA protecting the north-east of the area of Tierra de Campos; La Nava is also a Ramsar Site.

ORNITHOLOGICAL INTEREST

La Nava is an important site for wintering wildfowl and the thousands of Greylag Geese that flock here make

CASTILLA Y LEÓN

Laguna de La Nava — © Pedro Retamar

it the third best site for the species in the Iberian Peninsula after Doñana and Villafáfila. As well, winter sees the arrival of excellent numbers of dabbling duck and a 100-strong Marsh Harrier roost; summer brings breeding Grey Heron, Black-winged Stilt, Little Ringed Plover, Northern Lapwing, Black-headed Gull and Whiskered Tern. On passage, Eurasian Spoonbill and many waders are regular, along with enough Aquatic Warblers to make La Nava one of the most important migration stop-over points for this globally threatened warbler. In the surrounding arable fields, Montagu's Harrier, Great Bustard and Black-bellied Sandgrouse can be found.

ITINERARIES

1. 1 km

In the far south-east corner of the lagoon there is a car-park on the right of the road as you come from Fuentes de Nava (3.5 km). Cross the road and follow the path along the eastern side of the lagoon to the first boardwalk and then observation point (1) overlooking the lagoon. This area – El Prao – is good for passerines such as Reed Bunting and Savi's, Reed, Great Reed and Aquatic Warblers (the latter on post-breeding migration). Dusk sees the entry of many Marsh Harriers into their roosting sites.

Return to the path and you soon reach another observation point giving good views of the lagoon. A little further on you reach another boardwalk (2), from where the great flocks of Greylag Goose – with the occasional White-fronted or Barnacle Goose mixed in – and small groups of Eurasian Spoonbill during migration periods can be observed at pleasure.

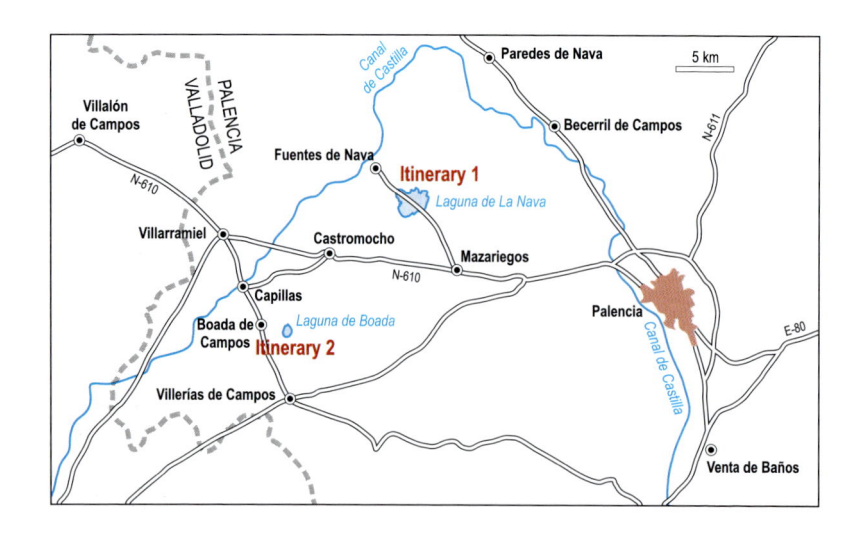

Returning to the car-park, take the road towards Fuente de Nava and park at the far end of the lagoon on the right-hand side of the road. From here walk to the hide overlooking the Corralillos area of the lagoon, best visited in the afternoon with the sun behind you. This is the best area for waders, which often include large groups of Black-tailed Godwit. Look here also for Black-necked Grebe and Purple Heron and in winter for Meadow and Water and Red-throated Pipits, the latter occasional during migration.

The surrounding arable land is home to easily observable flocks of Great Bustard and other steppe species such as Montagu's Harrier.

2. 1 km

Park in the village of Boada de Campos and follow the marked path towards the hide situated 200 m from the southern shore of the Laguna de Boada. In winter, almost all the Spanish species of dabbling duck and many Greylag Geese congregate here. Great Bustard and Peregrine Falcon frequent the area.

MOST REPRESENTATIVE SPECIES

Residents: Gadwall, Marsh Harrier, Great Bustard, Northern Lapwing, Black-headed Gull, Black-bellied Sandgrouse, Bearded Tit.

Summer visitors: Black-necked Grebe, Purple Heron, Montagu's Harrier, Black-winged Stilt, Avocet, Little Ringed and Kentish Plovers, Whiskered Tern, Savi's, Reed and Great Reed Warblers.

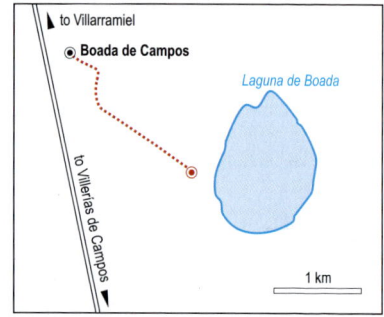

CASTILLA Y LEÓN

PRACTICAL INFORMATION

- **Access:** from Palencia, take the N-610 towards Benavente and at Mazariegos (25 km), turn north-west towards Fuentes de Nava. Boada de Campos (17 km from Fuentes de Nava) can be reached along minor roads passing through the villages of Castromocho and Capillas.
- **Accommodation:** in Fuentes de Nava, a study centre - Centro de Estudios Ambientales de Tierra de Campos (tel. 979 842 398) - has beds for 35 people. Also, try Paredes de Nava or the city of Palencia.
- **Visitor infrastructure:** the lagoon visitor centre in Fuentes de Nava (tel. 979 842 500) is worth visiting before you head for the lagoon. Open Wednesday to Friday in the afternoon, and all day at weekends.
- **Visiting tips:** best from the end of October to mid-March, coinciding with the greatest number of wildfowl. Migration periods are also of interest. Do not leave the marked paths; a telescope all but essential, as is mosquito repellent in summer.
- **Recommended:** Aside from being a hostal, the Centro de Estudios Ambientales de Tierra de Campos in Fuentes de Nava functions as an information point and organises work camps and guided excursions.
- **Further information:** maps 1:50,000, n°. 273, 310 and 311 (SGE). The web page of the Castilla y León government (www.jcyl.es, then go to 'Medio Ambiente/Espacios Naturales') provides details about the lagoon and a list of species present.

Winter visitors: White-fronted, Barnacle and Greylag Geese and other wildfowl, Red Kite, Hen Harrier, Peregrine Falcon, European Golden Plover, Dunlin, Common Snipe, Black-tailed Godwit, Water and Meadow Pipits, Bluethroat, Penduline Tit, Reed Bunting.

On migration: Eurasian Spoonbill, waders, Lesser Black-backed Gull, Black Tern, Red-throated Pipit, Grasshopper, Aquatic and Sedge Warblers.

56. AZUD DE RIOLOBOS

Located in the north of Salamanca province, the Azud de Riolobos is a recently built reservoir, designed to receive water from the river Tormes and then distribute it to the surrounding irrigation projects. When full, the Azud covers over 400 ha and, given the flat nature of the land it has flooded, is nowhere deep; it has yet to develop any aquatic or fringing vegetation. It lies amidst cereal plains dotted with small copses of pines and holm oaks, and lacks any type of legal protection.

ORNITHOLOGICAL INTEREST

Given that many of the steppe lagoons (*lavajos*) in the area have been drained or are exceedingly seasonal, the Azud has quickly become the best wetland for birds in Salamanca province, with notable numbers of birds (and species-richness) present throughout the year.

Its banks and islands hold hundreds of pairs of Black-winged Stilt, one of the biggest colonies in the centre of the Iberian Peninsula, and this is also one of the few sites in Castilla y León with breeding Shelduck. Winter sees the arrival of thousands of wildfowl, above all Greylag Goose and Mallard, and a Common Crane roost. The greatest diversity of birds occurs during migration periods, when good numbers of Eurasian Spoonbill and

Azud de Riolobos © Miguel Rouco

up to 30 species of waders put in an appearance. White-headed Duck numbers (100+) build up during post-breeding dispersion.

ITINERARY

8 km

A quiet road circumnavigates the Azud and permits good observations to be made over much of the lagoon. Begin at the houses at Pedrezuela de San Bricio (1) at the southern end of the Azud on the road from the village of Villar de Gallimazo.

Heading anti-clockwise along the eastern bank of the Azud, the only pinewood along this stretch (2) is also the best observation point. Numerous small bays and inlets make this area excellent for observing waders, and a good spot for Eurasian Spoonbill on passage and Greylag Geese and Common Crane in winter.

At the northern-most point of the Azud next to the buildings of La Confederación Hidrográfica del Duero (3), you can park and walk across the dam for views of Little, Black-necked and Great Crested Grebes, large groups of Eurasian Coots and duck (above all Northern Shoveler). Flowing out of the dam, a small stream (4) with a little

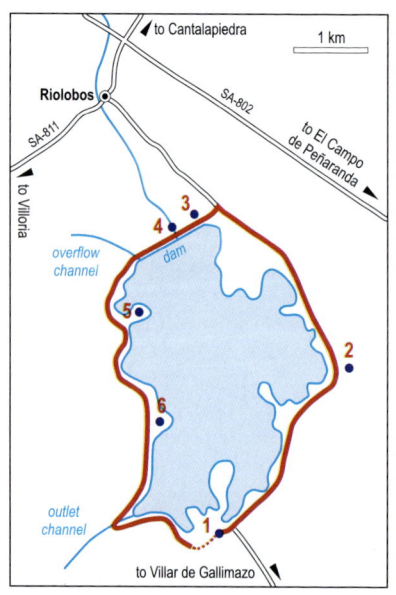

CASTILLA Y LEÓN

PRACTICAL INFORMATION

- **Access:** from Madrid and Ávila via the N-501 to Salamanca. Nine km after passing Peñaranda de Bracamonte (km point 64), turn right for Villar de Gallimazo (4 km). From there, a road will take you to Pedrezuela de San Bricio at the far end of the Azud.
- **Accommodation:** in Peñaranda de Bracamonte or the city of Salamanca (30 km).
- **Visitor infrastructure:** no visitor centre or infrastructure for birdwatchers.
- **Visiting tips:** December and January are the best months for wildfowl; from June onwards, White-headed Duck; between summer and autumn, post-breeding migration of Eurasian Spoonbill and waders. To avoid disturbing the birds, it is best not to leave the track or approach the shore. If you start at Pedrezuela de San Bricio in the morning, it is best to travel anti-clockwise because of the sun; if you start in the afternoon, travel clockwise.
- **Further information:** maps 1:50,000, n°. 453 and 479 (SGE); the local SEO/BirdLife group in Salamanca (e-mail: seo-salamanca@seo.org) and up-to-date information on http://members.fortunecity.es/riolobos

vegetation provides refuge for species unable to survive on the completely bare banks of the Azud.

Return to your car and drive across the dam and on to a small spit (5) frequented by waders, gulls and terns, from where you can approach the western side (6) of the lagoon. This is the best area for White-headed Duck, above all in the second half of the year, and diving duck such as Common Pochard and Tufted Duck. In Pedrezuela de San Bricio at the beginning of this itinerary and in the Alquería de Riolobos, 2 km north of the dam, there are colonies of Lesser Kestrel (in total almost 50 pairs). The surrounding plains – best to the east – are home to Montagu's Harrier, Little Bustard and Stone Curlew; with luck, you may also come across Great Bustard, Black-bellied and the occasional Pin-tailed Sandgrouse.

MOST REPRESENTATIVE SPECIES

Residents: Black-necked Grebe, Red Kite, Marsh and Hen Harriers, Eurasian Coot, Great Bustard, Little Bustard, Black-bellied and Pin-tailed Sandgrouse, Calandra Lark.
Summer visitors: Black Kite, Montagu's Harrier, Lesser Kestrel, Black-winged Stilt, Stone Curlew, Little Ringed and Kentish Plovers, Short-toed Lark, Yellow Wagtail.
Winter visitors: White-front, Greylag and Barnacle Geese, White-headed and other ducks, Peregrine Falcon, Merlin, Common Crane, Northern Lapwing, European Golden Plover, Short-eared Owl, Meadow Pipit.
On migration: Eurasian Spoonbill, Garganey, Osprey, Common Crane, Curlew Sandpiper, Black-tailed Godwit and other waders, Gull-billed, Whiskered and Black Terns, Sand Martin, Barn Swallow.

57. HOCES DEL DURATÓN

In the north-east of Segovia province stretches the stunning canyon (*hoces*) of the river Duratón, 25 km of limestone cliffs describing sinuous curves between the town of Sepúlveda and the dam at Burgomillodo.

Hoces del Duratón © Pedro Retamar

Reaching heights of up to 70 m in places, the cliffs separate two very different environments: the plateau on top of the cliffs, with patches of juniper woodland and thyme scrub, and the verdant fluvial woodland along the river banks. Most of these *hoces* are part of a natural park (over 5,000 ha) and an SPA.

ORNITHOLOGICAL INTEREST

Duratón is one of the best places for cliff-breeding raptors in central Spain. Over 500 pairs of Griffon Vultures breed along the length of the *hoces*, with the main concentrations on either side of the Villaseca bridge, and share space with Egyptian Vulture,

CASTILLA Y LEÓN

Golden Eagle, Peregrine Falcon and Eagle Owl. Here too the cliffs hold one of the best populations of Red-billed Chough in Castilla y León, Rufous-tailed and Blue Rock Thrushes and Black Wheatear. On the plateau a couple of hundred pairs of Dupont's Lark can be heard singing in spring, although numbers are declining.

ITINERARIES

1. 12 km

The first half of the *hoces* can be walked along a path in the base of the gorge that provides good views of Griffon and Egyptian Vultures, Red-billed Chough and other cliff-nesting species. Golden Orioles sing from the riverside poplars.

Begin in the car-park indicated off the road from Sepúlveda to Segovia. From the Roman Talcano bridge (1), you can walk as far as the next part of the river that is accessible by car, the Villaseca bridge (2). If you follow the whole itinerary, you will need another car to return from Villaseca, although, alternatively, return to the Talcano bridge having walked as far as you want.

Another option departs from the Romanesque church of La Virgen de la Peña (3) in the highest part of Scpúlveda. From here walk down to the Talcano bridge through a variety of habitats from top to bottom of the cliffs on a fairly gentle path.

2. 5 km

The chapel of Ermita de San Frutos perched atop a narrow rock promontory protruding into the *hoces* is a must: from Sepúlveda, take the road towards Segovia and after 1 km, turn off towards Villar de Sobrepeña. Once through this village and having crossed a bridge over the river Duratón, you reach the village of Villaseca. Here, take a broad track off to the left from next to the church and park when you reach the car-park. A gentle downhill path leads to the Ermita (4) on the edge of an enormous cliff.

This track passes through scrub frequented by Dupont's Lark and it is relatively easy to detect its characteristic call – emitted from the ground – during early spring. Other birds to look out for include Dartford

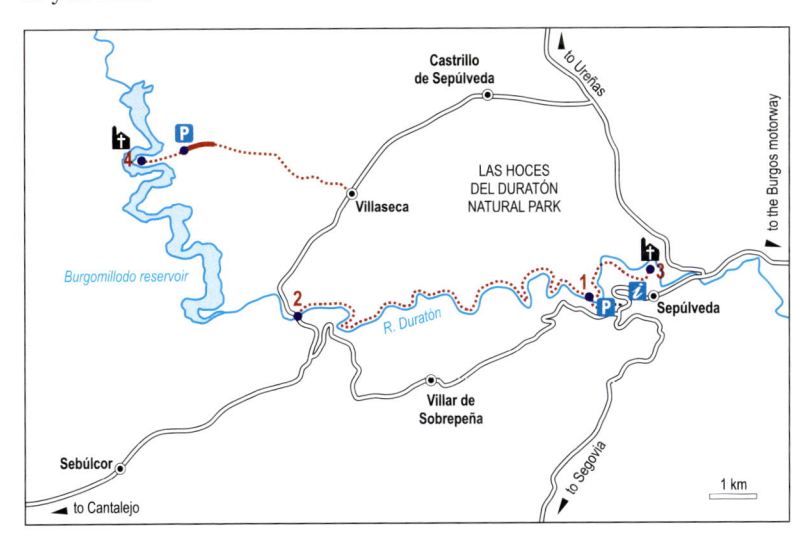

PRACTICAL INFORMATION

- **Access:** from Madrid take the Autovía de Burgos (N-I/E-5) and 20 km after the Puerto de Somosierra, turn off towards Sepúlveda.
- **Accommodation:** in Sepúlveda and Cantalejo; there is also a lot of rural accommodation available in nearby villages such as Sebúlcor, Burgomillodo and Carrascal del Río.
- **Visitor infrastructure:** there is a Natural Park interpretation centre in Sepúlveda located in a restored Romanesque church (tel. 921 540 586). Open every day from 10.00 to 18.00: on weekdays from October to June, it closes an hour earlier.
- **Visiting tips:** best in spring and autumn. Access is unrestricted, although from January to July, to avoid too much disturbance to the vulture colonies, a permit from the interpretation centre is needed to walk itinerary 1 and enter certain areas of the reserve.
- **Recommended:** a number of companies organise canoe trips – an original way to go birdwatching - along the Duratón.
- **Further information:** map 1:50,000, nº. 431 (SGE). Information about the natural heritage of the Hoces de Duratón can be found in the following books: *Cañón del Duratón, el lento trabajo del río* by Francisco Sánchez Aguado (Edilesa, 1995) and *Las Hoces del río Duratón. Guía de interpretación de la naturaleza* by Tomás Santamaría and Jorge A. Caballero (Proatur, 1998).

and Spectacled Warblers and, from the cliff top, cliff-breeding raptors and Red-billed Choughs.

MOST REPRESENTATIVE SPECIES

Residents: Red Kite, Griffon Vulture, Golden Eagle, Common Kestrel, Peregrine Falcon, Eagle Owl, Crag Martin, Dupont's and Thekla Larks, Black Wheatear, Blue Rock Thrush, Dartford Warbler, Southern Grey Shrike, Azure-winged Magpie, Red-billed Chough, Rock Sparrow, Rock Bunting.

Summer visitors: Black Kite, Egyptian Vulture, Short-toed and Booted Eagles, Common Sandpiper, European Bee-eater, Wryneck, Short-toed Lark, Northern and Black-eared Wheatears, Rufous-tailed Rock Thrush, Spectacled, Subalpine and Garden Warblers, Common Whitethroat, Golden Oriole, Woodchat Shrike.

58. MONTEJO DE LA VEGA

In the north-east of Segovia province near the provincial border with Burgos and Soria the mid-course of the river Riaza between the villages of Montejo de la Vega and the Linares reservoir has carved out a 12-km long canyon from the limestone plateau. Whilst the bottom of the valley is occupied by fluvial woodland and poplar plantations, the sides of the canyon and the surrounding plateau are covered by dry pastures and juniper scrub. In some areas well-constituted patches of holm oak forest remain and there are even a few stands of Lusitanian oak in the shadiest parts of the gorge.

This is a legendary site amongst Spanish birdwatchers due to the fact that since 1974, thanks to the persuasive powers of Félix Rodríguez de la Fuente (a pioneer Spanish naturalist), the western part of the canyon has been run by the WWF/Ade-

CASTILLA Y LEÓN

Raptor sanctuary at Montejo de la Vega © José Manuel Reyero

na as a raptor reserve (2,100 ha). At the eastern end, a further 200 ha are also protected, although this area is managed by the power company Confederación Hidrográfico del Duero. The whole site is an SPA and was declared a Natural Park in December 2004.

ORNITHOLOGICAL INTEREST

The gorge of the river Riaza holds one of Spain's largest Griffon Vulture colonies (400 pairs), as well as 10 pairs of Egyptian Vulture and other cliff-breeding species such as Golden Eagle, Peregrine Falcon and Eagle Owl. There are also frequent records of Black Vulture, wandering from their breeding grounds to the south, and in the surrounding plains, a few Dupont's Larks and Black-bellied Sandgrouse. The hedgerows and scrub are replete with passerines, and it is not difficult to find three species of shrike (including Red-backed), three of wheatear (including Black) and an abundance of warblers, thrushes and buntings.

ITINERARIES

1. 6,5 km

This itinerary takes you through the western end of the *hoces* and is ideal for observing passerines in the fields and fluvial woodland and the birds that haunt the cliffs and other rocky habitats.

Leave the village of Montejo de la Vega on foot from next to the cemetery (1) along a path passing through allotments and fields. Pass by a concrete bridge (2) on your left (don't cross) and continue upstream. As the gorge begins to narrow at the cliffs of La Calleja (3), the path splits: keep left alongside the river through a patch of fluvial woodland and out to a point opposite the spectacular cliffs of Peña Portillo (4) on the far bank.

Opposite the WWF/Adena information point in a field, cross the river on a footbridge (5) and return along the far bank at the base of the cliffs of Peña Portillo and Peña Rubia (6). Turn left at the road (7) and walk the short distance back to Montejo de la Vega.

2. 10 km 🚶 ☀ ↻

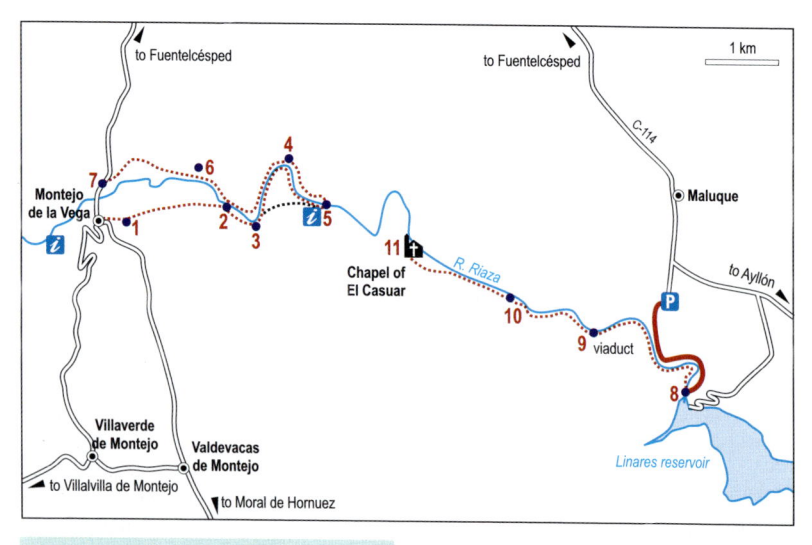

This walk departs from an abandoned quarry in the eastern sector of the gorge. Walk along the asphalted track that heads for the dam of the Linares reservoir; here cross the river on a footbridge (8) and pick up a footpath along the left-bank of the river. Pass under the railway viaduct (9) and reach two stone cairns indicating the limit of the Confederación Hidrográfico del Duero reserve.

Continue downstream as far as the Ermita del Casuar (11), from where you

PRACTICAL INFORMATION

- **Access:** at km point 134 of the Autovía de Burgos (NI/E5) there is a turn-off to Villalvilla de Montejo; from here, Montejo de la Vega is 10 km away (itinerary 1). Five kilometres from this village lies Fuentelcésped: here, take the C-114 towards Ayllón. After 10 km you will see a sign 'ADENA-CHD. Refugio de Rapaces', indicating a turn-off along an asphalted track that quickly reaches an abandoned quarry on the right, where you should park (itinerary 2).
- **Accommodation:** rural accommodation in Montejo de la Vega and Maderuelo. Hotels, hostals and rooms in Aranda del Duero.
- **Visitor infrastructure:** the interpretation centre (tel. 921 532 317) in Montejo de la Vega is open every day, morning and afternoon. WWF/Adena has an information point in the *hoces* themselves opposite Peña Portillo.
- **Visiting tips:** best in late spring and summer, although the autumn colours are spectacular. Do not leave the paths and during the breeding season in the first half of the year, do not climb the lower slopes up towards the cliffs or lean over the cliff-edges.
- **Recommended:** the WWF/Adena (tel. 913 540 578, info@wwf.es) organises activities and work camps all year: volunteers help the rangers and collaborate with management projects and bird counts, amongst other things.
- **Further information:** map 1:50,000, nº. 577 (SGE). *Guía del Refugio de Rapaces de Montejo de la Vega* by Jesús Cobo and Luis Suárez provides a detailed description of the wildlife of the site and itineraries.

CASTILLA Y LEÓN

should return along the same route. This route offers good opportunities to scan the cliffs, as well as to search for the passerines that haunt the river and the holm oak woodland.

MOST REPRESENTATIVE SPECIES

Residents: Griffon Vulture, Golden Eagle, Peregrine Falcon, Black-bellied Sandgrouse, Eagle Owl, Kingfisher, Dupont's, Calandra and Thekla Larks, Black Redstart, Black Wheatear, Blue Rock Thrush, Mistle Thrush, Dartford and Sardinian Warblers, Southern Grey Shrike, Azure-winged Magpie, Red-billed Chough, Rock Sparrow, Rock Bunting.
Summer visitors: Egyptian Vulture, Alpine Swift, Short-toed Lark, Crag Martin, Tawny Pipit, Northern and Black-eared Wheatears, Rufous-tailed Rock Thrush, Spectacled, Subalpine and Bonelli's Warblers, Red-backed and Woodchat Shrikes.
Winter visitors: Red Kite, Hen Harrier, Green Sandpiper, Alpine Accentor, Fieldfare, Redwing, Song Thrush.

59. PICOS DE EUROPA

The three awe-inspiring massifs of Los Picos de Europa, separated by equally startling river gorges, are the most spectacular part of the Cordillera Cantábrica. These limestone mountains – the most extensive in Atlantic Europe – have been moulded by severe folding and by a combination of glacial and karstic erosion, and today their landscapes are characterised by an impressive series of sheer cliffs, rocky outcrops, caves, limestone pavements and high-altitude lakes.

Los Picos are essentially part of the Atlantic bio-geographical region (although there are a few patches of more Mediterranean habitat) and at altitude pastures, scrub and bare rock dominate, grading into beech and oak forests at lower altitude. The highlights of the typical Cantabrian mammal fauna include European Brown Bear, Wolf, Izard and Castroviejo's Hare.

A National Park, shared between the provinces of León, Asturias and Cantabria, covers almost 65,000 ha of the three massifs. The autonomous government of Castilla y León has

Picos de Europa — Carlos Sánchez © nayadefilms.com

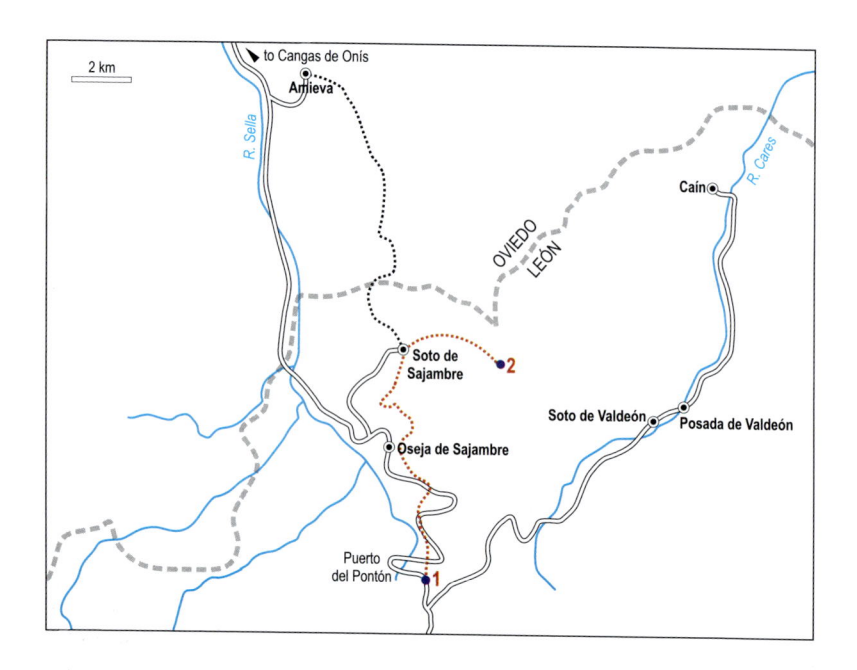

also protected their 'share' as part of a Regional Park of over 120,000 ha, which is also an SPA.

ORNITHOLOGICAL INTEREST

This is the best part of the Cordillera Cantábria for mid- and high-altitude birds such as Grey Partridge, Alpine Accentor, Wallcreeper, Alpine Chough and Snow Finch. In León province, a declining population of Capercaillie hangs on; beech forests have Black Woodpecker and oak forests Middle-spotted Woodpecker, and there are important communities of cliff-nesting raptors: Egyptian and Griffon Vulture (the latter in expansion) and Golden Eagle. Lammergeiers occasionally stray into the area.

ITINERARIES

1. 14 km

The valley of Sajambre (León province) in the south-west of the National Park can be explored along the Senda del Arcediano, an old mule trail connecting Castilla and Asturias. Begin at Puerto del Pontón (1) and follow the wide and comfortable path down the valley towards Oseja and then onto Soto de Sajambre, where you should have a second car waiting. From Soto de Sajambre it is worth detouring for 4 km along tracks to the mountain hut and pastures at Vegabaño (2).

Most of the itinerary passes through beech forest where it is not difficult to detect Black Woodpecker. A few Capercaillies still survive here, but probably a feather or excrement is all

CASTILLA Y LEÓN

that you will see of this shy bird. Look out too for Marsh and Crested Tits, Eurasian Treecreeper and Citril Finch, as well as Egyptian and Griffon Vultures and Golden Eagle from more open vantage points.

2. [7 km] 🚶 ☀ ⟳

From the village of Las Salas, near the Riaño reservoir in the eastern sector of the Regional Park, the long-distance footpath GR-1 heads up to the pass at Collado de Pando (3) and on to the village of Salamón. From here, return to Las Salas either back along the same path or by walking the peaceful road alongside the river Dueñas. Pass by the chapel of Virgin del Roblo (4) and come out on to the N-621.

This area is mainly covered by beech forests, although there are patches of oak, pastures and scrub. All the typical forest species breed here and open areas provide good views of most of the raptors breeding in these mountains. The heather and broom shrub formations are home to Grey Partridge, Ring Ouzel and Bluethroat and a quick ascent of a nearby peak may be rewarded by Alpine Accentor, Rufous-tailed Rock Thrush, Alpine and Red-billed Choughs and even Wallcreeper. The hedgerows and copses hold various species of warbler, Red-backed Shrike, Bullfinch and Yellowhammer.

MOST REPRESENTATIVE SPECIES

Residents: Griffon Vulture, Hen Harrier, Golden Eagle, Peregrine Falcon, Capercaillie, Grey Partridge, Eagle Owl, Black and Middle Spotted Woodpeckers, Crag Martin, Water Pipit, Dunnock, Alpine Accentor, Blue Rock Thrush, Ring Ouzel, Marsh and

PRACTICAL INFORMATION

- **Access:** from León, pass through Mansilla de las Mulas and take the N-625 to Cistierna. Then, 20 km further on, take the N-621 through Las Salas (itinerary 2) or continue to Cangas de Onís (Asturias) alongside the Riaño reservoir and Puerto de Pontón (itinerary 1).
- **Accommodation:** for itinerary 1, Soto de Sajambre, Oseja de Sajambre or Posada de Valdeón. Near itinerary 2, try Riaño, Las Salas or Boca de Huérgano.
- **Visitor infrastructure:** Casa Dago (tel. 985 848 614) in Cangas de Onís (Asturias) is the main information centre for the National Park. Open every day, morning and afternoon. In Posada de Valdeón in León there is another information point (tel. 987 740 549), open mornings from Monday to Friday and also in the afternoon in summer. The Regional Park has a visitor centre in Lario (tel. 987 742 215), open morning and afternoon from Wednesday to Sunday. There is another centre in Puebla de Lillo (tel. 987 731 091), some distance from the itineraries described here.
- **Visiting tips:** best in late spring and, despite the crowds, summer. Sudden changes in weather (rain and fog) can make leaving the marked tracks dangerous. Waterproofs essential.
- **Recommended:** in the summer the National Park organises guided walks within the protected area.
- **Further information:** maps 1:50,000, n°. 55, 80 and 105 (SGE). Information on the natural heritage and public access in the National Park from www.mma.es/parques/lared/picos.

Crested Tits, Wallcreeper, Eurasian Treecreeper, Alpine and Red-billed Choughs, Citril Finch, Bullfinch, Yellowhammer.
Summer visitors: Honey Buzzard, Egyptian Vulture, Short-toed Eagle, Alpine Swift, Bluethroat, Northern Wheatear, Rufous-tailed Rock Thrush, Common Whitethroat, Iberian Chiffchaff, Garden and Bonelli's Warblers, Red-backed Shrike, Ortolan Bunting.
Winter visitors: Redwing, Brambling.

60. HOCES DEL ALTO EBRO

In the province of Burgos the river Ebro flows through 145 km of magnificent gorges; the most attractive area is in the north-west where the river Ebro and its tributary the Rudrón have carved open the limestone rocks of the high plateau of La Lora. Karstic formations such as caves, pot-holes and underground rivers are also frequent in the area.

Straddling the divide between the Cordillera Cantábrica and the high plains of the Meseta, this area enjoys a combination of Atlantic and Mediterranean influences that is reflected in the atypical mix of beech and Lusitanian and holm oak forests. Add to this the well-conserved fluvial woodland, the extensive scrub formations on the plateaux and the small patches of cereal cultivation and you have a truly remarkable diversity of habitats. The only legal protection of the area is the Hoces del Alto Ebro y el Rudrón SPA (over 50,000 ha).

ORNITHOLOGICAL INTEREST

These *hoces* are exceptionally important as home to over 300 pairs of Griffon Vultures, excellent densities of Egyptian Vulture (declining in many parts of Spain), Golden Eagle and one (as of 2003) of the few remaining pairs of Bonelli's Eagles in the northern Meseta. Other more alpine species such as Alpine Chough (towards the southern limit of its distribution) and Alpine Accentor and Wallcreeper in winter can be seen here as well.

Many passerines frequent the mosaic of habitats surrounding the *hoces*. Both Hen and Montagu's Harriers fre-

Hoces del Alto Ebro © Pedro Retamar

CASTILLA Y LEÓN

quent the high *parameras* (plateaux) and in the scrub Mediterranean species such as Black-eared Wheatear, Spectacled and Orphean Warblers and Woodchat Shrike are found.

ITINERARIES

1. 8 km

This easy walk begins in the village of Valdelateja, where you should pick up a marked track from behind the church. Follow the right-bank of the river Rudrón to its confluence with the Ebro (1); continue alongside the Ebro as far as a footbridge near the small HEP station of El Porvenir (2). Cross the river and continue downstream along a narrower path as far as Pesquera de Ebro (with a car waiting in this attractive medieval village), or return to Valdelateja.

The fluvial woodland and cliffs are home to diverse bird colonies: along the rivers look out for Goshawk, Dipper, Marsh Tit, Red-backed Shrike, Bullfinch, Hawfinch and various species of bunting. The rock outcrops are home to Rufous-tailed Rock Thrush and the sky around the cliffs is airspace for myriad Griffon and Egyptian Vultures, Golden Eagle, Peregrine Falcon, Alpine Swift, Crag Martin and groups of Red-billed Chough (with a few Alpine Choughs mixed in).

2. 1 km

It is worth investigating the scrub-covered plateau of La Lora, dotted here and there with a few patches of holm and Lusitanian oak, for Hen and Montagu's Harriers, Thekla Lark, Black-eared Wheatear, Spectacled and Orphean Warblers, Red-backed, Southern Grey and Woodchat Shrikes and Ortolan Bunting.

Another good option is to take the N-623 northwards from Valdelateja. Ignore the right turn to Turzo (3) and at an old farm on the left (4), park and wander across the *páramo* as far as the edge of the Ebro gorge itself at Orbaneja del Castillo.

MOST REPRESENTATIVE SPECIES

Residents: Griffon Vulture, Hen Harrier, Goshawk, Golden and Bonelli's

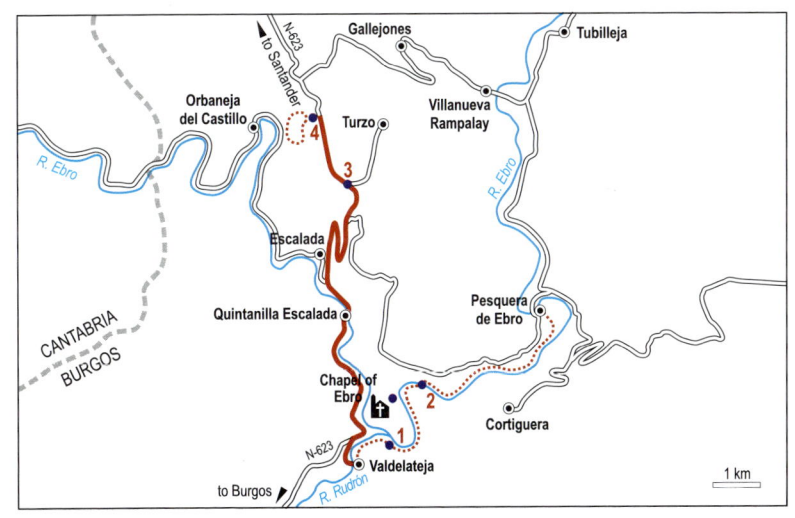

PRACTICAL INFORMATION

- **Access:** Valdelateja is 60 km from the city of Burgos along the N-623 towards Santander.
- **Accommodation:** in Valdelateja there is a spa near the river Rudrón. Also rural accommodation in Covanera, Orbaneja del Castillo, Pesquera de Ebro, San Felices and Sedano.
- **Visitor facilities:** no visitor centre or infrastructure for birdwatchers.
- **Visiting tips:** best in late spring and into summer.
- **Further information:** maps 1:50,000, nº. 109 and 135 (SGE). In Burgos try the offices of the Castilla y León Department of the Environment (tel. 947 281 500). Useful book: *Atlas de las aves nidificantes de la provincia de Burgos* by Jacinto Román *et al.* (Caja de Ahorros del Círculo Católico, 1996).

Eagles, Peregrine Falcon, Eagle Owl, Thekla Lark, Dipper, Dartford Warbler, Marsh Tit, Southern Grey Shrike, Red-billed and Alpine Choughs, Bullfinch, Hawfinch, Yellowhammer, Cirl and Rock Buntings.
Summer visitors: Honey Buzzard, Black Kite, Egyptian Vulture, Booted Eagle, Stone Curlew, Alpine Swift, Crag Martin, Common Redstart, Northern and Black-eared Wheatears, Rufous-tailed Rock Thrush, Bonelli's, Spectacled and Orphean Warblers, Iberian Chiffchaff, Spotted Flycatcher, Red-backed and Woodchat Shrikes, Ortolan Bunting.
Winter visitors: Alpine Accentor, Wallcreeper.

61. VALLE DEL ARLANZA

The rivers Arlanza and Mataviejas and their tributaries snake through the southern part of the Sistema Ibérico (Burgos province) between

Arlanza valley © Eduardo de Juana

CASTILLA Y LEÓN

high limestone cliffs such as Las Peñas de Cervera or in the incredibly narrow gorge of La Yecla near the famous monastery of Santo Domingo de Silos.

The most significant landscape features are the cliffs, fluvial woodland, the Spanish juniper and holm oak formations on *parameras* (high plateaus) and south-facing slopes, a few stands of Pyrenean Oak on north-facing slopes and the small fields in the valley bottoms. The only protection afforded to the area is that of the Sabinares del Arlanza SPA (almost 40,000 ha).

ORNITHOLOGICAL INTEREST

The cliffs of these valleys hold notable totals of over 500 pairs of Griffon Vulture and almost 20 pairs of Egyptian Vulture, along with Golden Eagle, Peregrine Falcon, Eagle Owl and one of the few pairs of Bonelli's Eagles left in Burgos province. Alpine Accentor and Wallcreeper frequent the area in winter.

The mixture of Atlantic and Mediterranean habitats provides for a superb range of passerines in the forest-agricultural mosaic and on the rocky outcrops. Both Rufous-tailed and Blue Rock Thrushes breed, as do Red-rumped Swallows, rare in the area. Other breeders include a few Black-eared Wheatears, many Azure-winged Magpies, both Stonechat and Whinchat, and Red-backed, Southern Grey and Woodchat Shrikes. In the stands of Spanish juniper and other forest formations look out for Common Redstart, Mistle and Song Thrushes, Orphean Warbler, Crested Tit and Ortolan Bunting.

ITINERARIES

The network of roads through the Arlanza (itinerary 1) and Mataviejas valleys (itinerary 2) between the villages of Covarrubias (west) and Salas de los Infantes (east) provide easy access to all the best birdwatching observation points and walks.

1. 35 km 🚗 🚶 ☀ ▬

At Covarrubias, you can explore the river and fluvial woodland from the bridge over the river Arlanza (1). From this village, continue towards Hortigüela and after the monastery of San

Pedro de Arlanza (between km points 14 and 15), park on the left (2) at the picnic site of El Torcón, from where with a telescope you will get good views of the Griffon Vulture colony (200+ pairs) on the other side of the river.

Halfway between Hortigüela and Salas de los Infantes, turn left in Barbadillo del Mercado towards Pinilla de los Moros. As you reach this village, a bridge (3) over the river Pedroso gives good access to the river bank and fluvial woodland downstream.

2. 50 km

The road from Salas de los Infantes to the monastery of Santo Domingo de Silos passes through the village of Carazo. From here, a 3-km path follows the river Mataviejas upstream as far as the chapel of La Virgin del Sol (4), located in an amphitheatre of cliffs. Return to the road and after Santo Domingo, turn left towards Espinoso de Cervera. Just after a tunnel, a signposted path takes you for 2 km along the narrow defile of La Yecla (5).

Return to the Santo Domingo road and continue on to Santibáñez del Val, from where an asphalted track leads to the hamlet of Barriosuso. Park and climb the mountain of Peña de Valdosa (6). Lastly, back in Santibáñez del Val, continue on towards Covarrubias and turn off towards Castroceniza where you can walk along the river Mataviejas for 4 km as far as the village of Ura.

MOST REPRESENTATIVE SPECIES

Residents: Griffon Vulture, Golden and Bonelli's Eagles, Peregrine Falcon, Eagle Owl, Kingfisher, Dipper, Blue Rock Thrush, Song and Mistle Thrushes, Dartford Warbler, Crested Tit, Southern Grey Shrike, Azure-winged Magpie, Red-billed Chough, Rock Sparrow, Yellowhammer, Cirl and Rock Buntings.

Summer visitors: Black Kite, Egyptian Vulture, Alpine Swift, Sand and Crag Martins, Red-rumped Swallow, Common Redstart, Whinchat, Northern and Black-eared Wheatears, Rufous-tailed Rock Thrush, Spectacled, Orphean and Subalpine Warblers, Common Whitethroat, Red-backed and Woodchat Shrikes, Ortolan Bunting.

Winter visitors: Red Kite, Alpine Accentor, Wallcreeper.

PRACTICAL INFORMATION

- **Access:** from Lerma, 35 km before Burgos along the N-I/E-5 from Madrid, take the C-110 west alongside the river Arlanza. After 24 km you reach Covarrubias, the start of itinerary 1 and end of itinerary 2. Salas de los Infantes links both itineraries.
- **Accommodation:** above all in Covarrubias, but also in Hortigüela and Salas de los Infantes.
- **Visitor infrastructure:** no visitor centre or infrastructure for birdwatchers.
- **Visiting tips:** late spring from May onwards and even in summer if you avoid midday.
- **Further information:** maps 1:50,000, n°. 276, 277, 314 and 315 (SGE). In Burgos try the offices of the Castilla y Leon Department of the Environment (tel. 947 281 500). Useful book: *Atlas de las aves nidificantes de la provincia de Burgos* by Jacinto Román *et al.* (Caja de Ahorros del Círculo Católico, 1996).

CASTILLA Y LEÓN

Sierra de Gredos © Pedro Retamar

62. SIERRA DE GREDOS

This imposing mountain chain – the highest in the Sistema Central – stretches east-west in the south of Ávila province for over 100 km and possesses various peaks topping 2,500 m. The north face, cool and poorly forested, descends gently towards the plains of the basin of the river Tormes. On the other hand, south-facing slopes are warmer and better vegetated and drop steeply to the valley of the river Tiétar. Almost 88,000 ha on both sides of the main ridge are protected by a Regional Park and an SPA.

The high granite uplands of the Gredos exhibit a wonderful variety of glacial features, from corries and glacial lakes, to hanging valleys and moraines. Up high, bare rock mixes with pastures and broom and dwarf juniper scrub, whilst at lower altitudes Pyrenean oak and sweet chestnut forests, pine plantations, holm oak formations (at the lowest altitudes) and abundant fluvial woodland provide the main forest cover. The best-known animal in these mountains is the Spanish Ibex.

ORNITHOLOGICAL INTEREST

This is a good site for both cliff-breeding and forest raptors, including Honey Buzzard, Black and Red Kites, Griffon and Black Vultures, Booted Eagle and Peregrine Falcon. Spanish Imperial Eagles sometimes stray from their breeding areas in and around the Tiétar valley. Look out too for forest and high mountain species such as Alpine Accentor, Bluethroat (the best Spanish populations) and Rufous-tailed Rock Thrush.

ITINERARIES

1. 7 km

From the car-park at La Plataforma (1,770 m) a well-marked path, paved at first, climbs between granite blocks with Black Redstart, Blue Rock Thrush and Dipper along streams to the pastures at Prado de las Pozas (1), frequented by Water Pipit. From here you enter the broom scrub, home to Dunnock, Bluethroat and Northern and Black-eared Wheatears; Golden

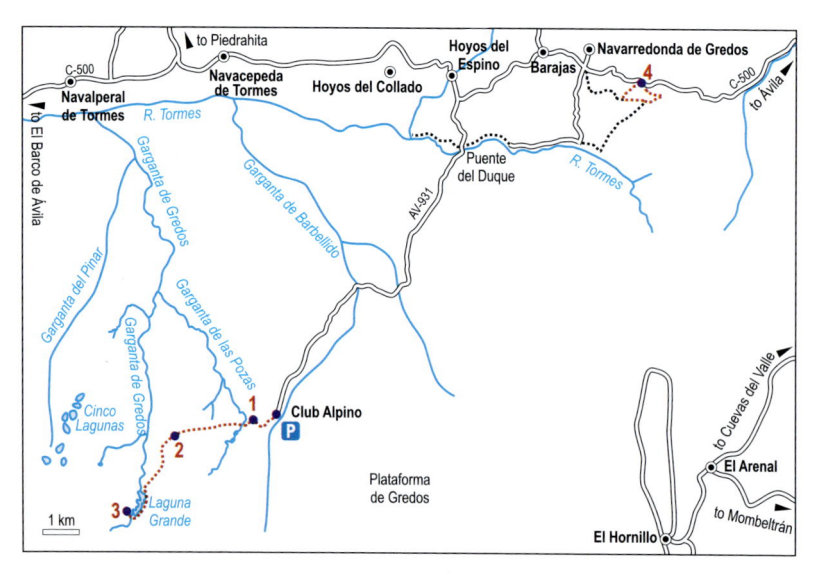

Eagle and Griffon and Black Vultures fly overhead.

After a somewhat steep climb, you reach Los Barrerones (2), from where the Laguna Grande comes into view. From here to Mirador del Circo the path is level, but then begins to drop a little to around 2,000 m and the Elola mountain hut (3), where you can spend the night. In and around the surrounding rocky outcrops and cliffs look for Crag Martin, Alpine Accentor, Red-billed Chough and, with luck, Rufous-tailed Rock Thrush.

This route from La Plataforma is the best known in the Gredos and can get very busy at weekends; alternatively and with the same ornithological possibilities, climb to the Laguna Grande via much quieter paths from the villages of Navalperal de Tormes and Bohoyo.

2. 4 km

This easy walk begins in the gardens of the Gredos Parador (4) and enters into the Navarredonda de Gredos pine forest. Likewise, from this village or from Hoyos del Espino other trails explore the well-conserved Scots pine forests on the north-face of Gredos that cover the slopes of the headwaters of the river Tormes. This is a good area for forest birds, including Black and Red Kites, Goshawk, Booted Eagle, Goldcrest, Firecrest, Citril Finch and Common Crossbill.

MOST REPRESENTATIVE SPECIES

Residents: Red Kite, Griffon and Black Vultures, Goshawk, Golden Eagle, Peregrine Falcon, Eagle Owl, Skylark, Water Pipit, Dipper, Dunnock, Alpine Accentor, Blue Rock Thrush, Mistle Thrush, Goldcrest, Firecrest, Crested and Coal Tits, Citril Finch, Eurasian Jay, Red-billed Chough, Common Crossbill, Rock Bunting.

Summer visitors: Honey Buzzard, Black Kite, Short-toed and Booted Eagles, Hobby, Crag Martin, Tawny Pipit, Bluethroat, Northern and Black-eared Wheatears, Rufous-tailed Rock Thrush, Bonelli's Warbler, Ortolan Bunting.

Winter visitors: Brambling, Siskin.

CASTILLA Y LEÓN

PRACTICAL INFORMATION

- **Access:** from Ávila take the N-110 towards Plasencia and after 7 km, turn off along the N-502 towards Arenas de San Pedro. After 50 km, turn right along the C-500 to Hoyos del Espino. From here, an asphalted track of over 10 km takes you to the car-park at La Plataforma (itinerary 1). The Gredos Parador is 7 km before Hoyos del Espino (itinerary 2).
- **Accommodation:** in all the villages and towns in the northern piedmont of the mountains, above all in Hoyos del Espino. Youth hostal, camp-site and Parador near Navarredonda de Gredos. The Elola mountain hut next to the Laguna de Gredos is permanently staffed (tel. 918 476 253).
- **Visitor infrastructure:** the visitor centre of the Sierra de Gredos Regional Park is in Hoyos del Espino.
- **Visiting tips:** once most of the snow has gone, best in spring from mid-April onwards. Despite the many visitors, summer is still worthwhile. Take care with ice in the morning at altitude; wear mountain gear – boots and waterproofs; crowds, especially on itinerary 1, at weekends and in summer.
- **Further information:** maps 1:50,000, n°. 555, 577 and 578 (SGE). In Ávila try the offices of the Castilla y Leon Department of the Environment (tel. 920 355 001). The book *Gredos, roca viva* by Tomás Santamaría *et al.* (Proatur, 1995) describes various nature walks on the northern side of the mountains and provides information about the birds of the area.

63. LA GRANJA AND SEGOVIA

The north face of the Sierra de Guadarrama between the towns of La Granja de San Ildefonso and Segovia exhibits an interesting succession of bird-rich habitats starting with rock outcrops up high, and continuing down through pastures, broom scrub, pine and oak forests and, in the valley bottoms, holm oak wood pastures (*dehesa*), rough grazing and a little cultivation. The area's streams and rivers are lined with relict stands of fluvial woodland. The eastern-most sector of the site is protected by the Sierra de Guadarrama SPA and the western by the Valles del Voltoya y Zorita SPA. A national park has been mooted for the area, which would also include the south-face of the Sierra de Guadarrama.

ORNITHOLOGICAL INTEREST

The 50 pairs of Black Vulture, the second-largest colony in Castilla y León after the Valle de Iruelas, and two Spanish Imperial Eagle territories are two of the main ornithological attractions to the area, although there are also good breeding populations of other raptors such as Red Kite, Golden Eagle, Booted Eagle and Lesser Kestrel. From the high mountains down to the arable plains, a fascinating variety of habitats unfolds and harbours a splendidly diverse avifauna, further enriched by the birds of the city of Segovia – 80 pairs of White Stork (but threatened by the closure of the municipal rubbish dump) and abundant Red-billed Chough.

ITINERARIES

1. 30 km

In La Granja de San Ildefonso wander through the Reales Jardines (1), beginning in the more manicured and popular north-west half before head-

Valsaín pine forests at la Granja and Segovia © José Manuel Reyero

ing for the woods in the south-eastern half. Look out for Crested and Coal Tits, Citril Finch and Common Crossbill, as well as overflying Black and Griffon Vultures and Golden and Spanish Imperial Eagles.

The CL-601 to Segovia passes by the Pontón reservoir (2), which can be reached after a short walk along the left-bank of the river Eresma: best in winter, with Great Cormorant, Grey Heron, duck and gulls. Back on the road, turn left (3) towards the village of Revenga, passing through excellent rough grazing land and Pyrenean and holm oak woodland with Black and Red Kites and Spanish Imperial Eagle not uncommon.

Some 9 km after the junction you reach a *soto* (fluvial woodland) near

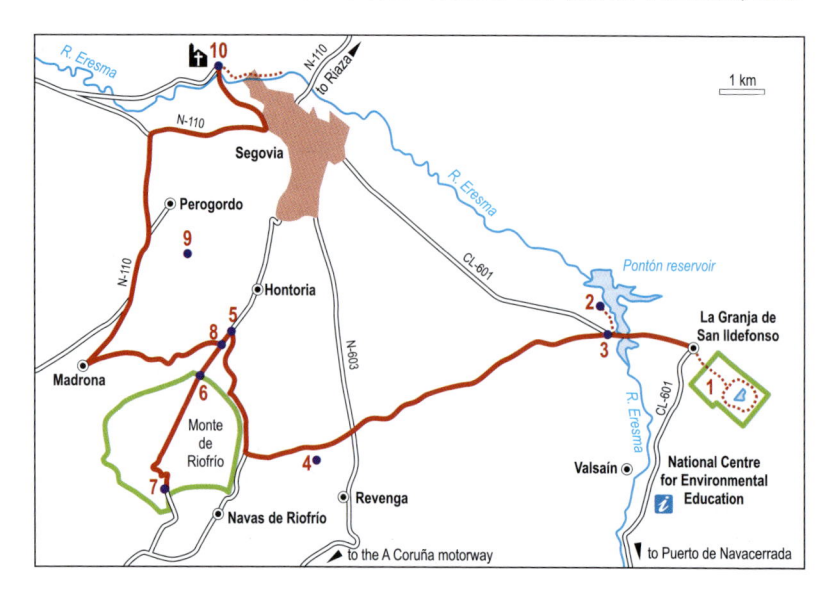

CASTILLA Y LEÓN

Revenga (4), a publicly owned but fenced-off area of *dehesas* and damp meadows. From the car the White Stork colony in the ash trees is quite visible. Further on, you come to the Riofrío palace and estate with a holm oak *dehesa* that was once a royal hunting estate. To gain access, first turn right (north) and continue as far as a junction (5) with the road leading from the palace to Segovia. Turn left here and enter the Riofrío estate through its northern (Hontoria) gate (6).

Once inside the estate, you are not allowed to park and get out of the car other then in the area near the palace (7); drive slowly and birdwatch from the vehicle, looking for Booted Eagle, European Bee-eater and Subalpine, Sardinian and Orphean Warblers.

Return to the Hontoria gate and head back north towards Segovia and after 1 km, turn left (8) into the Matamujeres valley. The church at Madrona (4 km) has a colony of Lesser Kestrels. From here arable cultivation (9) extends towards Segovia and is good for Calandra Lark, Skylark, Northern and Black-eared Wheatears and Tawny Pipit. Amongst the raptors, Montagu's Harrier is scarce here, unlike the kites and vultures that circle in search of the rich pickings from the area's pig farms.

2. 5 km 🚗 🚶 ☀ ➡

Birdwatchers should also find time to visit the city of Segovia itself, home to numerous White Storks that, once the breeding season is finished, roost on the spires of the city's cathedral. Here too there is a large Red-billed Chough roost (in the cathedral belltower) and a Common Swift colony in the famous Roman aqueduct, which attracts hunting Peregrine Falcon.

The sanctuary of Fuencisla (10) in the north of the city stands on a high crag and harbours Rock Dove, Red-billed Chough and Jackdaw. Another possibility is a walk upstream along the left-bank of the river Eresma for Kingfisher, Golden Oriole and other birds associated with riverside habitats.

MOST REPRESENTATIVE SPECIES

Residents: Red Kite, Griffon and Black Vultures, Golden and Spanish

---

PRACTICAL INFORMATION

- **Access:** take the CL-601 from the city of Segovia to La Granja de San Ildefonso (11 km). To reach Segovia from Madrid, take the Autovía de A Coruña (A-6) and then the N-603. Another worthwhile route is via the Puerto de Navacerrada, from where you can drop down to La Granja through Valsaín on the CL-601.
- **Accommodation:** in La Granja and, above all, in Segovia itself.
- **Visitor infrastructure:** the Centro Nacional de Educación Ambiental (tel. 921 471 711, ceneam@ceneam.mma.es) is in Valsaín, 3 km from La Granja: exhibitions, activities and four waymarked trails around the centre. Open weekends and public holidays, morning and afternoon.
- **Visiting tips:** best in late spring and summer for the breeding birds in Segovia. You have to pay to enter the state-run Reales Jardines de La Granja and Monte de Riofrío; open every day, morning and afternoon.
- **Further information:** maps 1:50,000, n° 483 (SGE).

Imperial Eagles, Peregrine Falcon, Kingfisher, Calandra Lark, Crag Martin, Dipper, Blue Rock Thrush, Dartford and Sardinian Warblers, Goldcrest, Crested and Coal Tits, Azure-winged Magpie, Red-billed Chough, Citril Finch, Common Crossbill.

Summer visitors: White Stork, Black Kite, Montagu's Harrier, Booted Eagle, Lesser Kestrel, Stone Curlew, European Nightjar, Common Swift, European Bee-eater, Short-toed Lark, Tawny Pipit, Northern and Black-eared Wheatears, Subalpine and Orphean Warblers, Golden Oriole.

Winter visitors: Great Cormorant, Grey Heron, Eurasian Wigeon, Black-headed Gull, Siskin, Bullfinch.

64. ARRIBES DEL DUERO

The great river Duero forms the Spanish-Portuguese on the western edge of the provinces of Salamanca and Zamora and, along with its tributaries the Tormes, Uces, Huebra and Água, has sculpted a deep, 100-km long gorge lined with majestic granite cliffs that plunge vertically hundreds of metres down to the river. Various habitats decorate the area: an undulating peneplain with *dehesas* and cereal cultivation, terraced olive and almond groves, hillside scrub consisting of juniper and holm oak on south-facing slopes and cork, and Pyrenean and Lusitanian oaks and even stands of European nettle tree on shadier north-facing slopes. Well-constituted fluvial woodland lines the rivers and a number of reservoirs store huge quantities of water. The Arribes are protected as a Natural Park (100,000+ ha) and as SPA.

ORNITHOLOGICAL INTEREST

Undoubtedly home to some of the best cliff-loving bird colonies in

Arribes de Duero · Carlos Sánchez © nayadefilms.com

CASTILLA Y LEÓN

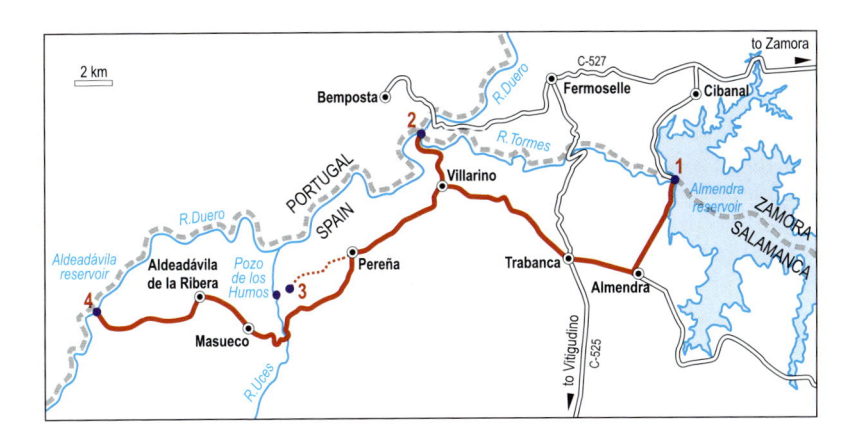

Spain, Los Arribes are vitally important for their 100+ pairs of Egyptian Vulture, 700+ pairs of Griffon Vulture and 20 pairs of Bonelli's Eagle, the greatest densities in Castilla y León. As well, there are good densities of Black Stork, Golden Eagle, Peregrine Falcon and Eagle Owl. Dupont's Larks have recently been found in Zamora province, a discovery that extends the species' known Iberian range much further westwards.

ITINERARIES

1. 70 km

This itinerary takes you to the main vantage points in the northern sector of the Arribes in Salamanca province. Begin at the enormous Almendra dam (1) between Zamora and Salamanca, with views over the reservoir to the east and the canyon of the river Tormes to the west. This is a good place for Black Stork, Egyptian and Griffon Vultures and other cliff-loving species such as Alpine Swift and Blue Rock Thrush.

From the dam, take the road through the villages of Almendra and Trabanca to Villarino, from where you can walk a 4-km long asphalted track down to Ambasaguas (2) at the confluence of the Duero and Tormes rivers. Bonelli's Eagles often fly over and the hillsides hold Thekla Lark and, with luck, Black Wheatear.

Return to Villarino and continue by road to the nearby village of Pereña. From here, a good dirt track can be driven to the viewpoint over the river Uces (3), one of the most-visited places in Los Arribes and the best site locally for Bonelli's Eagle (even from the car-park!). It is also worth walking down a narrow path to the base of the spectacular waterfall known as El Pozo de los Humos.

Lastly, from the village of Aldeadávila de la Ribera, take an asphalted track to the viewpoint of El Fraile (4), from where there are impressive views along the Duero and over the reservoir of Aldeadávila and good possibilities of observing Black Stork, Egyptian Vulture and Golden Eagle.

2. 4 km

In the extreme south of Los Arribes, the granite cliffs continue along the gorge of the river Águeda, one of the Duero's main tributaries. In San Felices de los Gallegos, an attractive medieval village, park and walk along the old drover's road to Puerto Seguro and then on to the river at Puente de Los Franceses (5). This is a good site for Black Stork, Griffon Vulture, Peregrine Falcon, Golden Eagle and, along the river, Kingfisher, Dipper and other river species.

MOST REPRESENTATIVE SPECIES

Residents: Grey Heron, Black-winged and Red Kites, Griffon Vulture, Golden and Bonelli's Eagles, Stone Curlew, Eagle Owl, Kingfisher, Lesser Spotted Woodpecker, Calandra and Thekla Larks, Dipper, Black Wheatear, Blue Rock Thrush, Dartford, Spectacled and Sardinian Warblers, Red-billed Chough, Rock Sparrow, Hawfinch, Rock Bunting.

Summer visitors: Black Stork, Black Kite, Egyptian Vulture, Montagu's Harrier, Alpine Swift, Roller, Crag Martin, Red-rumped Swallow, Northern and Black-eared Wheatears, Subalpine Warbler, Ortolan Bunting.
Winter visitors: Great Crested Grebe, Great Cormorant, Green and Common Sandpipers, Black-headed Gull, Water Pipit.

PRACTICAL INFORMATION

- **Access:** from Zamora take the C-527 and 10 km before Fermoselle turn left to Cibanal and the Almendra reservoir. From Salamanca, the quickest route is via Ledesma: before you reach Trabanca, turn right in Almendra to the reservoir. The C-517 from Salamanca passes through Vitigudino and then Lumbrales, from where the SA-324 towards Ciudad Rodrigo takes you to San Felices de los Gallegos in 10 km.
- **Accommodation:** hostals and rural accommodation in the larger villages in the area: Lumbrares, Villarino, Pereña, Masueco, Aldeadávila de la Ribera, Mieza, Vitigudino, Hinojosa de Duero and San Felices de los Gallegos.
- **Visitor infrastructure:** the visitor centre of the Arribes del Duero Natural Park is in Sobradillo (Salamanca). Open Wednesday to Sunday, morning and afternoon.
- **Visiting tips:** best in spring and autumn. Summer is too hot. Beware of disturbing breeding birds: many of the breeding failures in sensitive species such as Black Stork are caused by people approaching nest sites.
- **Recommended:** in Los Arribes in Zamora, there are a number of good sites such as the chapel of El Castillo near the edge of the Duero canyon, which can be reached by car from the village of Fariza.
- **Further information:** maps 1:50,000, n°. 422, 423, 449, 475 and 500 (SGE). In *Guía de la fauna vertebrada de los Arribes de Duero zamoranos y su entorno* by Joaquín Sanz-Zuasti and Jesús Fernández (Náyade, 2001) there is a complete discussion of all the bird species in the area.

CATALONIA

- 65. Ebro Delta
- 66. Aiguamolls de l'Empordà
- 67. Llobregat Delta
- 68. Aigüestortes
- 69. Cadí and Moixeró
- 70. Lleida Plains

Situated in the far north-east of the Iberian Peninsula, Catalonia is one of the most industrialised of all Spain's autonomous communities. Much of the population is concentrated in and around the Barcelona conurbation and along the immediate coastal strip; in the interior, however, human activity is less patent (above all in the extreme north and south), although certain factors such as the importance of tourism and the extensive road network clearly reflect the growth of the Catalan economy.

Outside the high Pyrenees and despite the changes imposed on most natural habitats, there is still a wonderful diversity of birdlife (and wildlife in general) to be found in Catalonia; the surprising diversity of species in this relatively small territory reflects the heterogeneous land-use, relief, climate and vegetation that is characteristic of the region.

The fact that an extraordinarily high number of bird species have been recorded here is indicative of the rare coincidence of habitats as different as coastal marshes, inland steppes and alpine forests that are found in such a small area. Furthermore, its position on the Mediterranean coastline in the north-east of the Iberian Peninsula, means that many birds migrating along the Mediterranean flyway will stop off in Catalonia on their travels.

The birdwatching highlight is undoubtedly the Ebro Delta (65). Aside from Doñana, no other site in the Peninsula can compare to the Ebro Delta, which in terms of gulls is actually the most important site in Spain and home to the world's largest breeding colony of Audouin's Gull. As well, it is of exceptional importance for breeding herons, Glossy Ibis, Greater Flamenco, ducks and waders, and for winter and passage wildfowl, waders and seabirds.

Two other coastal wetlands are musts for birdwatchers. In Aiguamolls de l'Empordà (66), management

Capercaillie © Juan Martín Simón

policies have been centred on restoring a complex of coastal lagoons and salt-marshes that were saved from destruction by popular protest in the 1980s: today these marshes have a wonderful network of itineraries and hides especially designed for birdwatchers. The Llobregat Delta (67), much loved by more than one generation of Catalan birdwatchers, has historically always been threatened by the expansion of the Barcelona conurbation and today survives as a small natural oasis in a sea of infrastructures and built-up areas.

For those who yearn for the solitude of nature at its majestic best, the high Pyrenees are the answer. Forming the northern border of Catalonia (and between Spain and France), with its highest peaks on the frontier with Aragon, the Pyrenees are replete with rocky summits, glacial lakes and a succession of true alpine and subalpine habitats that include vast forests of silver fir and pine.

Capercaillie and Tengmalm's Owl have their main Iberian strongholds in these coniferous forests, although both species are hard to find. Easier to locate are Black Woodpecker, Ring Ouzel, Citril Finch and other forest birds in areas such as the Aigüestortes National Park (68), well equipped with services, visitor centres and waymarked routes and highly recommendable for birdwatchers.

Outside the Pyrenees, most of Catalonia is occupied by rugged mid-altitude mountain ranges that in some cases even reach the coastline. The abrupt pre-Pyrenean ridges and gorges are home to Lammergeier and the Cadí-Moixeró (69), with its splendid mix of subalpine, Eurosiberian and even Mediterranean habitats, is one of the best such areas.

To the south in Lleida province lie the semi-steppes of the Ebro Depression, the continuation in Catalonia of the Aragonese Monegros. Cereal cultivation and a few areas of natural steppe predominate and are the last haunt in the Peninsula of Lesser Grey Shrike and the only site in Catalonia for Little Bustard, Black-bellied and Pin-tailed Sandgrouse and Dupont's Lark, although all face an uncertain future as irrigation projects take ever-increasing hold in the region.

CATALONIA

The heart of these plains – known as Els Secans de Lleida (70) – lies near the city of Lleida and offer birdwatchers the added attraction of two interesting wetlands, the reservoirs of Utxesa and the confluence of the rivers Segre and Cinca (Aiguabarreig).

65. EBRO DELTA

In the extreme south of Catalonia lies the Ebro Delta, a vast alluvial plain of some 36,000 ha lying at the mouth of the river Ebro. Arrow-shaped in outline and extending a full 20 km into the sea, the two halves of the Delta – with El Fangar to the north and La Banya to the south – are singular in their absolute flatness and chromatic fluctuations: from brown in winter, the paddy fields that occupy much of the Delta change colour in spring as they are flooded and then again in summer as the green rice thrusts up through the mud.

The best preserved parts of the Delta are the long empty beaches, the shallow interior bays behind the points of the 'arrowhead', the permanent saline lagoons and the salt-marshes. The salt-works, despite their artificiality, actually add to the species-richness of the Delta. These habits are protected by a Natural Park (almost 8,000 ha), including two *reserves naturals* and seven *reserves de fauna*, that have been declared a Ramsar Site and an SPA.

ORNITHOLOGICAL INTEREST

Second only to Doñana in its importance for seabirds and aquatic birds in Spain, the Ebro Delta is exceptional, above all, for its breeding gulls and terns: almost 12,000 pairs of Audouin's Gulls (by far the largest colony in the world and almost two-thirds of the total world population), the most important Spanish colonies of Slender-billed Gull and Sandwich and Common Terns, the only breeding Lesser Black-backed Gulls on the Mediterranean coast, the only Lesser Crested Terns in Spain (few but regular) and up to 30,000 wintering Mediterranean Gulls.

Furthermore, almost all the Iberian herons breed, including the scarce Great Bittern, the largest Spanish concentrations of Squacco Heron

Punta de la Banya, the southern tip of the Ebro delta © Anna Motis

(800 pairs) and a small breeding (larger in winter) population of Great Egret. The Ebro is also one of the few Spanish breeding sites for Glossy Ibis and boasts one of only two stable colonies of Greater Flamingo in Spain (the other breeding sites are Fuente de Piedra in Málaga and sometimes Doñana).

The delights for the birdwatcher continue: winter brings masses of wildfowl, including Greylag Geese. Shelduck and Red-crested Pochard have their strongest Iberian breeding concentrations here, Marsh Harriers winter in large numbers (although they do not breed) and Purple Swamp-hen have recently colonised the area. Wader-lovers will also do well: Spain's best populations of breeding Oystercatchers and excellent numbers of all the species that pass through or winter in the Iberian Peninsula occur here.

ITINERARIES

The Ebro Delta can be divided into two halves: the northern half on the left-bank of the river, and the southern on the right-bank.

The northern delta

1. 30 km

From the town of Deltebre, take the road towards L'Ampolla and after 8 km, turn right (1) along a track signposted to the beach at El Goleró (2). Park and walk north-west along the shore of El Fangar, one of the huge shallow interior bays, as far as the viewpoint at the Les Olles lagoon (3).

On the way you pass through paddy-fields that are flooded in April and attract many herons, waders, gulls and terns. In winter, the fields are dry and birds gather wherever there is a little water or where the specially designed tractors are ploughing in the rice stubble. Los Olles is one of the best places in the Delta for Osprey on passage, and also for the scarce Caspian Tern and groups of Slender-billed Gull.

Return to the car and drive along the shore of El Fangar south-eastwards. Just after passing a small fishing port, take a track on the right (4) alongside a canal and then turn left on to the road running from Deltebre to the beach of La Marquesa (5). From the beach, head out north-west past an observation point towards the sand-spit of El Fangar along a sandy track that can be flooded by storms (find out beforehand from the information centres).

In winter groups of Greater Flamingo and many waders grace the muddy shores of the bay behind the sand-spit, while its shallow waters are frequented by Black-necked Grebe and Red-breasted Merganser. Out to sea, Great Northern Diver, Balearic Shearwater and other seabirds fish and Mediterranean Gulls and, in summer, Audouin's Gulls ply back and forth.

The dunes of the sand-spit are home to breeding colonies of Oystercatcher, Kentish Plover, Slender-billed Gull and Gull-billed, Little, Common and Sandwich Terns (look amongst the latter for the odd Lesser Crested Tern). The drive to the lighthouse (6) provides views of all these species, but between May and August do not leave your cars or drive off the track to avoid disturbing the nesting birds.

2. 2 km

From Deltebre, the road to the Riumar holiday complex (7) passes through paddy-fields with similar birds to those described for the previous itinerary. From Riumar, head south-east along the shore of El Gar-

CATALONIA

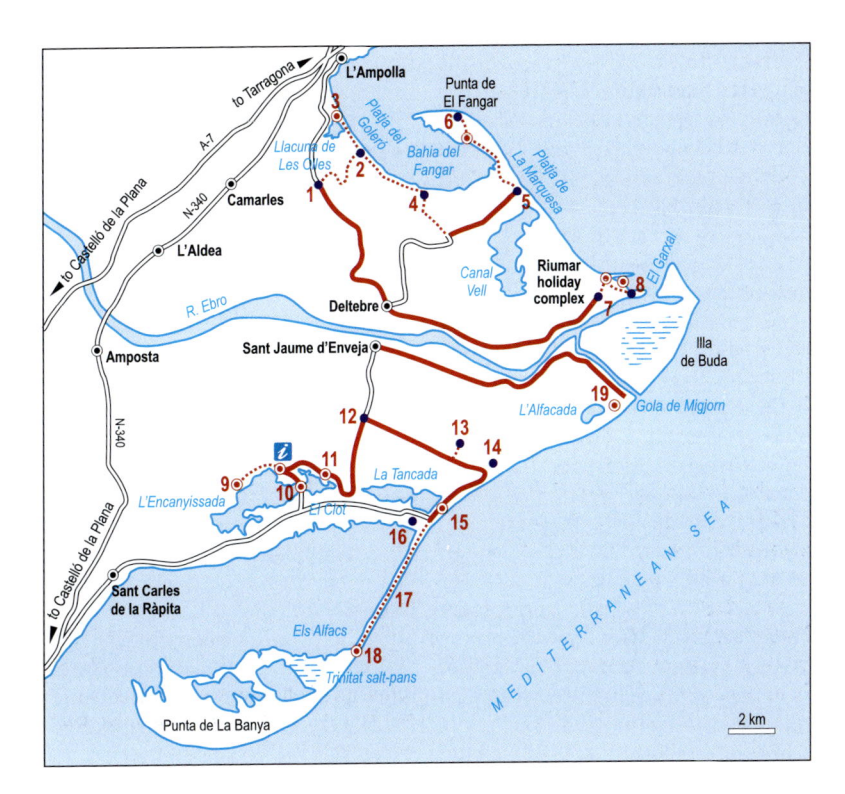

xal, another shallow coastal lagoon, as far as a lighthouse (8) next to the river mouth.

Two hides, one raised at the beginning of the walk, provide good views of the lagoon, which, when the rest of the Delta is dry at the beginning of spring or at the end of summer, becomes an alternative refuge for many birds. There are always good concentrations of Red-crested Pochard and of Garganey, waders and terns during migration periods. The dunes and salt-marshes hold Kentish Plover and Lesser Short-toed Lark.

The southern delta

1. 25 km

Next to the Casa de Fusta, an information centre, a raised hide overlooks L'Encanyissada, the Delta's largest lagoon; a path heads west from here for 1 km to another hide (9). Both viewing points overlook a magnificent reed-bed with Little Bittern and Purple Heron and, with luck, Great Bittern. Views of Greater Flamingo, Marsh Harrier (in winter) and large numbers of duck (above all in winter) are guaranteed. It is easy here to see Red-crested Pochard with young in the canals in the reed bed.

Heading east on the road along the north side of the lagoon you reach a junction, where you should turn right to a bridge (Pont del Través) and an observation tower (10), a good site for Kingfisher, Purple Swamp-hen, Reed Bunting and, in winter, Bluethroat. Return to the crossroads, turn right and pass by the raised viewpoint at L'Embut (11), next to the lagoon of El Clot, a natural extension of L'Encanyissada, before reaching another junction; turn left and head north.

After 3 km, turn right (12) on the road from Sant Jaume d'Enveja to Los Eucaliptos. After 5 km along this road (between km points 18 and 19), turn left to the reserve of Riet Vell (13), run by SEO/BirdLife. Aside from paddy-fields, this reserve has a small restored freshwater lagoon with a reed-bed and artificial islands where Collared Pratincole, Black-winged Stilt and Common Tern breed. The diversity of habitats here mean that most of the Delta's birds pass through at some time or other during the year.

Continue on towards Los Eucaliptos (14), but turn right along a road parallel to the beach before you reach the first houses, and pass through an area of saline pastures frequented by Collared Pratincole and Lesser Short-toed Lark. Soon you reach the eastern end of La Tancada, the other large lagoon in the southern part of the Delta, and a hide (15). From here you will see similar birds as on L'Encanyissada, along with Black-necked Grebe in winter.

A little further on, the road crosses through the abandoned Sant Antoni salt-pans (16) – good for waders and terns, including Caspian Tern – and meets the road from Sant Carles de la Ràpita. Turn left and then almost immediately turn right onto the sandy track along the narrow Trabucador sand-spit (17), over 5 km in length. Look out in winter for waders, including Oystercatcher, and seabirds out to sea and on the shallow waters of Els Alfacs, the bay behind the sand-spit.

This itinerary ends where La Punta de la Banya, a closed-off *reserva natural* with an extraordinary diversity of birdlife, begins. A tower hide here (18) provides views of the salt-pans of La Trinitat and part of La Punta and is the best place in the Delta for gulls and terns, including Audouin's and Slender-billed Gulls. Large flocks of Greater Flamingos (which breed on La Punta) are visible all year, and there are also breeding Shelduck and Avocet, good concentrations of duck and waders in winter and Eurasian Spoonbill on passage.

2. 10 km

La Gola de Migjorn, a secondary mouth of the river is also well-worth a visit. Leave Sant Jaume d'Enveja along the road parallel to the right-bank of the river and reach a high observation point (19) near the beach giving good views over L'Illa de Buda. There is no access to this island, but the tower is the best place in the Delta to see Glossy Ibis and Great Egret. Winter is good for Greylag Goose and waders and herons; the near by private salt-marshes of L'Alfacada add interest to the spot.

MOST REPRESENTATIVE SPECIES

Residents: Great Crested Grebe, Balearic Shearwater, Night Heron, Great Egret, Glossy Ibis, Greater Flamingo, Shelduck, Gadwall, Red-crested Pochard, Purple Swamp-hen, Oystercatcher, Avocet, Kentish Plover, Slender-billed, Audouin's and Lesser Black-backed Gulls, Sandwich and Whiskered Terns, Kingfisher, Lesser Short-toed Lark, Skylark, Moustached Warbler, Penduline Tit, Reed Bunting.

Summer visitors: Cory's Shearwater, Little Bittern, Squacco and Purple Herons, Black-winged Stilt, Collared Pratincole, Gull-billed, Lesser-crested, Common and Little Terns, Short-toed Lark, Yellow Wagtail.

Winter visitors: Great Northern Diver, Black-necked Grebe, Northern Gannet, Great Cormorant, Greylag Goose, Red-breasted Merganser and

CATALONIA

PRACTICAL INFORMATION

- **Access:** from the L'Aldea and Amposta junctions on the A-7 Barcelona-Valencia motorway. The N-340 runs along the eastern edge of the Delta and through L'Ampolla, Camarles, Amposta and Sant Carles de la Ràpita.
- **Accommodation:** in all the main towns in the Delta, but best in Deltebre and Sant Jaume d'Enveja.
- **Visitor infrastructure:** the main Natural Park information centre (tel. 977 489 679, infoecomuseu@wanadoo.es) is in Deltebre. Open every day, morning and afternoon; Sunday and public holidays, only in the afternoon. The other information centre, La Casa de Fusta (tel. 977 261 022), next to L'Encanyissada lagoon, keeps the same opening hours. Both centres programme guided visits.
- **Visiting tips:** good any time of year. From October onwards winter visitors arrive; between January and April there are great concentrations of birds in the bays behind the sand-spits and anywhere with water. May to July is the best time to see breeding birds and birds on spring passage. Find out about the state of the tracks along the coast, above all to El Fangar lighthouse, before attempting to drive them. Wear mosquito repellent and sun-cream; good maps available.
- **Recommended:** bike hire is a good way to see the Delta, above all from the bike lanes around L'Encanyissada and La Tancada. Tourist boat to the mouth of the river.
- **Further information:** maps 1:50,000, n°. 522, 523 and 547 (SGE) and from the Ebro Delta SEO/BirdLife office (in the town of Amposta; tel. 977 702 308, deltaebro@seo.org). They organise visits to the reserve at Riet Vell via the web site www.seo.org/rietvell. The book *Els Ocells del delta de l'Ebre* by Albert Martínez-Vilalta and Anna Motis (Lynx Edicions, 1991; text in Catalan) contains a complete description of all species and a number of itineraries.

other duck, Marsh Harrier, Osprey, Peregrine Falcon, Red-knobbed Coot, Grey Plover, Sanderling, Little Stint, Dunlin, Black-tailed and Bar-tailed Godwits, Eurasian Curlew, Arctic and Great Skuas, Mediterranean Gull, Razorbill, Short-eared Owl, Bluethroat.

On migration: Eurasian Spoonbill, Garganey, Red-necked Phalarope and other waders, Caspian Tern.

66. AIGUAMOLLS DE L'EMPORDÀ

This small area of marshland lies between the mouths of the rivers Muga and Fluvià behind the Bahia de Roses (Girona province) and consists of a series of permanent saline coastal (*llaunes*) and freshwater interior (*estanys*) lagoons associated with sandy beaches, salt-marshes, pastures (*closes*), arable land and fluvial woodland. After a popular campaign prevented the area from being built upon, a Natural Park was declared in 1983 (almost 5,000 ha) that since then has cleverly combined the joint priorities of conservation and public use. The Aiguamolls marshes are also a Ramsar Site and an SPA, and the most sensitive areas of the site are also *reserves naturals*.

ORNITHOLOGICAL INTEREST

Second in Catalonia only to the Ebro Delta in terms of water birds, the Aiguamolls are home to important heron colonies with some of the few Spanish breeding Great Bitterns and good numbers of Purple Heron; other

Aiguamolls de l'Empordà   © Fernando Barrio

breeders include Marsh Harrier, Gadwall, Black-winged Stilt, Kentish Plover and Roller. White Stork and Purple Swamp-hen have both been successfully reintroduced; this too is one of the few Iberian locations for Lesser Grey Shrike, although it has all but disappeared. Greater Flamingos do not breed but are present all year. Winter brings many duck and Glossy Ibis and Greylag Goose, while passage periods are good for Eurasian Spoonbill, Osprey, Red-footed Falcon and Audouin's Gull.

ITINERARIES

1. 9 km

The Empuriabrava holiday complex divides the park into two independent sectors: the southern sector includes the Reserva Natural Les Llaunes and protects six coastal lagoons. Begin the itinerary at the information centre at El Cortalet (1) next to an artificial freshwater lagoon with three hides. The route follows a canal (Rec Corredor) to La Massona, the first of the *llaunes*. The large reedbeds here are home to Little Bittern and Marsh Harrier and, in winter, to Moustached Warbler, Penduline Tit, Reed Bunting and, with luck, Bearded Tit.

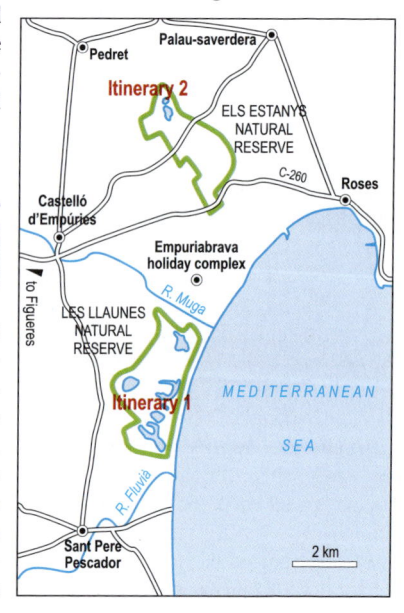

Three hides overlook La Massona: in between the first and the second, it is worth leaving the main route and turning right (west) past the former paddy-fields of El Matà (2), today restored as a semi-natural wetland. Black-winged Stilts breed here and both Purple Heron and Purple Swamp-hen come to feed; Garganey and thousands of waders pass through in spring and winter sees a contingent of Glossy Ibis present. At the top of the monumental nearby rice silos, a wonderful observation platform (3) gives excellent views over the whole area.

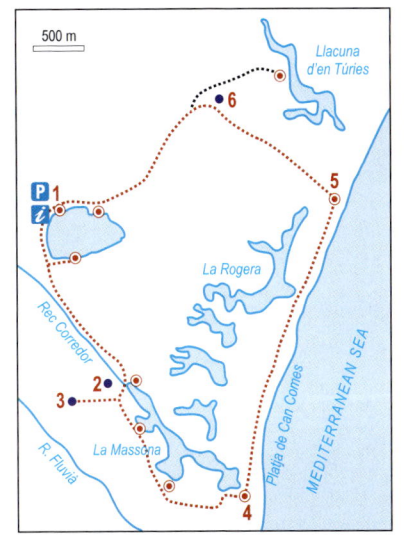

Return to the main route and head for the beach at Can Comes. Two viewing points (4, 5) at the beginning and end of this section are good for seawatching, with Black-throated Diver, Balearic Shearwater, Northern Gannet and Razorbill possible in winter. Kentish Plover run around the beach all year and other waders and Audouin's Gulls congregate here on migration and in winter. Halfway along the beach, there are good views over La Rogera, a saline lagoon frequented by Greater Flamingo.

A track off to the left heads inland through the saline pastures of Can Comes (6), where you might find Eurasian Dotterel and Richard's Pipit in autumn passage. From here, return to El Cortalet.

2. 2 km

The northern sector of the Aiguamolls is protected by the Reserva Natural Els Estanys and can be reached from Castelló d'Empúries along the road to Palau-Saverdera. Stop at Els Tres Ponts (7), a stretch of road that crosses three canals in quick succession and look for Great Bittern.

A little further on, park at the restaurant Aiguamolls (8), (6.5 km from Castelló d'Empúries) and walk back along the road until you find a track off to the right that passes through open fields with Roller and Red-footed Falcon (in May and June) and then takes you to the Vilaüt lagoon and a hide (9). This was once a breeding area of Lesser Grey Shrike.

MOST REPRESENTATIVE SPECIES

Residents: Balearic Shearwater, Great Bittern, White Stork, Greater Flamingo, Gadwall, Marsh Harrier, Purple Swamp-hen, Stone Curlew, Kingfisher, Moustached Warbler, Penduline Tit.

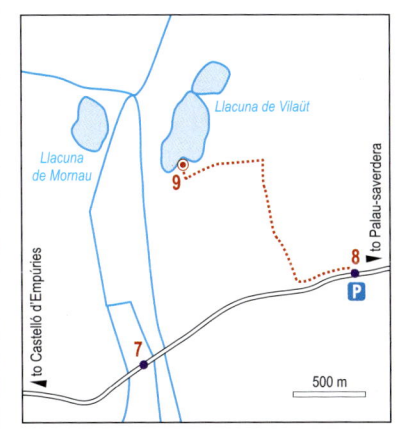

> **PRACTICAL INFORMATION**
>
> - **Access:** from Figueres, take the C-260 towards Roses. At Castelló d'Empúries (7 km), turn right at a roundabout towards Sant Pere Pescador and just after a petrol station on your right turn left to El Cortalet.
> - **Accommodation:** in Castelló d'Empúries, Roses or the Costa Brava.
> - **Visitor infrastructure:** The road to the information centre at El Cortalet is between km points 13 and 14 on the road from Castelló d'Empúries to Sant Pere Pescador. Open every day, morning and afternoon (tel. 972 454 222, pnaiguamolls.dmah@gencat.net).
> - **Visiting tips:** good all year, but above all in April and May when breeding and passage species coincide. The beach of Can Comes (itinerary 1) is closed to the public between April 1 and June 15 to protect Kentish Plovers and other birds.
> - **Recommended:** the naturalist groups Apnae (the. 972 454 672) and Iaeden (tel. 972 670 531) organise activities in the park.
> - **Further information:** map 1:50,000, n°. 258 (SGE). The book *Parc Natural dels Aiguamolls de l'Empordà. Guia d'itineraris* by Toni Llobet and Ponç Feliu (Edicions del Brau, 2001) gives more itineraries.

Summer visitors: Little Bittern, Purple Heron, Black-winged Stilt, Kentish Plover, Roller, Lesser Grey Shrike.
Winter visitors: Black-throated Diver, Black-necked Grebe, Northern Gannet, Great Cormorant, Glossy Ibis, Greylag Goose and other wildfowl, European Golden Plover, Northern Lapwing, Eurasian Curlew, Razorbill, Bearded Tit, Bluethroat.
On migration: Squacco Heron, Eurasian Spoonbill, Garganey, Osprey, Red-footed Falcon, Eurasian Dotterel and other waders, Little Tern, Slender-billed and Audouin's Gulls, Whiskered and White-winged Black Terns, Red-throated Pipit.

67. LLOBREGAT DELTA

A stone's throw from the city of Barcelona, the delta of the river Llobregat consists of a 10,000 ha alluvial plain that has been largely transformed into agricultural land and occupied by industrial and transport infrastructures such as the port of Barcelona and the airport of El Prat. Today a few natural enclaves such as saline lagoons with dense reed-beds, salt-marshes, stone pine woodland, dunes and beaches still exist.

The most valuable wetlands are included within two *reserves naturals* (Ricarda-Ca l'Arana and Remolar-Filipines), totalling around 500 ha, which are also an SPA. The immediate surroundings have been heavily affected by the development of large-scale infrastructures such as the airport and a high-speed train line.

ORNITHOLOGICAL INTEREST

In spite of its small size and its situation amidst dense industrial and urban development, the Llobregat Delta enjoys surprisingly high densities of birds and bird species: this fact can be put down to its strategic position *en route* between the Ebro Delta and the Aiguamolls de l'Empordà, the two most important Catalan wetlands, and its concentration of many different habitats in a small space. Many birdwatchers from Barcelona have cut their teeth here and the site has a long list of rarities to its name.

CATALONIA

Les Filipines salt-marsh in the Llobregat Delta © Jaume Orta

Above all of interest for its water birds, Little Bittern, Squacco Heron, Red-crested Pochard, Marsh Harrier and Purple Swamp-hen all breed. Kentish Plover nest on the beaches and in 2003 Red-knobbed Coot were introduced for the first time. Greater Flamingos are present almost all of the year and Audouin's Gulls are regular in spring and summer. Migration periods and winter are good for Balearic Shearwater, Great Bittern, Great Egret and many Mediterranean Gulls.

ITINERARY

3 km

The publicly owned Reserva Natural Remolar-Filipines is the most accessible area of the Llobregat Delta and encompasses in a small area habitats as diverse as lagoons, salt-marsh, pinewood and beach. Hunting was banned in 1984 and birds are thus fairly confiding; most of the hides have good wheelchair access.

From the Autovia de Castelldefels (C-31) there is access to the road that runs alongside La Vidala (1), an arm of the much larger Remolar lagoon. Common species such as Common Moorhen, Eurasian Coot and ducks swim here, Marsh Harriers fly over and the reed beds are home to Moustached Warbler and Penduline Tit (above all in winter). About halfway along La Vidala (heading towards the sea) there is an observation platform (2) that offers good views of the whole reserve and the chance to spot Little Bittern or Purple Heron in flight.

Further on, opposite a hide that is normally closed, park at a junction and keep right heading towards the sea. On the right you have the Riera de Sant Climent, and on the beach the observation point of Cal Francès (4), a good place for seawatching for Balearic Shearwater, Northern Gannet, Mediterranean Gull and Sandwich Tern. From April onwards, Audouin's Gulls frequent the area, and Kentish Plover breed in the dunes.

Return to the junction and continue alongside La Vidala as far as a boardwalk that provides access to the area of Les Filipines. From the information centre near the entrance, two short itineraries take you to hides overlooking the permanent lagoon of Bassa dels Pollancres and other more temporary pools. Most of the typical birds of the Delta can be seen from here: on artificial islands there are colonies of Black-winged Stilt and Little Ringed Plover, winter sees numerous Great Cormorants, herons and ducks resting here, Great Bittern (above all in winter), Red-crested Pochard, waders, Kingfisher and reed-bed passerines such as Bluethroat, Penduline Tit and Reed Bunting.

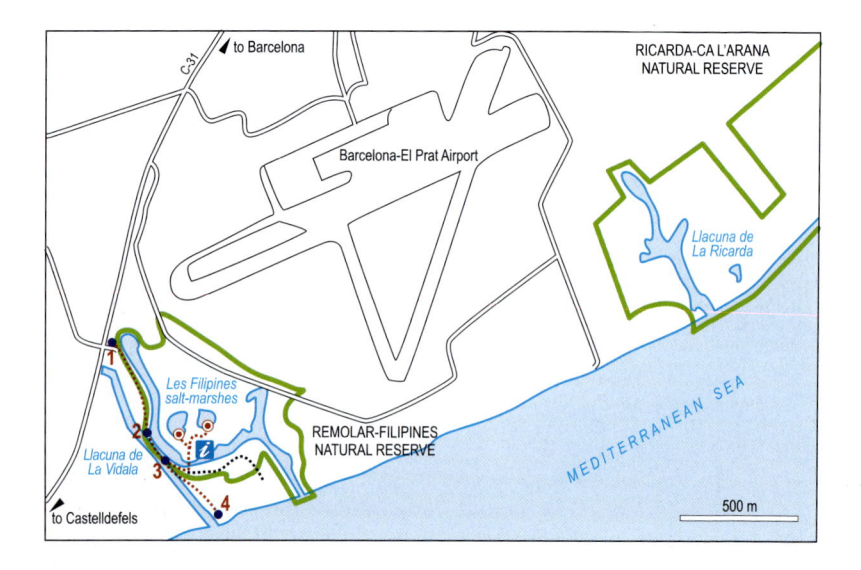

MOST REPRESENTATIVE SPECIES

Residents: Greater Flamingo, Gadwall, Northern Shoveler, Common Pochard, Marsh Harrier, Purple Swamp-hen,

PRACTICAL INFORMATION

- **Access:** from Barcelona take the Autovia de Castelldefels (C-31) and at junction 13 (Gavà Mar), change direction by leaving the dual-carriageway and retracing your steps for 2 km until you see on your right a sign for 'Reserva Natural Remolar-Filipines'.
- **Accommodation:** in Barcelona and Castelldefels, as well as various camp-sites in the vicinity.
- **Visitor infrastructure:** there is an information centre (tel. 936 586 761, rndeltallobregat.dmah@gencat.net), which doubles as a field station, at the entrance to the reserve. Open every day, morning and afternoon. Guided visits are organised.
- **Visiting tips:** all year, but fewest birds in June and July. Except for the road that goes to the beach, cars are not allowed in the reserve and it is forbidden to leave the marked paths. Some of the beaches have restricted access from March 15 to August 1 to protect the nesting Kentish Plovers.
- **Further information:** map 1:50,000, n°. 448 (SGE). Try the official website www.gencat.net/mediamb/rndelta or the book (in Catalan) *Els Ocells del Delta del Llobregat* by Ricard Gutiérrez *et al.* (Lynx Edicions, 1995).

Red-knobbed Coot, Kentish Plover, Penduline Tit.
Summer visitors: Little Bittern, Night, Purple and Squacco Herons, Black-winged Stilt, Little Ringed Plover, Audouin's Gull.
Winter visitors: Balearic Shearwater, Northern Gannet, Great Cormorant, Great Bittern, Great Egret, duck, European Golden Plover, Mediterranean Gull, Sandwich Tern, Razorbill, Kingfisher, Bluethroat, Moustached Warbler, Reed Bunting.
On migration: Eurasian Spoonbill, Osprey, Collared Pratincole, Temminck's Stint, Marsh Sandpiper and other waders, Little and Slender-billed Gulls, Caspian, Whiskered and Black Terns.

68. AIGÜESTORTES

The Aigüestortes i Estany de Sant Maurici National Park (14,000+ ha; also SPA) is located in the north-west of Lleida province and consists of a rugged collection of abrupt granite and slate peaks accompanied by *aigüestortes* (high-altitude river meanders) and the largest concentration of *estanys* (glacial lakes) anywhere in the Iberian Peninsula. Alpine pastures and silver fir and, above all, mountain pine forests predominate up high and merge into Scots pine and mixed deciduous forests at lower altitude. To the north extends the Parc Natural de l'Alt Pirineu (almost 70,000 ha), the largest in Catalonia and visited occasionally by European Brown Bears from the reintroduced French population.

ORNITHOLOGICAL INTEREST

Aigüestortes holds important communities of upland forest birds such as Capercaillie (almost 500 males in the Catalan Pyrenees), Tengmalm's Owl, Black Woodpecker, Ring Ouzel and Citril Finch, while above the tree line, Ptarmigan, Grey Partridge, Alpine Accentor, Wallcreeper and Alpine Chough eke out a living. Lammergeier are not uncommon (20 pairs in the Catalan Pyrenees) and Golden Eagle and Griffon Vulture, the latter roaming from its breeding colonies in the gorges of the pre-Pyrenees, are also fairly visible.

Estany Llong, Aigüestortes i Estany de Sant Maurici National Park © Pedro Retamar

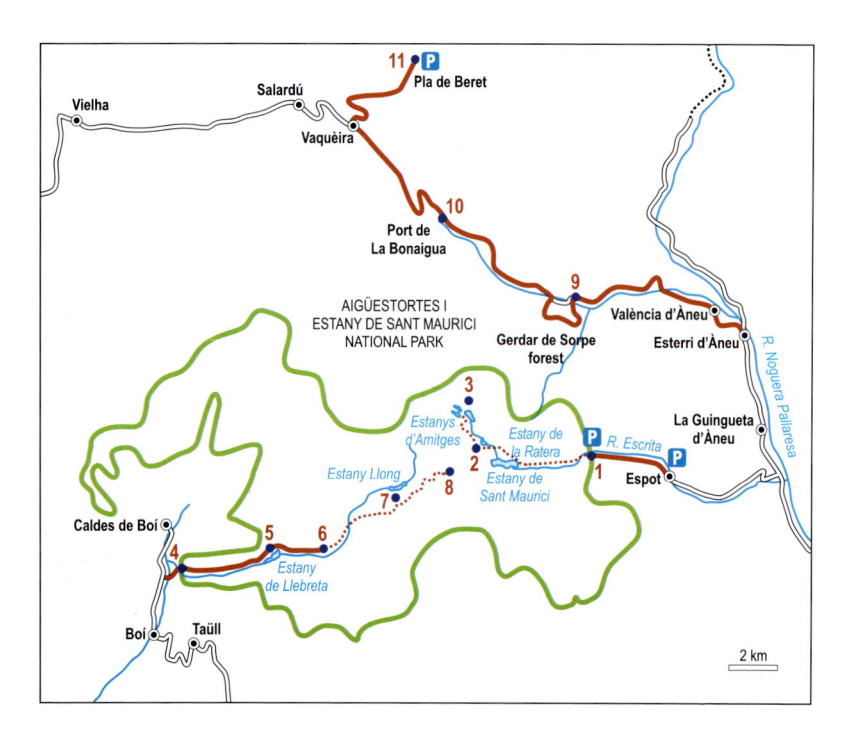

ITINERARIES

The two best-known itineraries in the National Park follow two east-west running valleys that penetrate the Park from the east (Sant Maurici) and west (Aigüestortes) and meet head on at the high pass of Portarró d'Espot. Both pass through subalpine European silver fir and mountain pine forests with passerines such as Crested Tit, Citril Finch and Common Crossbill, as well as open areas that provide the chance to see Lammergeier, Golden Eagle and Griffon Vulture in flight. Capercaillie and Tengmalm's Owl also frequent these woods, but the chances of seeing these shy birds are remote.

1. 8 km

From Espot take the road up to the entrance to the park at Prat de Pierró (1), where you have to park. From here the path runs up the right-bank of the river Escrita, crosses over and then works its way up the opposite bank to Estany de Sant Maurici, lying in the shadow of the magnificent twin peaks of Els Encantats. If you continue as far as Estany de la Ratera (2), there is a good chance of seeing Ring Ouzel in summer, while for typical alpine species such as Water Pipit, Alpine Accentor and Alpine Chough, you should continue even further up the steep path to Els Estanys d'Amitges (3) at 2,400 m. Wallcreepers breed here but are very hard to locate.

2. 12 km

Halfway between Boí and the spa at Caldes de Boí a track off to the right takes you in 1 km to another entrance (4) to the Parc Nacional, where you have to park. From here the route follows the Sant Nicolau valley to

Estany Llebreta (5) and then Planell d'Aigüestortes (6), a flat bog created by a myriad of river meanders. Continue up the valley, past riverside aspen woods where Black Woodpeckers call from March to June and, after the damp meadows of Aiguadassi, reach Estany Llong (7). A good option for seeing alpine species is to continue on to the saddle at Portarró d'Espot (8).

magnificent silver fir forest of the Mata de Valencia and are good for forest birds such as Black Woodpecker. In some years, Siskin breed.

After passing over Bonaigua, in the Baqueira complex turn right towards Pla de Beret (11) and its car-park, above the tree line. A walk in the surrounding area can produce interesting alpine birds, including Ptarmigan and Grey Partridge.

3. 36 km 🚗 🚶 ☀️ ➡️

Outside the National Park to the north, the road between Esterri d'Aneu and Vielha passes through magnificent pine and fir forests. A good idea is to take the left turn (9) to Gerdar de Sorpe, 10.5 km before reaching to top of the Bonaigua (10) pass. The forests here are part of the

MOST REPRESENTATIVE SPECIES

Residents: Lammergeier, Griffon Vulture, Goshawk, Golden Eagle, Ptarmigan, Capercaillie, Grey Partridge, Tengmalm's Owl, Black Woodpecker, Dipper, Mistle Thrush, Goldcrest, Marsh and Crested Tits, Wallcreeper, Eurasian Treecreeper, Red-billed and

PRACTICAL INFORMATION

- **Access:** the eastern sector of the park (Sant Maurici) is reached from Lleida via Tremp and then Sort. Continue on the C-13 to just before La Guingueta d'Àneu, where you should turn left to Espot or continue straight on to Esterri d'Àneu, the start of itinerary 3. The western sector (Aigüestortes) is accessible from Lleida along the N-230 to El Pont de Suert: 3 km past this town, turn right to Boí and Caldes de Boí.
- **Accommodation:** in Boí, Espot, Esterri d'Àneu and other nearby villages. There is a network of permanently staffed mountain huts in the National Park (best to make a reservation in advance).
- **Visitor infrastructure:** there are two visitor centres: the main one in Boí (tel. 973 696 189) and the other in Espot (tel. 973 624 036). Both open Monday to Saturday, morning and afternoon; Sunday and public holidays not in summer, mornings only. Book guided visits here.
- **Visiting tips:** good all year, although bear in mind the snow in winter. Snow and ice may close some roads. The best way to get up high quickly on the itineraries described here is to take a 4-wheel drive taxi from Espot to Estany de Sant Maurici or from Boí to Planell d'Aigüestortes.
- **Recommended:** near the village of Son, 10 km from Esterri d'Àneu, visit the environmental education centre of Les Planes de Son run by Fundació Territori i Paisatge (tel. 902 400 973, fundtip@fundtip.com)
- **Further information:** maps 1:50,000, nº. 149 and 181 (SGE). Pirineu Viu (tel. 973 622 126, info@pirineuviu.com) is a municipal association in the Pallars Sobirà dedicated to the promotion of ecotourism. Also, www.mma.es/parques/lared/aigues.

Alpine Choughs, Citril Finch, Siskin, Common Crossbill.
Summer visitors: Woodcock, Water Pipit, Dunnock, Alpine Accentor, Rufous-tailed Rock Thrush, Whinchat, Northern Wheatear, Ring Ouzel, Red-backed Shrike, Bullfinch, Yellowhammer, Rock Bunting.

69. CADÍ AND MOIXERÓ

Straddling the border between Barcelona, Girona and Lleida provinces, the imposing Cadí and Moixeró ridges (running east-west for 30+ km and peaking at over 2,500 m) are the highest pre-Pyrenean mountains in Catalonia. Both of these limestone ridges are well-known for their sheer cliffs, on the north-face of the Cadí and south-face of the Moixeró.

A wide diversity of habitats is on display here, from sub-Mediterranean holm oak forests to subalpine pastures. Much of the area is covered by Scots pine forests, although in higher areas mountain pine formations come into their own. Some of the most southerly silver forests in the Iberian Peninsula thrive here, along with mixed deciduous forests with beech. The two ridges are part of a Natural Park of over 41,000 ha, which is also an SPA.

ORNITHOLOGICAL INTEREST

High-altitude forest birds are well-represented here and include one of best Spanish non-Pyrenean populations of Capercaillie (50 males) and Tengmalm's Owl. Black Woodpecker, the symbol of the Natural Park, is abundant, as are Ring Ouzel and Citril Finch. Above the tree line, there are Grey Partridge, Alpine Accentor, Rufous-tailed Rock Thrush, Wallcreeper, Alpine Chough and, in winter, Snow Finch. Golden Eagle and two pairs of Lammergeier breed on cliffs in the Park (two others nearby) and Griffon Vultures are regularly seen but do not breed.

Due to the meeting of Mediterranean, Eurosiberian and subalpine habitats, there is in general a broad diversity of bird species in the area, especially noticeable amongst the passerines. Raptor migration, includ-

Gòsol valley in El Cadí i Moixeró Natural Park © Pedro Retamar

ing good numbers of Honey Buzzards, kites and harriers, is also good.

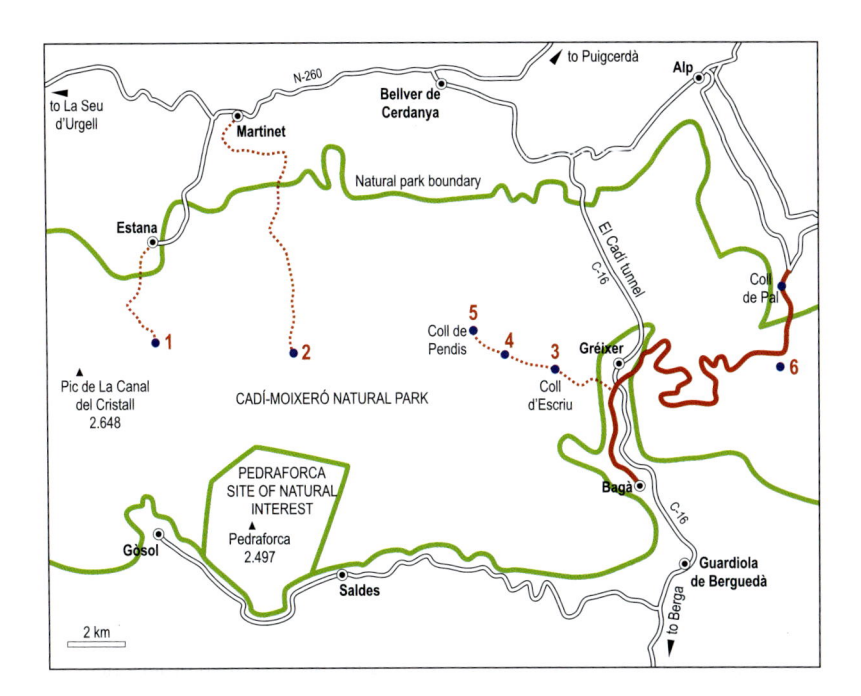

ITINERARIES

1. 3 km 🚶 ☀ ➔

On the north-face of the Cadí ridge, the well-marked walk from the village of Estana up to Prat de Cadí (1), a large pasture amidst the pine forest, passes successively through oakwoods, montane pastures, Scots and then mountain pine forests, and finally subalpine pastures. Alternatively, in Martinet take the driveable track (10 km) up to the mountain hut at Prat d'Aguiló (2).

Black Woodpecker, Ring Ouzel, Coal Tit, Citril Finch and Common Crossbill frequent the pine forests, while above the tree line there is plenty of habitat for Water Pipit, Alpine Accentor and Alpine and Red-billed Chough. Both Capercaillie and Tengmalms' Owls live in the mountain pine forests but are exceedingly hard to detect.

2. 8 km 🚶 ☀ ➔

On the south side of the Moixeró ridge, follow a gentle track closed to traffic which heads up to the left from the road to Coll de Pal 4 km from Bagà. This itinerary passes through Coll d'Escriu (3) on its way to the mountain hut at Sant Jordi (4), and then climbs to Coll de Pendís (5). Successively, you pass through a small oakwood, an extensive beech forest, Scots and then mountain pine forests and, finally enter a zone of subalpine pastures. Alternatively, drive the 20 km road between Bagà and Coll de Pal.

The birdlife on this itinerary is varied. Red-backed Shrike, Dunnock and Bullfinch breed around the lower pastures and hedgerows, with Black Woodpecker in the beech forest. Cliff-breeding raptors such as Lammergeier, Golden Eagle and Griffon

PRACTICAL INFORMATION

- **Access:** from Barcelona, the C-17 passes through Vic and Ripoll and joins the N-260 that runs from Puigcerdà to La Seu d'Urgell along the northern boundary of the park. To the south, the C-16 heads north through Manresa, Berga and Bagà before entering the park via the Túnel del Cadí.
- **Accommodation:** on the north side, Bellver de Cerdanya, Alp and Martinet; in the south, Bagà, Guardiola, La Pobla de Lillet, Castellar de n'Hug and Saldes. Outside the park, also Berga, La Seu d'Urgell and Puigcerdà.
- **Visitor infrastructure:** the main visitor centre is in Bagà (tel. 938 244 151, pncadimoixero.dmah@gencat.net). Open Monday to Saturday, morning and afternoon; Sunday only mornings. The park runs guided visits, courses and other activities for visitors. Secondary interpretation centres are located in Saldes (devoted to the mountain of Pedraforca), Bellver de Cerdanya (forest fauna) and Martinet (river ecosystems).
- **Visiting tips:** spring and summer are ideal moments for observing breeding birds. As a mountainous area, be sure to carry the correct equipment given the likelihood of rain or snow.
- **Recommended:** to the south-west between the Serra del Cadí and the Segre Valley (outside the park) lies the Reserva de la Muntanya d'Alinyà, a protected area managed by the Fundació Territori i Paisatge (tel. 902 400 973, fundtip@fundtip.com), a foundation run by the savings bank, Caixa Catalunya. A network of footpaths offers visitors the opportunity to see Lammergeier, Griffon Vulture and Golden Eagle.
- **Further information:** maps 1:50,000, n°. 254 and 255 (SGE). Interesting descriptions in the book (in Catalan) *La fauna del Parc Natural del Cadí-Moixeró* by Jordi Garcia-Petit (Lynx Edicions, 1997).

Vulture are best searched for up high. Around the mountain hut of Coll de Pal (6), Snow Finches are regular when there is snow on the ground. All the mountain passes are generally good for migrating raptors.

MOST REPRESENTATIVE SPECIES

Residents: Lammergeier, Griffon Vulture, Golden Eagle, Peregrine Falcon, Capercaillie, Grey Partridge, Tengmalm's and Eagle Owls, Black Woodpecker, Crag Martin, Dipper, Dunnock, Alpine Accentor, Blue Rock Thrush, Coal Tit, Wallcreeper, Red-billed and Alpine Choughs, Citril Finch, Common Crossbill, Bullfinch.
Summer visitors: Short-toed Eagle, Woodcock, Alpine Swift, European Bee-eater, Water Pipit, Northern Wheatear, Rufous-tailed Rock Thrush, Ring Ouzel, Subalpine Warbler, Red-backed Shrike, Bullfinch.
Winter visitor: Snow Finch.
On migration: Honey Buzzard, Black and Red Kites, Marsh, Hen and Montagu's Harriers.

70. LLEIDA PLAINS

These plains (around 40,000 ha) are Catalonia's main area of steppe and lie near the city of Lleida in the north of the Ebro Depression. Cereal cultivation is predominant, although a few patches of natural steppe such as the well-known thyme scrub at Alfés or the Reserva Natural de Mas de Melons still exist. The site is enriched by a large number of artificial wetlands

CATALONIA

(paddy-fields, lagoons and reservoirs), rivers with fluvial woodland and to the south, Mediterranean scrub with pinewoods and olive and almond groves.

ORNITHOLOGICAL INTEREST

This site harbours the only Catalan Little Bustards (almost 1,000 males), Black-bellied and Pin-tailed Sandgrouse and Dupont's Larks, the latter confined in Catalonia to the scrub at Alfés. All these species are threatened by the progressive implantation of irrigated farming and the building of major infrastructures, although there has been an encouraging increase in Lesser Kestrel and Montagu's Harrier numbers thanks to recovery programmes launched after their disappearance from the area as breeding birds. This is also the main Iberian stronghold of Lesser Grey Shrike.

The Utxesa reservoir holds the best breeding populations of Marsh Harrier in Catalonia, as well as a colony of Little Bittern and Night and Purple Herons. The same species breed at the confluence of the rivers Segre and Cinca (Aiguabarreig), along with Squacco Heron.

ITINERARIES

1. 50 km

From Lleida take the N-11a towards Zaragoza. Just over 1 km beyond Alcarràs, turn left at a roundabout to Torres de Segre (home to an important colony of White Storks). From here, a road takes you to the reservoir of Utxesa, skirting its western bank and reaching its outlet canal (1). The thick reed-beds hold most of Spain's breeding herons, Marsh Harrier and Reed Bunting. In winter, duck, Great Cormorants, egrets, gulls and starlings roost here in huge numbers.

The track alongside the canal takes you to a junction (2), where you should turn right to Aitona, and once, there, left towards the Aiguabarreig (3), the confluence of the rivers Segre and Cinca. The best option is to head for the towns of Massalcoreig and La Granja d'Escarp, from where there are boardwalks through the fluvial woodland and reed-beds of the area. Just before passing the provincial boundary between Lleida and Zaragoza, the road from Granja d'Escarp

Mas de Melons, Lleida plains © Anna Motis

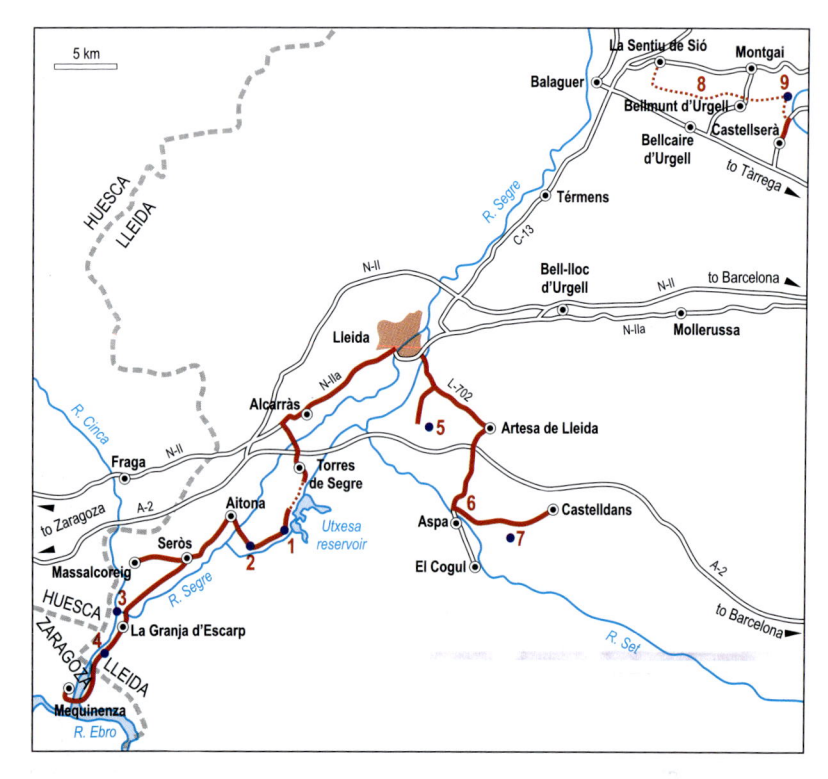

to Mequinenza passes next to a lookout point on the left, well-placed to view the islands in the Aiguabarreig, including the heronry on Illa dels Martinets.

2. 30 km 🚗 ☀️ ➡️

Leaving Lleida on the road to Artesa de Lleida, after 5 km turn right along an asphalted track to the small aerodrome (5) that occupies much of the Timoneda d´Alfés, the patch of natural thyme scrub that holds Catalonia's only Dupont's Larks. Park and walk around looking for Stone Curlew, Pin-tailed Sandgrouse, Great Spotted Cuckoo and Calandra and Short-toed Larks. Lesser Kestrels and Lesser Grey Shrike come to hunt in the scrub, Eurasian Dotterel appear on passage and Dupont's Larks can be heard singing from February onwards.

Return to the road and head for Artesa; once here, turn south towards Aspa. Just before this town, turn left (6) along a track (asphalted at first) that heads for Castelldans. At first you follow the river Set, with Black Wheatear and Blue Rock Thrush on the small cliffs overlooking the river. On passage, Fieldfare, Redwing and Hawfinch frequent the riverside woods. Further ahead, the track enters and crosses the Reserva Natural de Mas de Melons (7), a mosaic of arable land, patches of steppe vegetation and low hills frequented by Pin-tailed Sandgrouse, Roller and dispersing juvenile Golden and Bonelli's Eagles.

3. 14 km 🚗 ☀️ ➡️

The narrow corridor of cereal plains near the village of Bellmunt run

CATALONIA

east-west between low gypsum hills. Pick up a farm track (8) in La Sentiu de Sió and continue straight across the road between Montgai and Bellmunt d'Urgell. Once at the Canal d'Urgell (9), keep right alongside the canal until you reach the road between Preixens and Castellserà.

This is a good area for Little Bustard, best between April and May, Montagu's Harrier, Stone Curlew and, on migration, Red-footed Falcon.

MOST REPRESENTATIVE SPECIES

Residents: Marsh Harrier, Golden and Bonelli's Eagles, Little Bustard, Stone Curlew, Black-bellied and Pin-tailed Sandgrouse, Dupont's, Lesser Short-toed and Calandra Larks, Blue Rock Thrush, Black Wheatear, Dartford Warbler, Reed Bunting.

Summer visitors: Little Bittern, Night, Squacco and Purple Herons, Montagu's Harrier, Lesser Kestrel, Black-winged Stilt, Great Spotted Cuckoo, Roller, Short-toed Lark, Reed, Great Reed and Spectacled Warblers, Lesser Grey Shrike.

Winter visitors: Great Cormorant, duck, Hen Harrier, Peregrine Falcon, European Golden Plover, Black-headed Gull, Fieldfare, Redwing, Hawfinch.

On migration: Garganey, Osprey, Red-footed Falcon, Eurasian Dotterel and other waders.

PRACTICAL INFORMATION

- **Access:** from Lleida take the N-II towards Alcarràs (itinerary 1) or the L-702 to Artesa de Lleida (itinerary 2). For itinerary 3, take the C-13 to Balaguer (30 km) and then La Sentiu de Sió (7 km).
- **Accommodation:** in Lleida.
- **Visitor infrastructure:** the Estació Biològica de l'Aiguabarreig (tel. 974 464 435, info@aiguabarreig.net) in Mequinenza (Zaragoza province) runs an interpretation centre in Massalcoreig. Open weekends and public holidays.
- **Visiting tips:** early spring is best for steppe birds. After heavy rain, the tracks can become impassable.
- **Recommended:** the naturalist group Egrell (egrell@egrell.org) organises activities for birdwatchers. More information on www.egrell.org.
- **Further information:** maps 1:50,000, n°. 360, 388, 415 and 416 (SGE).

COMUNITAT VALENCIANA

71. L'Albufera
72. Marjal del Moro
73. El Hondo
74. Salt-pans of southern Alicante

The Comunitat Valenciana, made up of the provinces of Castellón, Valencia and Alicante, occupies a coastal strip between Catalonia and Murcia of over 300 km and is characterised by a line of vast sandy beaches separating the Mediterranean from a densely populated hinterland with a thriving economy based on industry, tourism and intensive agriculture. Further inland, however, the plains give way to a poorly populated region, characterised by abrupt mountains and plateaux that differ wildly from the bright, wide-open horizons that are wont to be used as a symbol of this autonomous region.

The limestone mountains in Castellón and Valencia represent the last flourish of the Sistema Ibérico, while in Alicante to the south, they are the continuation of the Sierras Béticas; on reaching the coast they form impressive cliffs (for example, El Montgó).

The mountains of the Comunitat Valenciana are dry – especially in Alicante – and covered by large stands of Aleppo pine. Forest fires have destroyed many hectares of forest and forest bird communities are somewhat impoverished. More significant are the cliff-breeding raptors that include good populations of Golden and Bonelli's Eagles, the latter in decline.

Nevertheless, most birdwatchers are attracted to this region by the coastal wetlands, despite the fact that most have been all but completely denaturalised. The few remaining semi-natural sites hold the best numbers of birds and most bird species in the region and areas such as the Illes Columbretes off the coast of Castellón are very important for breeding seabirds.

The few remaining coastal wetlands are scattered along the coast and are a reminder of what must have been once a continuous line of *marjals*

Bonelli's Eagle — Carlos Sánchez © nayadefilms.com

(marshes) and *albuferes* (lagoons) from north to south. Despite the protection they are afforded, many sites are still under threat from pollution and drainage schemes aimed at providing land for more intensive agriculture and urban development. Hunting is very popular in the region and has led to some of the highest levels of lead-poisoning in wildfowl (caused by the ingestion of lead pellets) ever reported anywhere.

L'Albufera (71) near the city of Valencia is the best-known birdwatching site in the region and the rice-paddies that surround this shallow lagoon provide refuge to vast numbers of waders, ducks and a myriad of other birds. Breeding birds include some of the most important heron and tern colonies in the Iberian Peninsula and there are huge numbers of wintering wildfowl (above all Red-crested Pochard) that are only surpassed in number in Spain by Doñana and in the Ebro Delta.

Further north near the port of Sagunt lies El Marjal del Moro (72), once an area of rice-paddies and now a natural wetland that is producing a string of highly interesting sightings. As well, this site has begun to be used as a testing ground for the reintroduction of species such as Purple Swamp-hen and Red-knobbed Coot that had become extinct as breeders in the Comunitat Valenciana.

El Hondo (73) in the south of Alicante province consists of two reed-choked irrigation reservoirs, both part of the once immense Albufera d'Elx, that are of vital importance as the main European refuge of the globally threatened Marbled Duck. Furthermore, this site also holds in some years the largest number of breeding White-headed Duck, another threatened duck, in the Iberian Peninsula.

Many of the birds from El Hondo feed and rest in areas connected to some of the region's working salt-pans (74). This is the case of the salt-pans at Santa Pola, with its permanent groups of Greater Flamingos, and the Laguna de La Mata, which acts as a warming pool for the commercial salt-pans of Torrevieja. Despite being surrounded by heavily built-up tourist resorts, both of these sites are home to notable concentrations of breeding and migrant birds.

71. L'ALBUFERA

This wetland of over 20,000 ha, lying to the south of the city of Valencia between the rivers Túria and Júcar, consists of a large freshwater coastal lagoon of 3,000 ha surrounded by extensive *marjals* (marsh), largely occupied today by rice-paddies. Masses of aquatic vegetation, including islands in the lagoon known as *mates*, survive in what is today a very humanised environment, characterised by poor-quality water and intense hunting activity. Such levels of human activity inevitably pose many legal problems for the Parc Natural de l'Albufera (an SPA), which also includes the lengthy sand-bar that separates the wetland from the sea.

ORNITHOLOGICAL INTEREST

This is the most important wetland in the Comunitat Valenciana and home to all of Spain's colonial herons, including some of the country's best Squacco (almost 100 pairs) and Grey Heron (over 500 pairs) and Little Egret (almost 3,000 pairs) colonies. Of great interest too are the gull and tern colonies, with over 2,000 pairs of Common Tern accompanied by Gull-billed, Sandwich and Little Terns, and Slender-billed and Audouin's Gulls, both recent colonisers. Other breeders include Marbled Duck, Purple Swamphen, Black-winged Stilt, Avocet and Bearded Tit; Red-knobbed Coots have recently been reintroduced.

The winter flooding of the rice-paddies brings huge numbers of wintering ducks, including almost a third of all Iberian Red-crested Pochards, as well as numerous herons, Great Cormorants, Marsh Harriers, waders and gulls, to the site.

ITINERARIES

1. 2 km

In the visitor centre of Racó de l'Olla an observation tower giving good views over L'Albufera and La Mata del Fang (1) is well placed for birdwatchers to enjoy the constant transit of herons and, in winter, Great Cormorants and ducks. Nearby an

El Palmar, L'Albufera © José Manuel Reyero

COMUNITAT VALENCIANA

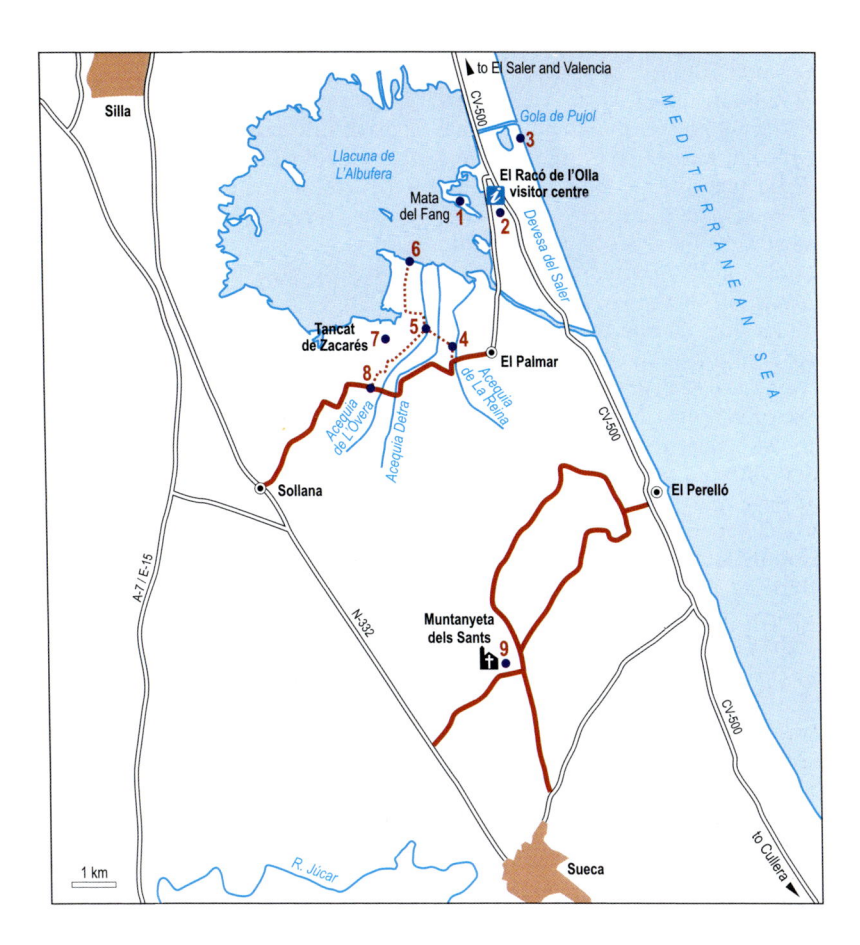

artificial lagoon (2) is good for Marbled Duck and is home to hundreds of pairs of Common Tern. A path runs along the eastern bank of the lagoon to a hide, from where there are spectacular views of the tern colony.

The Devesa del Saler is the best-preserved stretch of the sand-bar separating L'Albufera from the sea. A walk along the beach on the other side of the CV-500 can produce views of Balearic Shearwater in winter and Northern Gannet in autumn. The beach (3) south of the Gola de Pujol is a good area for Sandwich Tern.

2. 40 km

From the village of El Palmar, there is access to the rice-paddies to the south of L'Albufera from the road to

Sollana. After a long straight section of road, turn right along a track next to the bridge over the canal La Reina. At the first junction (4), turn left and cross two other canals (Dreta and L'Overa). At the next junction (5), turn right towards the southern retention wall of L'Albufera (6) (flooded in winter) or left towards El Tancat de Zacarés (7) and then back to the road (8) to El Palmar or Sollana.

Another good way of exploring the *marjals* is to drive the agricultural tracks between the towns of El Perelló and Sueca and up the small hill of Muntanyeta dels Sants (9), some of which may be closed off during the hunting season.

In winter, most herons, ducks and waders, as well as Marsh Harrier and, with luck, Great Egret, concentrate in the flooded rice fields, provided that they are not disturbed by hunters. In spring and summer Squacco and Grey Herons, Black-winged Stilt, Audouin's Gull and often numerous groups of Whiskered Tern congregate here too; fallow land and salt-marshes are the best place to look for Collared Pratincole.

Look in the reed-beds around the lagoon for Purple Swamp-hen and Bearded Tit and along the drainage ditches for Little Bittern.

MOST REPRESENTATIVE SPECIES

Residents: Little Egret, Marbled Duck, Red-crested Pochard, Purple Swamp-hen, Black-winged Stilt, Avocet, Kentish Plover, Black-headed Gull, Sandwich Tern, Kingfisher, Moustached Warbler, Bearded Tit.

Summer visitors: Little Bittern, Night, Squacco and Purple Herons, Collared Pratincole, Slender-billed and Audouin's Gulls, Gull-billed, Common and Little Terns, Savi's, Reed and Great Reed Warblers.

Winter visitors: Black-necked Grebe, Balearic Shearwater, Northern Gannet, Great Cormorant, Great Egret, Northern Shoveler and other duck, Marsh Harrier, Osprey, Euro-

PRACTICAL INFORMATION

- **Access:** from Valencia take the V-15 motorway to El Saler and then the CV-500 coast road to Cullera. The entrance to El Racó de l'Olla is just before the turn-off to El Palmar.
- **Accommodation:** in Valencia (10 km) or Cullera.
- **Visitor infrastructure:** El Racó de l'Olla houses the Natural Park visitor centre (tel. 961 627 345, raco.olla@gva.es). Open every day in the morning; Tuesday and Thursday, also in the afternoon.
- **Visiting tips:** all year is good, although hunters can make birdwatching in the rice-paddies a little fraught at weekends in autumn and winter. Always respect the no entry signs.
- **Recommended:** SEO/BirdLife organises a programme of activities for birdwatchers and volunteers in the Estació Ornitològica de l'Albufera (tel. 961 627 389, valencia@seo.org) in El Saler. A boat trip on L'Albufera from El Palmar or along the canals from Silla or Catarroja is a different way of birdwatching.
- **Further information:** maps 1:50,000, nº. 722 and 747 (SGE). The book *Las aves de l'Albufera de Valencia* by Bosco Dies and José Ignacio Dies *et al.* (Vaersa, 1999) provides a lot of information about the birds of the site. Also, the official web page http://parquesnaturales.gva.es.

COMUNITAT VALENCIANA

Marjal del Moro © Eduardo de Juana

pean Golden Plover, Black-tailed Godwit, Bluethroat, Penduline Tit, Reed Bunting.
On migration: Glossy Ibis, Garganey, Eleonora's Falcon, Temminck's Stint and other waders, Whiskered and Black Terns.

72. MARJAL DEL MORO

This small wetland near the town of Sagunt covers little more than 300 ha, but can still boast a succession of habitats that range from pebble beaches, flooded salt-marshes, freshwater lagoons – former rice-paddies that still maintain their structure – to fields liable to flood. Stands of reed and bulrush dominate the vegetation, although there are also areas of *Salicornia* and other halophyte scrub. In recognition of its value for wildlife, the Valencian government has purchased the site and now runs it as a protected area.

ORNITHOLOGICAL INTEREST

Despite its size, this is an extremely important wetland, as the five breeding pairs of Marbled Duck prove. Other breeders include Little Bittern, Purple Heron, Red-crested Pochard, Montagu's Harrier and good densities of Purple Swamp-hen. There are also colonies of Black-winged Stilt, Collared Pratincole and Whiskered Tern; Red-knobbed Coot have recently been introduced.

Winter sees the arrival of Marsh Harrier and many other birds, including many species of heron (the reed-beds hold a Cattle Egret roost of at times over 2,000 birds), duck, gulls and passerines. Migration periods are always lively, with a variety of waders present and the globally threatened Aquatic Warbler turning up regularly.

ITINERARY

5 km

From the Centre d'Educació Ambiental de la Comunitat Valenciana take the track known as Pas de les Egües towards the sea and stop after a few metres at a small hillock (1) with an observation point that provides good views of the whole area. Halfway to

Head south along a sea-wall, a good site for seawatching from October onwards with constant movement of seabirds and groups of Balearic Shearwater and Common Scoter. Inland, the flooded salt-marsh are frequented by herons, duck and other water birds such as Black-necked Grebe, Greater Flamingo and even Marbled Duck. The viewing platform (4) halfway along this coastal stretch of the itinerary is the best observation point.

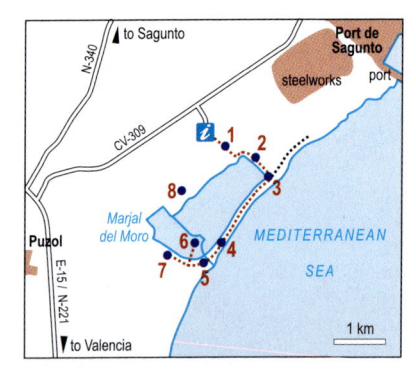

the sea there is another viewpoint (2): from here, look and listen for Savi's, Moustached, Reed and Great Reed Warblers in spring. In winter, Bluethroat, Penduline Tit and Reed Bunting frequent these reed-beds, Marsh Harrier fly over and Purple Swamphens are abundant. Migration periods see Aquatic Warbler here and huge roosts of Sand Martin, Barn Swallow and Yellow Wagtail.

The track merges with another running parallel to the coast: on the left, there are a number of restored saline lagoons (3) with islands which have been colonised by Black-winged Stilt, Collared Pratincole, Kentish Plover and Common and Little Terns. Be careful not to disturb the birds here.

The itinerary crosses the outlet canal from the *marjal* and passes a bar, L'Estany (5), behind which there is a asphalted road and a number of houses. A little further on, continue where the asphalt finishes along a track closed to traffic that skirts the southern edge of the reserve. Within a few metres, a boardwalk heads right into the *marjal* and provides good views over this shallow sector of the marsh, choked with reeds, that is good for many birds including Red-knobbed Coot.

Passage migrants and winter visitors on the last lagoon are visible from a final viewing platform (7), from where you can either retrace your steps or return to the Centre d'Edu-

PRACTICAL INFORMATION

- **Access:** from Valencia, take the A7C/N-221 motorway and leave at junction 18 (Sagunt) along the road to Port de Sagunt. After 2 km, a asphalted track off to the right takes you to Marjal del Moro.
- **Accommodation:** in Port de Sagunt, Sagunt or Valencia (20 km).
- **Visitor infrastructure:** the Centre d'Educació Ambiental de la Comunitat Valenciana (tel. 962 680 000) is housed in an old restored farm and has a live video link to the *marjal*. Open Monday to Thursday, morning and afternoon; Friday and weekends, only mornings.
- **Visiting tips:** April, when the last winter visitors, spring migrants and first breeding birds coincide, sees the greatest diversity of birds on the site. Heed all signposts and do not enter closed-off areas, above all during the breeding season.
- **Recommended:** The ornithologist Marcial Yuste (tel. 667 047 369, marjaldelmoro@terra.es) is a specialist local bird guide.
- **Further information:** map 1:50,000, n°. 696-II (SGE).

COMUNITAT VALENCIANA

ció Ambiental by continuing along farm tracks around the eastern (8) sector of the *marjal*. When flooded in migration periods, the pastures here are frequented by good numbers of Black-winged Stilt, European Golden Plover, Black-tailed Godwit, Eurasian Curlew and other waders.

MOST REPRESENTATIVE SPECIES

Residents: Cattle Egret, Red-crested Pochard, Purple Swamp-hen, Red-knobbed Coot, Moustached Warbler.
Summer visitors: Little Egret, Purple Heron, Marbled Duck, Montagu's Harrier, Black-winged Stilt, Collared Pratincole, Kentish Plover, Common and Whiskered Terns, Yellow Wagtail, Savi's, Reed and Great Reed Warblers.
Winter visitors: Black-necked Grebe, Balearic Shearwater, Shelduck, Common Scoter and other duck, Marsh Harrier, European Golden Plover, Northern Lapwing, Common and Jack Snipes, Mediterranean Gull, Sandwich Tern, Water Pipit, Bluethroat, Penduline Tit, Reed Bunting.
On migration: Squacco Heron, Glossy Ibis, Eurasian Spoonbill, Greater Flamingo, Garganey, Osprey, waders, Audouin's Gull, Sand Martin, Barn Swallow, Aquatic Warbler.

73. EL HONDO

Along with the nearby Salinas de Santa Pola and the lagoons of La Mata and Torrevieja, the Parc Natural d'El Hondo (also Ramsar Site and an SPA) is one of the most important wetlands on the west coast of the Mediterranean. Lying in the south of Alicante province, El Hondo consists essentially of two enormous reservoirs built on a large tract of marshland that irrigate 40,000 ha of land with water from the basin of the river Segura. Reeds and other helophytes occupy vast areas of these shallow reservoirs, although around the edges there are still some areas of salt-marsh and a few palm plantations.

El Hondo © Pedro Retamar

ORNITHOLOGICAL INTEREST

El Hondo is an extraordinarily important site for the globally threatened Marbled Duck and, in some years, for the likewise threatened White-headed Duck. Its interesting breeding herons include Little Bittern and mixed colonies of Squacco and Purple Herons. Other breeders include Black-necked Grebe (one of the most important Spanish breeding colonies), Red-crested Pochard, Purple Swamp-hen, Black-winged Stilt, Blacked-headed Gull and Little and Whiskered Terns, as well as Montagu's Harrier and Collared Pratincole in the surrounding salt-marshes and fields.

Greater Flamingo de not breed regularly, but are present all year; winter brings duck (above all Northern Shoveler), Marsh Harrier and Booted Eagle; the greatest wader diversity occurs during migration periods.

ITINERARIES

1. 3 km

La Vereda de Sendres is the road that marks the northern limit to the Parc Natural and is the best way of reaching the so-called Segunda Elevación (1), the start of the itinerary that follows the central canal of El Hondo. Both of the reservoirs, Poniente (western) and Levante (eastern), belong to the Comunidad de Riegos de Levante and so permission is needed to enter: contact the visitor centre at least 24 hours before you intend to visit.

The path followed by the itinerary follows the embankment around the western reservoir and visits three hides, the first a tower hide (2).When water levels are high, the sheer abundance of birds here is remarkable: be prepared for multitudes of Black-necked Grebe, Squacco and Purple Herons, Greater Flamingo, Marbled and White-headed Ducks, Marsh Harrier, Purple Swamp-hen and Whiskered Tern, amongst other birds.

The last stretch of the embankment leaves the western reservoir and penetrates the Charco del Norte (3) in the eastern reservoir. The viewing platform of La Roseta (4) provides winter views of wonderful concentrations of ducks, above all of Northern Shoveler and Common Pochard, and huge flocks of Common Starling. The final hide – Peu Verd (5) – sits amongst the reeds and is ideal for viewing reed-bed passerines such as Bearded Tit.

2. 7 km

Returning to the start of itinerary 1, continue along La Vereda de Sendres towards San Felipe Neri (westwards) and after 4 km on the left you will come to the visitor centre. A nearby trail (6) passes through a small area

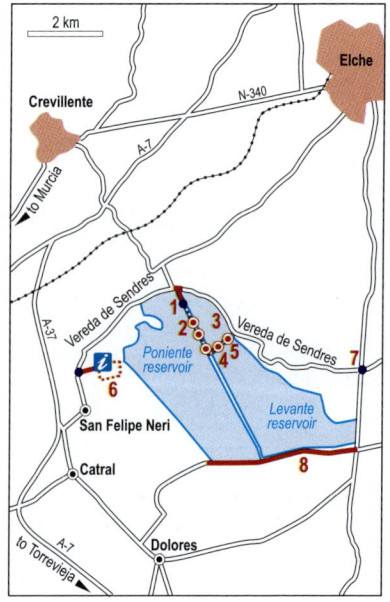

COMUNITAT VALENCIANA

> ### PRACTICAL INFORMATION
>
> - **Access:** From Elx (Elche), take the A-7 motorway towards Murcia and at junction 76 head for Crevillente railway station. A couple of kilometres after passing over the railway line, turn off towards San Felipe Neri and the northern entrance to El Hondo (itinerary 1).
> - **Accommodation:** in Elx and the tourist resorts of Santa Pola, La Marina and Guardamar.
> - **Visitor infrastructure:** the Natural Park visitor centre (tel. 966 678 515) near San Felipe Neri is open every day in the morning, except Monday; Tuesday and Thursday also in the afternoons.
> - **Visiting tips:** good all year, although summer is hot and birds become very inactive.
> - **Further information:** maps 1:50,000, n°. 893 and 894 (SGE). *Las aves de los humedales del sur de Alicante y su entorno* by Antonio Jacobo Ramos and Luis Fidel Sarmiento (Editorial Club Universitario, 1999) is worth a read. Also the official web site http://parquesnaturales.gva.es.

of salt-marsh and by a small pool, giving good views of Marbled Duck, Black-winged Stilt, Avocet and Collared Pratincole.

Alternatively, head east along La Vereda de Sendres to the junction with the Elx-Dolores road. Turn right (south) and then after just over 2 km, turn right again along Carretera de Vistabella (8) running along the southern edge of the Park Natural.

Summer brings Montagu's Harrier and Collared Pratincole to the area and, with luck, a small group of Little Bustard. There is also a surprising variety of waders to be found in the recently irrigated fields and flooded salt-marshes, including Eurasian Dotterel on passage. Winter is the time to look for Greylag Geese (scarce), European Golden Plover and a variety of raptors such as Common Buzzard and Booted Eagle.

MOST REPRESENTATIVE SPECIES

Residents: Black-necked Grebe, Greater Flamingo, Marbled and White-headed Ducks, Red-crested and Common Pochards, Purple Swamp-hen, Black-winged Stilt, Lesser Short-toed Lark, Moustached Warbler, Bearded Tit.

Summer visitors: Little Bittern, Night, Squacco and Purple Herons, Montagu's Harrier, Collared Pratincole, Common and Whiskered Terns, Yellow Wagtail, Savi's, Reed and Great Reed Warblers.

Winter visitors: Great Cormorant, Greylag Goose, Shelduck, Northern Shoveler and other duck, Marsh Harrier, Common Buzzard, Booted Eagle, Osprey, European Golden Plover, Black-tailed Godwit, Bluethroat, Penduline Tit, Common Starling, Reed Bunting.

On migration: Garganey, Little Bustard, Eurasian Dotterel and other waders, Black Tern.

74. SALT-PANS OF SOUTHERN ALICANTE

This large group of wetlands lies today on the site of the once huge Albufera d'Elx, which was drained in the nineteenth century. The working Santa Pola salt-pans (Natural Park, Ramsar Site and SPA) always have water, while the endorheic lagoons of

Laguna de Torrevieja, Salt-pans of southern Alicante © Pedro Retamar

Clot de Galvany and El Fondet de la Senieta are both seasonal. Further south near the tourist resorts of La Mata and Torrevieja, two large lagoons containing commercial salt-pans have also been declared a Natural Park, Ramsar Site and an SPA.

ORNITHOLOGICAL INTEREST

Most of the Spanish herons breed here, including Squacco and Purple Herons, as well as the globally threatened Marbled Duck and one of Spain's largest populations of Shelduck. The salt-pans are also home to colonies of Slender-billed Gull (only surpassed in number in Spain by the Ebro Delta), Avocet, Kentish Plover, Black-headed Gull, Common, Little and Whiskered Terns and, in recent years, Glossy Ibis and Purple Swamp-hen.

Greater Flamingo are abundant and breed in some years. Good numbers of Black-necked Grebe, Marsh Harrier and Osprey winter, and passage periods see many groups of waders passing through.

ITINERARIES

1. 1 km

Ten kilometres south of Alicante along the N-332 towards Santa Pola, turn into the Arenales del Sol holiday complex (1). At the end of the coastal road, a track leads off to the right to the nature school at Clot de Galvany. Walk along the paths to the hides overlooking a small artificial lagoon (2), with White-headed Duck in the post-breeding period, and the main pool (3), home to breeding Red-crested Pochard and Marbled Duck.

2. 10 km

From the town of Santa Pola and once beyond the crossroads (4) with the road to Elx, the N-332 passes through the Santa Pola salt-pans. On a small artificial lagoon next to the visitor centre (on the left just after the crossroads) and on the

COMUNITAT VALENCIANA

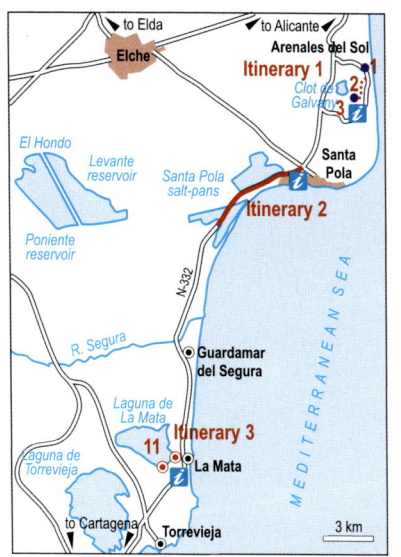

salt-pans (Braç del Port) (5) next to the first stretch of road, Greater Flamingo, duck, waders and gulls are always obvious.

Five kilometres further on, park next to the Roman look-out tower of Tamarit (6) on the right. Both Slender-billed and Audouin's Gulls frequent this sector of the salt-pans, along with Great Cormorant, Greater Flamingo, Shelduck, Red-crested Pochard and, with luck, a Glossy Ibis or two. In winter, Eurasian Spoonbills are often seen here.

Just 300 m further on, a lay-by (7) on the right opposite the pools of Canalets and Cuadretas on the other side of the road is a good spot for Marbled Duck, especially in the drainage ditches, and Great Crested Grebe, Marsh Harrier and Common, Little and Whiskered Terns. On the far side of the lagoons, Black-headed Gulls breed.

The following stop is at the junction (8) with a road to the Bonmatí salt-pans on the left. Aside from groups of Greater Flamingo, many species of waders (*Tringa* and *Calidris* sps. and godwits, for example) frequent these pools, and in winter Ospreys sit on the posts. The nearby beaches of El Tamarit (9) and El Pinet (10), accessible along roads off the N-332, are good for waders in passage and, with an onshore wind, seabirds.

3. 1 km

On the N-332 in the town of La Mata there is a turn-off to a visitor centre, from where an itinerary passes through salt-marshes to a hide and then a raised observation platform next to a pine plantation. From here, there are good views of the south-east corner of the lagoon of La Mata (11). Greater Flamingo and, in winter, Black-necked Grebe are regular. Look out too for Shelduck, Marsh and Hen Harriers, Black-winged Stilt, Avocet and Slender-billed Gull.

MOST REPRESENTATIVE SPECIES

Residents: Great Crested Grebe, Glossy Ibis, Greater Flamingo, Shelduck, Marbled Duck, Red-crested Pochard, Purple Swamp-hen, Avocet, Kentish Plover, Black-headed and Slender-billed Gulls, Lesser Short-toed Lark.

Summer visitors: Little Bittern, Night, Squacco and Purple Herons,

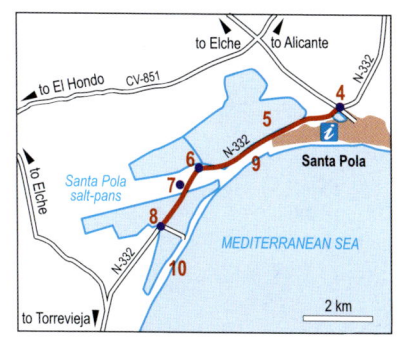

Montagu's Harrier, Black-winged Stilt, Collared Pratincole, Common, Little and Whiskered Terns.
Winter visitors: Black-necked Grebe, Balearic Shearwater, Northern Gannet, Great Cormorant, Eurasian Spoonbill, Marsh Harrier, Osprey, European Golden and Grey Plovers, Black-tailed Godwit, Mediterranean and Audouin's Gulls, Sandwich Tern.
On migration: Garganey, White-headed Duck, Curlew Sandpiper and other waders, Caspian Tern.

PRACTICAL INFORMATION

- **Access:** the N-332 between Alicante and Torrevieja provides access to all the itineraries.
- **Accommodation:** lots of hotels all along the coast.
- **Visitor infrastructure:** the visitor centres in the natural parks of Salines de Santa Pola (tel. 966 693 546, parque.santapola@gva.es) and Lagunas de La Mata y Torrevieja (tel. 966 920 404) are open every day in the morning, except Monday; Tuesday and Thursday also in the afternoons. The nature school at El Clot de Galvany is run by the Elx City Council (tel. 966 658 028, mediambient@ayto-elche.es).
- **Visiting tips:** good all year, but especially during migration periods and winter. To avoid robberies, be sure to lock cars and do not leave optical equipment in sight in cars.
- **Recommended:** the local SEO/BirdLife group in Alicante (the. 620 809 786, seo-alicante@seo.org) and Amigos de los Humedales del Sur de Alicante (tel. 966 796 464, ahsaahsa@worldonline.es) organise outings, courses and activities for birdwatchers.
- **Further information:** maps 1:50,000, n°. 893 and 914 (SGE). *Las aves de los humedales del sur de Alicante y su entorno* by Antonio Jacobo Ramos and Luis Fidel Sarmiento (Editorial Club Universitario, 1999) and *Las Aves del Clot de Galvany* (Ayuntamiento de Elche, 2002) are worth a read. Also the official web site http://parquesnaturales.gva.es.

WHERE TO WATCH BIRDS IN SPAIN. THE 100 BEST SITES

EXTREMADURA

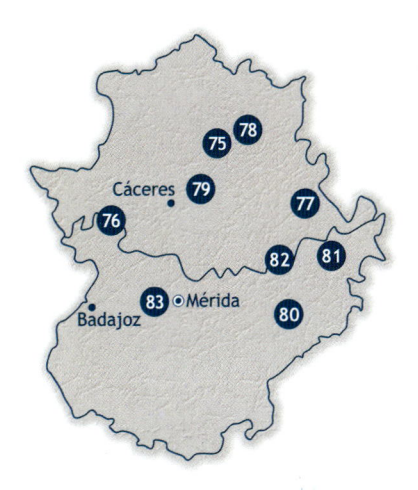

75. Monfragüe
76. Sierra de San Pedro
77. Las Villuercas
78. Campo Arañuelo
79. Cáceres Plains
80. La Serena
81. Puerto Peña and Orellana
82. Vegas Altas del Guadiana
83. Mérida

According to a report published by SEO/BirdLife in 1999, no less than three-quarters of Extremadura is of international importance for birds. Low population density, a less developed economy than other parts of Spain and a predominance of traditional, low-impact agricultural and economic activities such as stock-raising, hunting and cork production, explain why this autonomous community heads the list of Spanish and European regions when it comes to bird-rich square kilometres.

Many birdwatchers from the north of Europe make the pilgrimage to south-west Spain in search of the large tracts of well-preserved land – with birdlife to match – that their countries lack; Monfragüe (75) on the banks of the river Tajo is the first stop for most of these wildlife tourists.

Monfragüe's long parallel mountain ridges, covered in dense Mediterranean forest formations, and its high quartzite outcrops dropping vertiginously down to the Tajo are home to concentrations of raptors unmatched anywhere in Europe. Moreover, the network of hides in the Natural Park makes the observation of the main breeding species – for example, the world's largest colony of Black Vultures, Spanish Imperial Eagle and a high density of Black Storks – relatively easy.

Another exceptional site for these three jewels of Mediterranean birdlife is the somewhat less known Sierra de San Pedro (76), most of which is occupied by private hunting and stock-raising estates that greatly limit access. Nevertheless, a drive through the area is often more than sufficient to enjoy this much under-appreciated and sin-

Black-winged Kite © Gabriel Sierra

gular relict of nature at its most natural.

Like Las Villuercas (77), an abrupt mountain range all but bordering on Castilla-La Mancha, San Pedro (almost on the border with Portugal) can be thought of as a continuation of Los Montes de Toledo. Both these sites, which are part of an east-west running chain separating the Tajo and Guadiana watersheds, are of vital importance for Bonelli's Eagle and other cliff-breeding raptors.

The land between these wonderful mountain ranges is occupied by plains, of which many are covered by one of the most characteristic sights in Extremadura – vast tracts of *dehesas* (wood pasture). The rational use of natural resources in these Mediterranean forests has led to the creation of a balanced landscape in which the predominant tree species, usually cork and holm oaks, are sufficiently spaced out to allow pastures to form on the forest floor.

Many thousands of hectares are covered by *dehesas*, an excellent example of sustainable economy and one of motors of the regional economy. At the same time, they are home to many of the most characteristic species of Mediterranean biomes and witness the mass arrival in winter of Common Cranes and Wood Pigeons in search of nutritious acorns. Campo Arañuelo (78) north of Monfragüe is a good example of a wildlife-rich working *dehesa*, with the added attraction of the Arrocampo reservoir.

Wherever pressure from grazing and agriculture is more intense, the *dehesas* have given way to treeless, steppe-like plains occupied by cereal cultivation and natural grasslands. Los Llanos de Cáceres (79) and La Serena (80) are superb examples of these steppe-like habitats and harbour some of Spain's best Great Bustard populations and species such as Little Bustard and Black-bellied and Pin-tailed Sandgrouse. Cities such as Trujillo and Cáceres and all the small towns scattered around these plains boast some of Europe's most important colonies of Lesser Kestrel and White Stork.

A sizeable part of La Serena has been flooded by the largest reservoir in Europe, one of a number of such

EXTREMADURA

reservoirs in Extremadura. This is the case of Puerto Peña and Orellana (81) on the river Guadiana, where mountains, forests and agricultural land mix in an excellent site for raptors and steppe birds.

To the north of this fluvial barrier, the area of Vegas Altas del Guadiana (82) includes large areas of *dehesa* and dry plains, but also large areas that have been transformed into rice-paddies. These artificial wetlands, along with the many small reservoirs and artificial pools that feed them, have become the best wetland in Extremadura and the main wintering quarters in Europe of the Common Crane. Downstream, the river Guadiana continues to be of interest, including the stretch of the river that flows through the city itself.

75. MONFRAGÜE

Located at the confluence of the dammed rivers Tajo and Tiétar, Monfragüe (Cáceres province, Natural Park and SPA) is the best-known natural area in Extremadura. A succession of quartzite ridges are clothed by some of the best stands of Mediterranean forest in Spain, exhibiting markedly different south- and north-facing slopes: the former are covered by holm oak forests and gum cistus, while the latter are clothed by cork and Lusitanian oaks and tree heath and strawberry-tree scrub. In the piedmont, extensive *dehesas* stretch away as far as the eye can see.

ORNITHOLOGICAL INTEREST

Home to the best concentrations in the world of large Mediterranean raptors, Monfragüe boasts the highest known densities of Black Vultures (250 pairs) and Spanish Imperial Eagle (12 pairs), along with over 500 pairs of Griffon Vulture and good numbers of Egyptian Vulture and Golden and Short-toed Eagles. No other site in Spain has so many cliff-breeding Black Storks.

Salto del Gitano, Monfragüe © Pedro Retamar

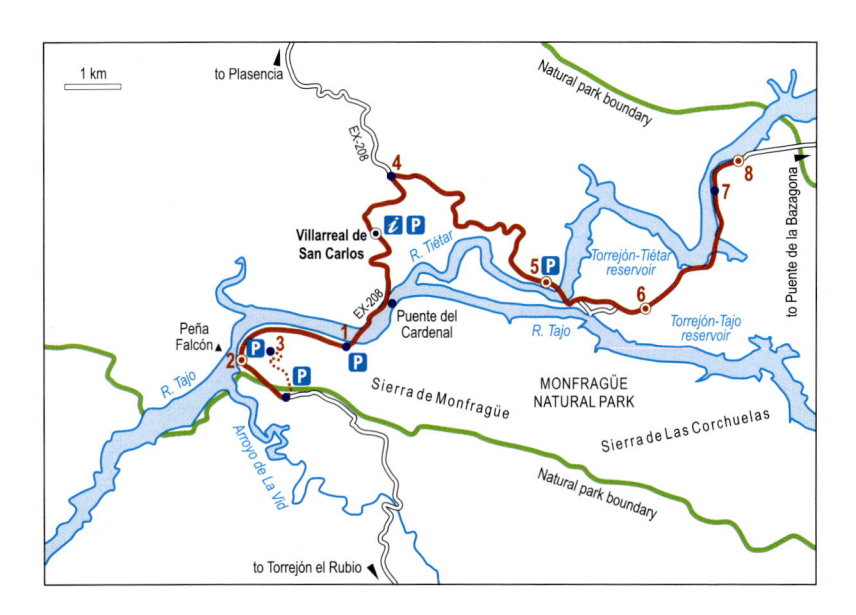

ITINERARIES

1. ⟨9 km⟩ 🚗 🚶 ☀️ ▬

The ease with which so many rare species can be seen explain why Monfragüe is a place of pilgrimage for many birdwatchers. Leave Villarreal de San Carlos along the EX-208 towards Trujillo and once across the bridge over the Tajo, park at La Fuente del Francés (1). Check the road bridge you have just crossed for Alpine Swifts amongst the many House Martins.

The road continues alongside the river as far as the famous Salto del Gitano (2), a viewpoint opposite the imposing quartzite cliff of Peña Falcón. From here the Griffon Vulture nests are perfectly visible; Egyptian Vulture and Common Kestrel also breed on the cliffs, along with smaller rock-loving birds such as Crag Martin, Red-rumped Swallow, Blue Rock Thrush and a few Black Wheatears. In winter, Alpine Accentor appear. Perhaps, however, the most outstanding aspect of the spot are the two Black Stork nests (in some years three) that are perfectly visible from the road just to the right of the viewpoint.

Just over 1 km from El Salto del Gitano a car-park marks the beginning of a walk up the south-face of the Sierra de las Corchuelas. Cars but not coaches can be left at a second car-park a little higher up. Begin the walk up through wild olive trees (frequented by Hawfinch in winter) and finish on a long set of stone steps that take you up to El Castillo de Monfragüe (3).

From the summit Griffon Vultures and abundant Black Kites fly past at arm's length. In summer and until the beginning of autumn, this is a good site for White-rumped Swift. A few pairs of Red-billed Choughs breed nearby and Black Stork and numerous large raptors including Spanish Imperial Eagle often fly over.

2. ⟨14 km⟩ 🚗 ☀️ ▬

Just 1.5 km from Villarreal de San Carlos, turn left (4) along the EX-208 towards Plasencia on a road passing

EXTREMADURA

through a deforested area, home to Thekla Lark and Black-eared Wheatear, which was once a eucalyptus plantation. Five kilometres further on you reach the viewpoint of La Tajadilla (5), with a hide overlooking the cliffs on the other side of the river Tiétar that hold a few Griffon Vulture nests and a single pair of Egyptian Vultures. Black Kites nest in the nearby pines and Bonelli's Eagles regularly fly over the area.

The road drops down to the Torrejón-Tiétar reservoir and then 1.5 km later reaches the small open area known as La Báscula (6). With a telescope, scan the north-facing slopes of the Sierra de Corchuelas for the Black Vulture nests and a single distant Spanish Imperial Eagle nest belonging to a pair of birds that often fly over the area.

From here, the road heads northwards and drops down to the Tiétar again. Just after crossing over a small tributary, a small lay-by (7) on the left gives views over to the far bank where there is a Black Stork's nest, occupied in some years by a pair of Griffon Vultures.

This itinerary ends 300 m further on at the viewpoint of Portilla del Tiétar (8), a must for all birdwatchers; there is no parking place here and so take care where you leave your cars. The rock outcrop on the far bank has a Griffon Vulture colony, and is commonly used as a perching site by Black Stork, Egyptian and Black Vultures and Bonelli's and, occasionally, Spanish Imperial Eagle. Eagle Owl can heard well in winter and is easily seen in spring when the young have just left the nest. Continue along the same road to reach the EX-108 – soon to be replaced by a dual-carriageway – at the bridge at La Bazagona.*

MOST REPRESENTATIVE SPECIES

Residents: Red Kite, Griffon and Black Vultures, Spanish Imperial, Golden and Bonelli's Eagles, Peregrine Falcon, Common Kestrel, Eagle Owl, Kingfisher, Lesser Spotted Woodpecker, Thekla Lark, Crag Martin, Black Wheatear, Blue Rock

---

PRACTICAL INFORMATION

- **Access:** Navalmoral de la Mata and Plasencia are both on the EX-108 (soon to be dual-carriageway); from this road take the EX-208 to Trujillo via Villarreal de San Carlos.
- **Accommodation:** in Plasencia, Malpartida de Plasencia, Villarreal de San Carlos and Torrejón El Rubio.
- **Visitor infrastructure:** the Natural Park visitor centre is in Villarreal de San Carlos (tel. 927 199 134). Open every day, morning and afternoon. Find out here about the companies and experts that work as guides in the park.
- **Visiting tips:** spring is best, providing you avoid Easter and long weekends, when Monfragüe is rather overcrowded. Do not leave the marked trails as most of the park is closed off to the public.
- **Further information:** maps 1:50,000, n° 623 and 651 (SGE). Many books have been published on the park. Also, www.juntaex.es/consejerias/mut/monfrague.

---

* Observing the nests as indicated in the texts of the itineraries will not disturb nesting birds and is approved by the authorities of the Natural Park.

Thrush, Dartford and Sardinian Warblers, Azure-winged Magpie, Red-billed Chough, Rock Sparrow, Rock Bunting.
Summer visitors: Black Stork, Black Kite, Egyptian Vulture, Short-toed and Booted Eagles, Alpine and White-rumped Swifts, Red-rumped Swallow, Black-eared Wheatear, Subalpine and Orphean Warblers, Golden Oriole.
Winter visitors: Alpine Accentor, Fieldfare, Redwing, Hawfinch.

76. SIERRA DE SAN PEDRO

In truth the Sierra de San Pedro is not one, but a collection of various low quartzite ridges that run parallel (north-west to south-east) for around 70 km along the border between Cáceres and Badajoz provinces near the frontier with Portugal. Excellent oak forests – holm oak on south-facing and cork oak on north-facing slopes – cover mountain sides, with *dehesas* in valleys. Up to recently Pardel Lynx and Wolf were present, although both are possibly now extinct. Large private hunting and stock-raising estates covering most of the area have maintained much of the natural habitat intact. An SPA of 120,000 ha protects part of the area.

ORNITHOLOGICAL INTEREST

The cliffs, forests and *dehesas* of San Pedro hold high densities of raptors (only topped by those of Monfragüe) that include over 20 pairs of Spanish Imperial Eagle (one of the world's most important populations), around 200 pairs of Black Vulture, Egyptian Vulture, Griffon Vulture and a vital population of almost 20 pairs of Bonelli's Eagle. As well as many other Mediterranean raptors, around 15 pairs of Black Stork nest and the surrounding plains hold Black-winged Kite, Lesser Kestrel and steppe birds such as Great Bustard. Winter sees the arrival of vast flocks of Wood Pigeon that are unrivalled in size anywhere in Europe.

El Torrico in Sierra de San Pedro © Pedro Retamar

EXTREMADURA

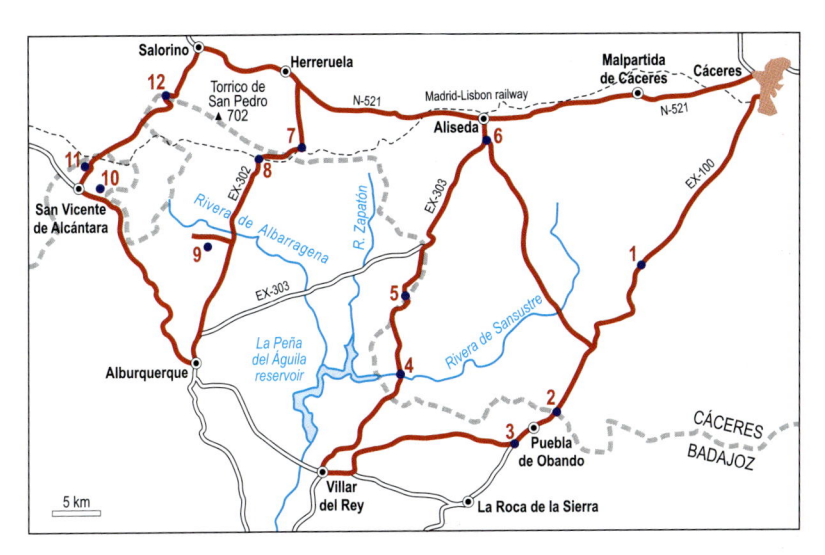

ITINERARIES

1. 160 km 🚗 ☀ 🔄

Between Cáceres and Badajoz the EX-100 passes through *dehesas* and plains before entering the typical forests of the Sierra de San Pedro. Here Black Vulture and Spanish Imperial Eagle breed and can sometimes be seen in flight from any of the mountain passes on the road such Alto de Clavín (1), Alto de La Covacha (2) or, after Puebla de Obando, Puerto de Zángano (3), the latter a magnificent viewpoint over the surrounding *dehesas*.

Just after Puerto de Zángano, turn right to Villar del Rey, from where another road heads north to Aliseda through the eastern sector of San Pedro. The tracts of *dehesas* with colourful scrub are broken only by the wooded slopes of ridges such as Sierra de Alpotreque, with Egyptian Vulture and Golden and Bonelli's Eagles. A good stopping point is the bridge (4) over Rivera de Sansustre (also called Rivera de El Saltillo) where Eagle Owl can be heard calling in winter.

A little further on the same road the Sierra de Mercadores (5) with magnificent north-facing forests breaks the hegemony of the endless *dehesas*: look out for Black Vulture, Black Stork and Spanish Imperial Eagle. Soon the road joins the EX-303, where you can turn right through a shady area to Aliseda and the main N-512 to Cáceres.

Alternatively, turn right before Aliseda along a narrow and somewhat bumpy (6) road that passes through one of the wildest sectors of San Pedro, with all the species previously mentioned present. In 20 km, you return to the EX-100 and can head back to Cáceres.

2. 210 km 🚗 ☀ 🔄

From Cáceres, the N-521 passes through Aliseda and when you are almost in Herreruela, turn left along the EX-302 to Alburquerque. After 7 km, a road off to the left takes you to the abandoned station of Herreruela (7), a wonderful place for viewing the north-facing forests at the heart of San Pedro and for seeing Spanish Imperial Eagle. In winter, immense

PRACTICAL INFORMATION

- **Access:** Cáceres is the best starting point for exploring San Pedro. The N-521 towards Valencia de Alcántara and the EX-100 to Badajoz allow you to explore the area from the north and south, respectively.
- **Accommodation:** in Cáceres, Alburquerque, San Vicente de Alcántara and Valencia de Alcántara.
- **Visitor infrastructure:** in San Vicente de Alcántara there is a wildlife interpretation centre dedicated to the Sierra de San Pedro, which is open Monday to Friday, mornings and afternoon.
- **Visiting tips:** spring is the best time for breeding birds and the flowering of the *dehesa*. Most of the estates are private and so it is best not to leave the roads.
- **Further information:** maps 1:50,000, nº. 702, 703, 727, 728 and 751 (SGE).

groups of hundreds of thousands of Wood Pigeons darken the sky here.

The road continues on across San Pedro and at the bridge over the railway line (8) it is worth stopping to scan the south-facing slopes for Egyptian and Black Vultures and for perched Golden Eagles. An interesting alternative is to take the asphalted track on the right just after passing over the river Albarragena that penetrates the *dehesa* of Piedrabuena: pass close to a Grey Heron and White Stork colony (on holm oaks) with a private castle (9) rising up in the background.

Near Alburquerque the landscape opens out and a mosaic of pastures, Spanish broom scrub and dry fields, dotted here and there with huge granite boulders, extends as far as San Vicente de Alcántara. It is worth investigating some of the plains (10, 11) for Great Bustard, Black-winged Kite, Lesser Kestrel and Roller.

A road northwards from San Vicente de Alcántara crosses San Pedro once again and at Puerto de Élice (12) stop for one of the best panoramas anywhere in the peninsula of a unbroken, natural Mediterranean forest. Continue on through *dehesas* to the village of Salorino, and then onto the N-521 back to Cáceres.

MOST REPRESENTATIVE SPECIES

Residents: Grey Heron, Black-winged Kite, Griffon and Black Vultures, Spanish Imperial, Golden and Bonelli's Eagles, Little and Great Bustards, Stone Curlew, Wood Pigeon, Eagle Owl, Kingfisher, Crested and Thekla Larks, Blue Rock Thrush, Eurasian Jay, Azure-winged Magpie, Spanish Sparrow.

Summer visitors: Black and White Storks, Black Kite, Egyptian Vulture, Short-toed and Booted Eagles, Lesser Kestrel, Roller, Red-rumped Swallow, Subalpine and Orphean Warblers, Golden Oriole.

Winter visitors: Red Kite, European Golden Plover, Redwing, Bullfinch.

77. LAS VILLUERCAS

The abrupt quartzite ridges of the parallel ridges of Las Villuercas rise knife-like in the south-east of Cáceres province. Mountain slopes are covered by a combination of Mediterranean forest, scrub and Atlantic forests of Pyrenean oak and sweet chestnut, with a number of interesting plant communities such as relict Portugal laurel formations found in some of the wild-

EXTREMADURA

est gorges in the area. Pine and eucalyptus plantations exist and around the villages small olive groves fields and market gardens add variety to the landscape. The only protection is an SPA of almost 70,000 ha.

ORNITHOLOGICAL INTEREST

Cliff-breeding raptors, without being common anywhere, are widespread throughout Las Villuercas and most suitable cliff-faces are occupied by one or more species. Almost 100 Griffon Vultures breed, along with Black Stork, Egyptian Vulture, Golden and Bonelli's Eagles and high densities of Peregrine Falcon. Surrounding forests hold a few Black Vultures and Spanish Imperial Eagles, as well as a good mixture of Mediterranean and Atlantic passerines.

ITINERARIES

1. 115 km

From Deleitosa, head through the northern part of Las Villuercas to Robledollano and then on towards Castañar de Ibor. At the river Viejas, take a path along the river towards a fish-farm (1): look out for Dipper, Kingfisher and Golden Oriole. Back on the road, pass the cliffs of El Frontón (2) with its vulture colony on your left and follow the river Viejas until it meets the river Ibor. Continue over the bridge and head for Castañar de Ibor, where you should take the EX-118 towards Navalvillar de Ibor.

Some 8 km beyond Navalvillar, turn left (3) towards Navatrasierra. After passing over Puerto del Hospital del Obispo, the road enters a Pyrenean oak forest and at a picnic site with a spring (4) you can wander through the woods in search of Lesser Spotted Woodpecker, Bonelli's Warbler and Eurasian Jay, amongst other birds.

Ten kilometres beyond the previous junction on the road to Guadalupe, opposite the chapel of El Humilladero (5), an asphalted track heads up to Pico Villuercas (1,600 m), the highest peak in the area. Pass through Pyrenean oak and sweet chestnut

Las Villuercas © Pedro Retamar

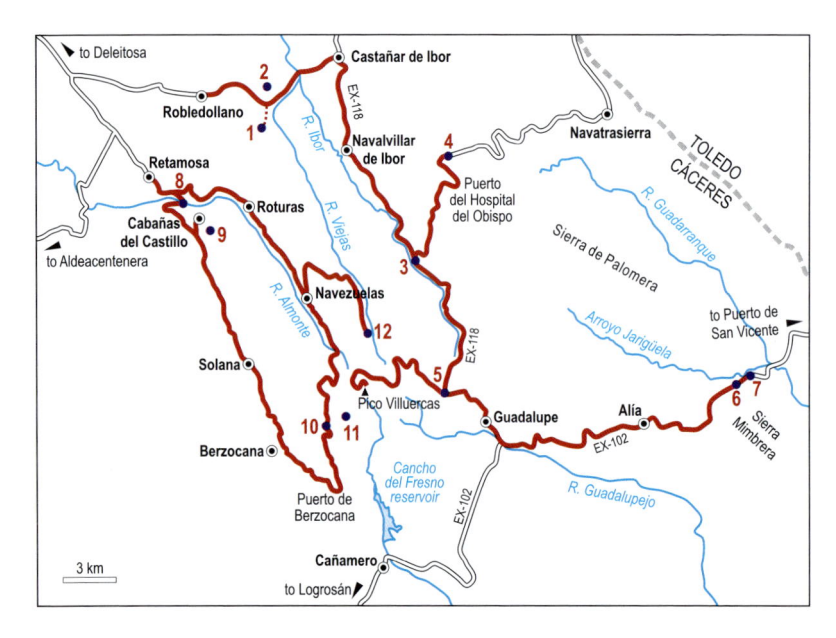

woodland (with Honey Buzzard relatively common) and up to the abandoned radar base, from where there is a splendid panorama and, amongst the birds, Golden Eagle, Red-billed Chough and reports of Rufous-tailed Rock Thrush.

From Guadalupe continue along the EX-102 through Alía. Eight kilometres further on, a stream, Arroyo Jarigüela, has carved open a spectacular gorge – Estrecho de La Peña – between the twin ridges of La Palomera and Mimbrera near its confluence with the river Guadarranque. There are two viewpoints (6, 7) next to the road from where the nests of the small Griffon Vulture colony are obvious. Egyptian Vulture, Bonelli's Eagle, Peregrine Falcon and Eagle Owl also breed here. Continue along the road towards Puerto de San Vicente to head for Toledo province.

2. 80 km

From Deleitosa head for Retamosa and 2 km afterwards, turn right and cross almost immediately the river Almonte. Just after the bridge (8), Griffon and Egyptian Vultures and Bonelli's Eagle frequent the nearby cliffs here and at Peña Buitrera (9), on the right of the road to the village of Cabañas del Castillo. This village, perched in between the cliffs that overlook the valley of the river Almonte or its castle are excellent observation points for raptors and other cliff-loving birds.

Return to the road from Retamosa, and head for Berzocana and, without entering this village, continue on to Puerto de Berzocana, where you should turn left and head north. The road climbs to Collado del Brazo (10) with good views over the Barrera de los Peñones (11) frequented by Golden Eagle. This eagle and other raptors can also be seen around the cliffs overlooking the river Almonte that are visible from the road through Navezuelas and Roturas that takes you back to Retamosa.

From Navezuelas a track off to the right at the far end of the village drops down to the river Viejas and follows

PRACTICAL INFORMATION

- **Access:** head for Navalmoral de la Mata along the Autovía de Extremadura (N-V), and then from here to Castañar de Ibor, Guadalupe and Deleitosa along the EX-118. The EX-102 through Logrosán and Cañamero will take you to Guadalupe.
- **Accommodation:** in Guadalupe, Castañar de Ibor, Cañamero and Logrosán, or La Coraja, rural accommodation for birdwatchers (tel. 927 314 129, lacoraja@terra.es), in Aldeacentenera. More information in www.lacoraja.com.
- **Visitor infrastructure:** no visitor centre or infrastructure for birdwatchers. A visitor centre is to be opened in Cañamero.
- **Visiting tips:** late spring is best to coincide with the peak flowering period. The autumn colours are splendid.
- **Recommended:** west of Las Villuercas between Aldeacentenera and Torrecillas de la Tiesa a road heads off to the village of Belén, near Trujillo. This road is excellent for Great Bustard and other steppe birds, as well as for large raptors such as Spanish Imperial Eagle.
- **Further information:** maps 1:50,000, nº. 680, 681, 707 and 708 (SGE).

it upstream along its left-bank. This valley possesses a variety of habitats, from quartzite cliffs on either side of the river and small market gardens, to sweet chestnuts and cherry woods that offer good prospects for birdwatchers. It is best to turn around once the tarmac ends (12).

MOST REPRESENTATIVE SPECIES

Residents: Red Kite, Griffon and Black Vultures, Golden and Bonelli's Eagles, Peregrine Falcon, Common Kestrel, Eagle Owl, Kingfisher, Lesser Spotted Woodpecker, Thekla Lark, Crag Martin, Dipper, Robin, Black Wheatear, Blue Rock Thrush, Eurasian Nuthatch, Eurasian Jay, Azure-winged Magpie, Red-billed Chough, Cirl and Rock Buntings.
Summer visitors: Black Stork, Honey Buzzard, Black Kite, Egyptian Vulture, Short-toed and Booted Eagles, Alpine Swift, Red-rumped Swallow, Common Redstart, Orphean and Bonelli's Warblers, Golden Oriole.
Winter visitors: Alpine Accentor.
On migration: Bluethroat.

78. CAMPO ARAÑUELO

This plain, lying between the rivers Tiétar and Tajo in the north-east of Cáceres province, is covered in the main by magnificent *dehesas* that eastwards give way to treeless pastures and southwards connect with the Sierras de Monfragüe and Villuercas; to the north around the river Tiétar, the *dehesas* have largely been replaced by irrigated agriculture. This site also includes two reservoirs, the enormous Valdecañas on the river Tajo, with granite outcrops at both ends, and the much smaller Arrocampo, used to cool the Almaraz nuclear power station. No legal protection of the site exists.

ORNITHOLOGICAL INTEREST

This site includes a series of reservoirs and irrigation pools that provide favourable habitat for many water birds: the reservoir of Arrocampo has constant water levels and large stands of bulrushes that are home to many species of bird infrequent in Extremadura. Most Iberian herons including Purple (20 pairs) and Squacco

Arrocampo reservoir, Campo Arañuelo © José Enrique Capilla

Herons, Marsh Harrier and Purple Swamp-hen breed. Winter sees the arrival of good numbers of Great Cormorant and a few Great Egrets.

Black Stork, Black Vulture, Spanish Imperial Eagle and other Mediterranean raptors disperse into the *dehesas* here - also home to wintering Common Cranes - from their breeding quarters in Monfragüe and Las Villuercas. The treeless pastures are the best place to look for Black-winged Kite and a number of towns and villages have colonies of Lesser Kestrel.

ITINERARIES

1. 10 km

From the town of Almaraz, best-known for its nuclear power station, take the road towards Saucedilla along the east bank of the Arrocampo reservoir. After 2 km, cross one of the arms of the reservoir on a dirt ford (1): park on the left and view the reservoir.

To the west there are good views of the water and the partially submerged wall that channels water for cooling the nuclear reactor, on which an extraordinary number of White Storks have built their nests and Great Cormorants and gulls rest. To the east there is a large stand of bulrushes frequented by Purple Swamphen, Purple and other herons; Gull-billed Terns sometimes wander in from their colonies on nearby Valdecañas reservoir, and passerines such as Savi's, Reed and Great Reed Warblers and Red Avadavat breed.

This is also a good site year-round for Marsh Harrier, although winter is better and also brings Great Cormorants, duck, gulls and Great Egret to the reservoir. At times numerous groups of Eurasian Spoonbill pass through on migration, along with Osprey and a number of waders.

Saucedilla is a good point of departure for exploring northwards from the reservoir. Take the asphalted track (2) that reaches the service road for the interesting Balsa de Cerro Alto (3). Keep left for views of this small reservoir and another similar site 2 km further on (4). Turn right to reach the town of Casatejada.

EXTREMADURA

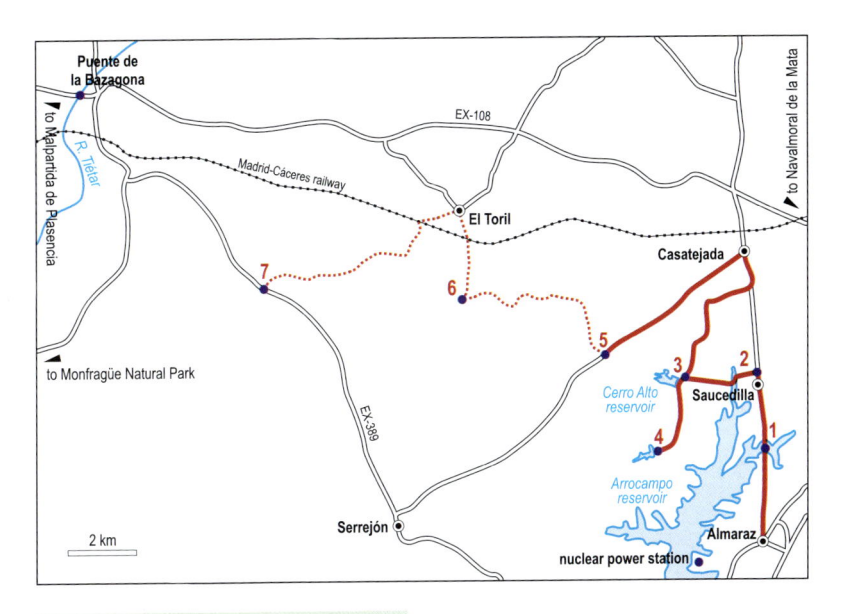

2. [20 km] 🚗 ☀ ➡

From Casatejada take the road towards Serrejón and after 5 km, turn right along a public track (5) that heads into the El Pizarral estate. At the *cortijo* (farm) of Los Calles (6), turn north towards the hamlet of El Toril and pass through excellent *dehesas* used by Black Vulture, Spanish Imperial Eagle and other raptors. Check too the many small pools in the area for Black Stork.

On winter, with luck, you may come across a group of Common Cranes.

From El Toril, another track leads to the EX-389, the road from Serrejón to Puente de La Bazagona on the river Tiétar. From here you can complete the circuit by following the EX-108 to Navalmoral de la Mata via Casatejada, the point of departure of this itinerary. This section visits more bird-rich *dehesas*, some rather more open and thus better for Black-winged Kite (above all in winter).

PRACTICAL INFORMATION

- **Access:** at km point 193 of the Autovía de Extremadura (N-V), turn off to Almaraz (itinerary 1). At km point 185, just after Navalmoral de la Mata, turn off along the EX-108 towards Malpartida de Plasencia and then after 10 km, head for Casatejada (itinerary 2). The EX-108 is to be converted into a dual-carriageway and some accesses may change.
- **Accommodation:** in Navalmoral de La Mata. In Serrejón, the rural hotel El Alcaudón (tel. 927 547 600, info@alcaudon.net) is much used by birdwatchers from Spain and from further afield.
- **Visitor infrastructure:** no visitor centre or infrastructure for birdwatchers.
- **Visiting tips:** all year is good apart from high summer; spring for breeding species and winter for concentrations of water birds. Once off road only use public tracks as most of this area is private property.
- **Further information:** maps 1:50,000, n°. 623, 624 and 652 (SGE).

MOST REPRESENTATIVE SPECIES

Residents: Little Grebe, Little Bittern, Cattle and Little Egrets, Grey Heron, White Stork, Black-winged Kite, Griffon and Black Vultures, Marsh Harrier, Spanish Imperial and Bonelli's Eagles, Water Rail, Purple Swamphen, Wood Pigeon, Kingfisher, Spanish Sparrow, Red Avadavat.
Summer visitors: Night, Squacco and Purple Herons, Black Stork, Egyptian Vulture, Short-toed and Booted Eagles, Lesser Kestrel, Gull-billed Tern, Roller, Red-rumped Swallow, Savi's, Reed and Great Reed Warblers.
Winter visitors: Great Cormorant, Great Egret, Common Crane, Black-headed and Lesser Black-backed Gulls, Bluethroat, Penduline Tit.
On migration: Eurasian Spoonbill, Osprey, Whiskered Tern, Sedge Warbler.

79. CÁCERES PLAINS

Over 250,000 ha of rolling cereal cultivation, dry grazing and Mediterranean scrub, interspersed by numerous huge granite boulders and interrupted by the occasional river (for example, Almonte) or quartzite outcrop (Sierra de Fuentes), surround the city of Cáceres. The monotony of these great plains as they merge into tracts of *dehesa* is also broken by a number of large reservoirs and smaller cattle pools. An SPA of over 70,000 ha protects part of the site.

ORNITHOLOGICAL INTEREST

Along with La Serena, the great Cáceres plains constitute the best site in Extremadura for steppe birds, amongst them almost 2,000 Great Bustard and good numbers of Montagu's Harrier, Little Bustard and Black-bellied and Pin-tailed Sandgrouse. The towns of Cáceres and Trujillo and surrounding areas are home to one of Europe's best populations of White Stork and Lesser Kestrel, while the numerous reservoirs and pools in the area are frequented by many aquatic birds. Black Storks and, to a lesser extent, Spanish Imperial Eagle are seen, and in winter there are a number of Common Crane roosts.

Belvís de Monroy in the Cáceres plains © Eduardo de Juana

EXTREMADURA

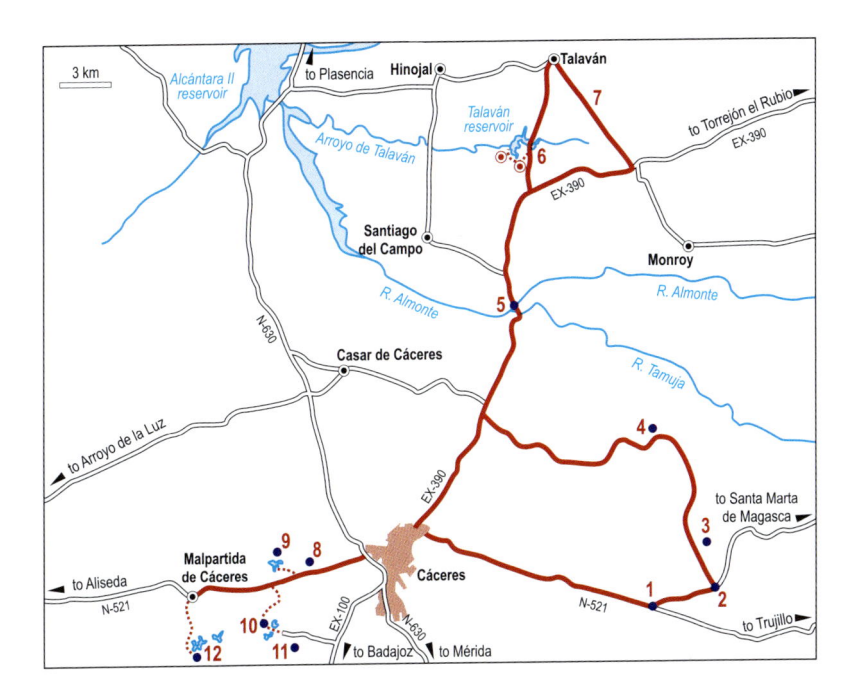

ITINERARIES

1. 50 km

At around 15 km from Cáceres along the N-521 towards Trujillo, turn left towards Santa Marta de Magasca (1) and then left again (2) on a bumpy road running through a mosaic of pastures, arable fields and fallows that hold excellent populations of steppe birds. Stop wherever there are good views, parking off-road or at the entrance to a track (without blocking it).

The first stretch of the road heading north passes through the cereal fields of the La Pulgosa estate (3), with abundant Montagu's Harrier. After a broad curve (4), the road suddenly turns west and enters an area of pastures and rough grazing frequented by Little and Great Bustards, Black-bellied and Pin-tailed Sandgrouse and Calandra Lark, with Lesser Kestrels in some of the old farm buildings. At the junction with the EX-390, turn left to return to Cáceres or right to link up with itinerary 2.

2. 70 km

The first stretch of EX-390 from Cáceres to Talaván passes through an area of rolling country where Rollers and Common Kestrel use the nest-boxes fixed to the pylons of the electric line running parallel to the road. Fifteen kilometres from Cáceres you reach the river Almonte at its confluence with Tamuja: look here for Common, Alpine and Pallid Swifts, Crag Martin and Red-rumped Swifts, all nesting in the road bridge (5).

Ten kilometres further on from the bridge you come to Talaván reservoir. Just before crossing the river flowing into the tail-end of the reservoir, there is a hide on the left and a track (6) that leads to the dam and another hide. This wetland is a good place in winter to view Common Crane coming in to roost from the *dehesa* extending away to the east; also in winter, look out for Black-necked Grebe, Greylag Goose and groups of Great Cormorants, duck and gulls. Post-breeding groups of Black Stork, as well as Eurasian Spoonbill and Osprey pass through on migration.

Continue for five kilometres beyond the reservoir to the village of Talaván, where you should pick up a road (7) heading south-east through *dehesa* – with Black-winged Kite and Common Crane (in winter) – as far as the EX-390. Alternatively, return to the Cáceres road from Talaván by heading west to Hinojal and Santiago del Campo through plains with Little Bustard, Stone Curlew, Black-bellied Sandgrouse and Roller.

3. 20 km

Five kilometres west of Cáceres on the N-512 west towards Aliseda a house (Los Arenales) (8) has an eye-catching White Stork colony of over 70 pairs. Just after the house, but before the road crosses the Madrid-Lisbon railway line, turn right along a track to Charca del Majón (9), a good site for water birds.

Return to the main road and just after a small petrol station, turn left towards Las Charcas del Millar (10), a group of gravel pits with similar birdlife to the previous stop. Continue along the same track to the municipal rubbish dump (11), popular with massive flocks of Cattle Egret, White Stork, Black and Red Kites and Lesser Black-backed and Black-headed Gulls.

From Malpartida de Cáceres (12 km from Cáceres), take a road heading south to the Monumento Natural de los Barruecos (12), an area of huge granite boulders crowned with White Stork nests.

MOST REPRESENTATIVE SPECIES

Residents: Cattle Egret, Black-winged and Red Kites, Griffon and Black Vultures, Spanish Imperial Eagle, Common Kestrel, Little and Great Bustards, Stone Curlew, Black-bellied and Pin-tailed Sandgrouse, Calandra Lark, Crag Martin, Kingfisher, Spanish Sparrow.

Summer visitors: Black and White Storks, Egyptian Vulture, Short-toed and Booted Eagles, Montagu's Harrier, Lesser Kestrel, Great Spotted Cuckoo, Alpine and Pallid Swifts, Roller, Short-toed Lark, Red-rumped Swallow, Black-eared Wheatear.

PRACTICAL INFORMATION

- **Access:** the Autovía de Extremadura (N-V) passes close to Trujillo, from where the N-521 connects to Cáceres and the N-630, the north-south road that runs through the whole of western Spain.
- **Accommodation:** many hotels in Cáceres.
- **Visitor infrastructure:** except for the hides on the Talaván reservoir, no visitor centre or infrastructure for birdwatchers.
- **Visiting tips:** all year apart from mid-summer (too hot) is good. Early spring is an ideal time to see groups of male Great Bustards displaying. Do not leave roads and tracks to avoid disturbing birds and cattle in the (mostly) private estates.
- **Further information:** maps 1:50,000, n°. 650, 678, 679 and 704 (SGE).

EXTREMADURA

Winter visitors: Great Cormorant, duck, Hen Harrier, European Golden Plover, Northern Lapwing, Lesser Black-backed and Black-headed Gulls.
On migration: Eurasian Spoonbill, Osprey.

80. LA SERENA

This steppe landscape in the east of Badajoz province, stretching north to the river Zújar and two massive reservoirs, Zújar and La Serena, is the largest tract of uncultivated land in Western Europe (100,000 ha). To the south a line of low ridges acts as the frontier of this treeless plain, which mainly consists of rough grazing along with a few scattered patches of arable land and scrub. Heavy grazing pressure from large herds of Merino sheep is being rectified by joint sustainable development projects run by the government of Extremadura and SEO/BirdLife with support from the EU. The only legal protection is an SPA.

ORNITHOLOGICAL INTEREST

This is possibly the best site in Spain for steppe birds such as Great (almost 500 birds) and Little Bustards, Montagu's Harrier, Collared Pratincole, Stone Curlew, Pin-tailed Sandgrouse and Calandra Lark. Black-bellied Sandgrouse are restricted to certain areas and Lesser Kestrels breed in villages and farm buildings. Many of the large raptors such as Egyptian and Griffon Vultures and Golden and Bonelli's Eagles that nest on the surrounding quartzite ridges, as well as Black Stork, come to feed in the area. There are also half a dozen Common Crane roosts in winter with up to 5,000 birds.

ITINERARIES

1. 50 km

From Castuera, the EX-103 towards Herrera del Duque runs through the midst of a typically desolate part of La Serena, with populations of birds such as Little Bustard and Stone Curlew.

La Serena © Anna Motis

The electric pylons are used as perches by Short-toed Eagle. Many also have nest-boxes occupied by Roller and Common Kestrel, while Lesser Kestrels nest in some of the farmhouses and even on occasions in piles of stones. The endless pastures are hunted over by large raptors such as Golden Eagle and summer sees large groups of White Stork and Black Kite feeding on the myriad of grasshoppers. In winter, groups of European Golden Plover and Northern Lapwing are present.

It is worth investigating some of the tracks and drover's roads that head off east and west of the road: try El Cordel Transversal, heading left 5 km north of Castuera to the abandoned Miraflores (1) mine (with a colony of Red-billed Choughs). Another alternative is a track off to the right 6 km further on that crosses the whole of this steppe area, passing by another abandoned mine, Peña Lobosa (2).

Some 8 km beyond the previous turn-off, an asphalted track (3) heads off to the right, crosses an arm of the Zújar reservoir and then joins the so-called Carretera de La Golondrina. Heading south-east to Cabeza del Buey, you will have a good chance of seeing Great and Little Bustards and Pin-tailed and, with luck, Black-bellied Sandgrouse.

Once you have driven 6.5 km from the beginning of this road, turn left along a 2-km track to the Pavorosa estate (4) through one of the best areas for Great Bustard in the site. Return to the road and 4 km further on, turn left again (5) along a disused but made-up road that crosses an excellent area of natural steppe and cereal cultivation with Montagu's Harrier and Collared Pratincole.

2. 32 km

From Castuera, the EX-104 runs along the Sierra de Tiros, the southern border of La Serena. Head for the Castillo de Benquerencia de la Serena for Black Stork and large raptors, as well as other rock-loving species such as Alpine Swift, Crag Martin, Black Wheatear, Blue Rock Thrush and Alpine Accentor (in winter).

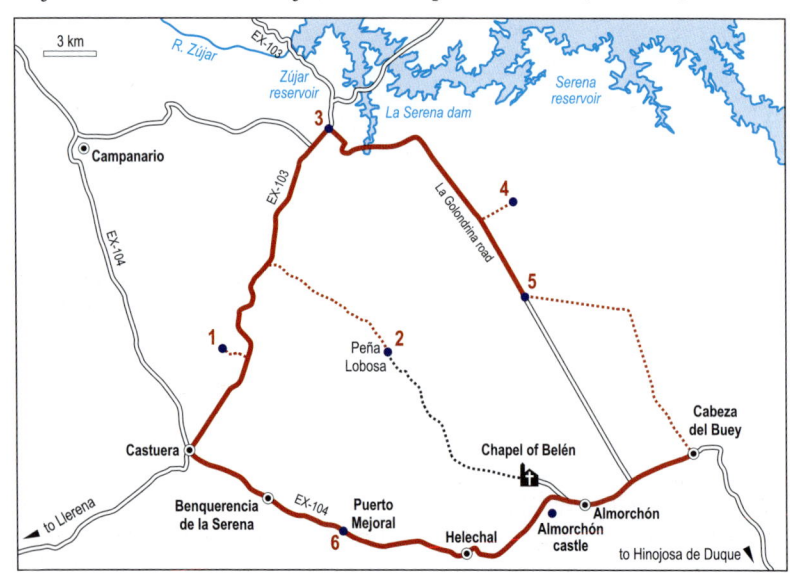

EXTREMADURA

PRACTICAL INFORMATION

- **Access:** at km point 287 of the Autovía de Extremadura (N-V), turn-off to Miajadas and Don Benito, and then continue on to Villanueva de La Serena and, finally, along the EX-104 to Castuera.
- **Accommodation:** in Castuera. The tourist complex of Isla del Zújar (tel. 924 146 010, sertur@laserena.org) in the heart of La Serena and near the Zújar reservoir has apartments and rural accommodation and organises wildlife activities.
- **Visitor infrastructure:** the nature school with hostal at Puerto Mejoral run by the wildlife protection NGO Adenex (tel. 924 371 202, adenex@bme.es) is devoted to the Common Crane.
- **Visiting tips:** all year apart from mid-summer (too hot) is good. Early spring is an ideal time to see groups of male Great Bustards displaying. After heavy rain, some tracks may be impassable.
- **Recommended:** specialist bird guides from Anser (tel. 924 773 869, anser-la-serena@teleline.es).
- **Further information:** maps 1:50,000, n°. 779, 780, 805, 806 and 807 (SGE).

Nevertheless, the main interest here are the groups of Common Crane that in winter feed in the *dehesas* to the south of the ridge and then cross the ridge to roost in the steppes to the north. The best point to wait at dusk is Puerto Mejoral (6).

MOST REPRESENTATIVE SPECIES

Residents: Cattle Egret, White Stork, Griffon Vulture, Marsh Harrier, Golden and Bonelli's Eagles, Common Kestrel, Little and Great Bustards, Stone Curlew, Black-bellied and Pin-tailed Sandgrouse, Eagle and Little Owls, Calandra, Crested and Thekla Larks, Crag Martin, Black Wheatear, Blue Rock Thrush, Red-billed Chough, Spanish Sparrow.
Summer visitors: Black Stork, Black Kite, Egyptian Vulture, Short-toed Eagle, Montagu's Harrier, Lesser Kestrel, Collared Pratincole, Great Spotted Cuckoo, Alpine Swift, Roller, Short-toed Lark, Black-eared Wheatear, Spectacled Warbler.
Winter visitors: Great Cormorant, Hen Harrier, Merlin, Common Crane, European Golden Plover, Northern Lapwing, Alpine Accentor.

81. PUERTO PEÑA AND ORELLANA

The reservoirs of Puerto Peña – also known as García Sola – and Orellana in the north-east of Badajoz province form part of a succession of large dams on the Guadiana river that together retain the largest amount of water in any reservoir system in Spain. Although all but touching, these two reservoirs are in fact very different in type: the former twists between abrupt cliffs and mountain sides covered in pine and eucalyptus plantations and Mediterranean forest, while the latter spreads out over a flat, undulating peneplain and is only interrupted by the odd exposed ridge such as the Sierra de Pela. Both are part of an SPA.

ORNITHOLOGICAL INTEREST

The quartzite cliffs that surround the reservoir of Puerto Peña are home to a community of breeding birds that

Orellana reservoir   © Pedro Retamar

includes important numbers of Griffon Vultures and Black Storks, as well as Egyptian Vulture, Golden and Bonelli's Eagles and Eagle Owl. Orellana is traditionally more important for water birds such as wintering Common Crane and has been declared a Ramsar Site. In recent years, however, many of the species that bred or wintered in Orellana have moved to the rice-paddies in Las Vegas Altas del Guadiana and the reservoir has lost some of its importance. Nevertheless, the surrounding steppe still has good numbers of Great Bustard and is hunted over by large raptors.

ITINERARIES

1. 65 km

From Herrera del Duque, with its ruined convent and church occupied by a colony of Lesser Kestrels, a road heads for the village of Peloche and then follows the eastern bank of Puerto Peña reservoir with the Sierra de los Golondrinos (1) on you left. View the typical cliff-loving birds from next to the dam. Just after crossing the dam, pick up the N-430 towards Mérida and then in 400 m, turn left to a wonderful viewpoint (2) overlooking a spectacular cliff face with the largest Griffon Vulture colony (almost 50 pairs) in Badajoz Province. Here and in the neighbouring ridges of El Escorial (east) and La Chimenea (west) you have chances of seeing Black Stork, Egyptian Vulture and Golden and Bonelli's Eagles.

Return to the road and head along the western border of the reservoir towards Guadalupe and 7 km after Valdecaballeros, turn right (3) to the never-finished and now dismantled nuclear power station of the same name. Continue along a track when the asphalt finishes and reach a earth dyke (4), which you can drive northwards for 3 km as far as a place you can turn your vehicle. From the dyke view the flooded gravel pits from where material was extracted to build the

EXTREMADURA

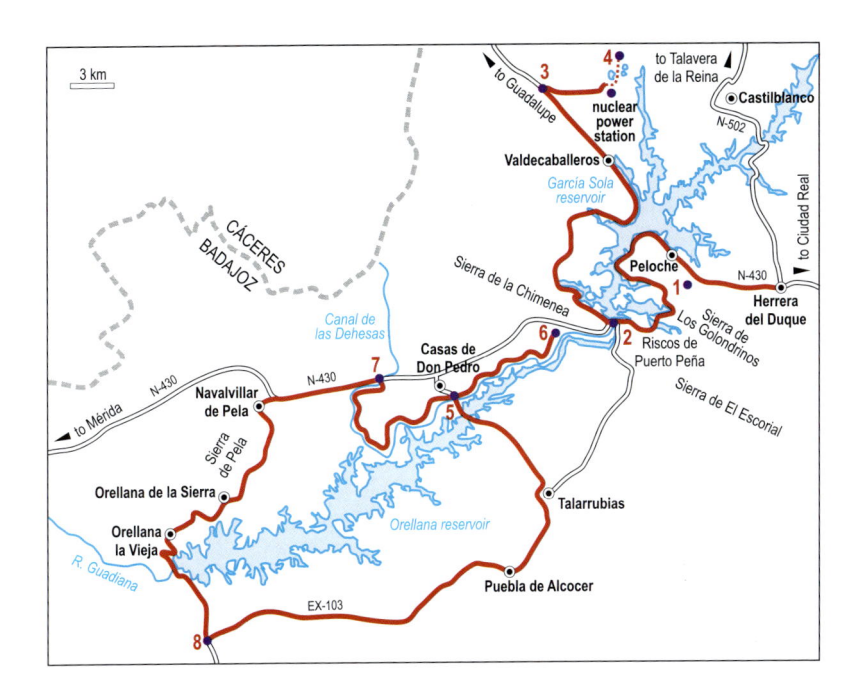

power station and which today hold Little Bittern, the odd Purple Heron, post-breeding groups of Black Stork and groups of wildfowl in winter.

2. 80 km

In Puebla de Alcocer, with its colony of over 100 pairs of Lesser Kestrel in an abandoned convent, it is worth climbing up the castle perched on a crag to look for Crag Martin, Black Wheatear, Blue Rock Thrush and Red-billed Chough.

From here head to Talarrubias, whose church is home to a spectacular White Stork colony. Continue north towards Casas de Don Pedro and after crossing Orellana reservoir, reach a road that runs alongside the Canal de las Dehesas (5), an irrigation canal that in turn follows the northern shore of the reservoir. Turn right towards the dam of Puerto Peña and in just over 10 km in a group of tamarisks reach an immense colony (6) of over 2,000 pairs of Cattle Egret. In the same area at the tail-end of the reservoir, post-breeding groups of Black Stork form.

Return to the previous junction and continue west alongside Canal de las Dehesas through *dehesas* with Black-winged Kite that are used as hunting grounds by large raptors and, in winter, by flocks of Common Cranes. The canal road runs into the road (7) from Casa de Don Pedro; turn left towards Navalvillar de Pela and then Orellana de la Sierra and Orellana la Vieja, passing near the interesting rocky outcrops of Sierra de Pela.

Roughly 6 km after crossing the Orellana dam, take the EX-103 (8) through an area of pastures and cereal cultivation with Great and Little Bustards and Black-bellied and a few Pin-tailed Sandgrouse and return to Puebla de Alcocer. This area holds the best densities of Montagu's Harriers in Extremadura and there are plenty of Lesser Kestrels in the farms and other rural buildings.

PRACTICAL INFORMATION

- **Access:** take the Autovía de Extremadura (N-V) to Talavera de la Reina, where you pick up the N-502 to Herrera del Duque (itinerary 1). From this road there are a number of ways of reaching Puebla de Alcocer (itinerary 2).
- **Accommodation:** in Herrera del Duque and Talarrubias. The Puerto Peña camp-site (tel. 924 631 411) is at the foot of the impressive cliffs that give it its name.
- **Visitor infrastructure:** except for the viewpoint at Puerto Peña, no visitor centre or infrastructure for birdwatchers.
- **Visiting tips:** all year except high summer (too hot); spring for steppe birds and winter for Common Crane.
- **Further information:** maps 1:50,000, n°. 732, 733, 755 and 756 (SGE).

MOST REPRESENTATIVE SPECIES

Residents: Cattle Egret, White Stork, Black-winged Kite, Griffon Vulture, Golden and Bonelli's Eagles, Little and Great Bustards, Stone Curlew, Black-bellied and Pin-tailed Sandgrouse, Eagle Owl, Calandra Lark, Crag Martin, Black Wheatear, Blue Rock Thrush, Red-billed Chough, Spanish Sparrow.
Summer visitors: Little Bittern, Purple Heron, Black Stork, Egyptian Vulture, Short-toed and Booted Eagles, Montagu's Harrier, Lesser Kestrel, Black-winged Stilt, Roller.
Winter visitors: Great Cormorant, duck, Common Crane, European Golden Plover, Northern Lapwing, Black-headed and Lesser Black-backed Gulls.

82. VEGAS ALTAS DEL GUADIANA

Between Cáceres and Badajoz a large plain extends southwards from the foothills of the Sierra de las Villuercas to the river Guadiana. Crossed by important tributaries – rivers Ruecas, Cubilar and Gargáligas – running north-east to south-west, this plain is given over to cereal cultivation, grazing land and *dehesas*, although increasingly rice-paddies supplied with water by a network of small reservoirs and canals are taking over. This area has no legal protection.

ORNITHOLOGICAL INTEREST

This site is by far the most important area for wintering Common Crane in Europe (between 20,000 and 30,000 birds), most of which roost in the rice-paddies. These artificial wetlands and the numerous pools and reservoirs form one of the most interesting sites for water birds in Extremadura, with excellent numbers of wintering duck and Greylag Goose and groups of thousands of Black-tailed Godwits amongst other waders passing through on passage. Colonies of escaped passerines such as Red Avadavat and Common Waxbill are well established.

Although declining as a result of habitat loss, there are still good densities of Great Bustard and other steppe birds in the area, as well as colonies of Lesser Kestrel (for example, the 40 pairs in the village of Acedera). Black-winged Kites are abundant in open *dehesas* and in general the area is used by large raptors to hunt and by Black Storks on passage (a few winter).

EXTREMADURA

Dehesas de Madrigalejo in the Vegas Altas del Guadiana © José Enrique Capilla

ITINERARIES

1. 25 km 🚗 ☀️ ➡️

The new irrigation projects and all the associated hydraulic infrastructure produce changes locally that may mean that the best birdwatching sites change from year to year. Bearing this in mind, one of the best possible birdwatching itineraries begins in Navalvillar de Pela, where you should take the N-430 towards Mérida and then after 3 km, turn along the EX-116 towards Guadalupe.

In winter there is constant movement of Common Cranes all along this road. Eight kilometres from the start of the route, a pool – the Charca de Gorbea or Moheda Alta – appears on your right: just past it, enter a track (1) on the right and park. This is a perfect spot to wait for the cranes to come in and roost in the rice-paddies behind the pool and thus to witness one of the greatest birdwatching sights in Europe. From mid-December onwards, groups of up to 12,000 cranes enter to roost, accompanied also by Greylag Geese and Marsh Harriers.

The EX-116 continues through areas of rice-paddies with waders on passage and open *dehesa* with Black-winged Kites, as well as groups of feeding cranes in winter.

Fifteen kilometres from the start of the itinerary, turn left along an asphalted track running alongside El Canal de las Dehesas. In 2.5 km, turn right on a road that crosses the canal (km point 57 of the canal is a good reference point) and approach the Cubilar reservoir. From the dam (2), view the open water for Black-necked Grebe and Common Crane in winter and wildfowl such as Greylag Goose and Red-crested Pochard.

2. 60 km 🚗 ☀️ ➡️

From Navalvillar de Pela, take the N-430 towards Mérida and after 7 km, pick up the EX-355 to the right towards Madrigalejo. After just over 4 km, turn right to the village of Vegas Altas and its cemetery (3), some distance from the village. This stretch of road passes through rice-paddies used by Common Crane, duck, herons, gulls and large groups of Black-tailed Godwit in January and February.

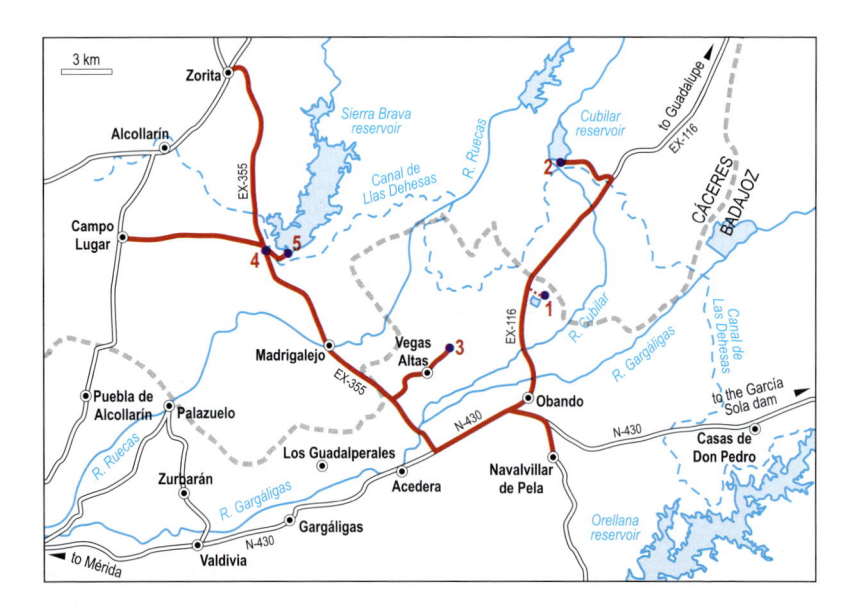

Back on the EX-355, continue north-west and turn right in 12 km (after passing through the village of Madrigalejo) along an asphalted track (4) to the dam of the Sierra Brava reservoir (5). This is a roosting site for tens of thousands of wildfowl, including Greylag Goose, in winter; as well, the two islands in the reservoir hold a large colony of Gull-billed Tern (up to 500 pairs, the highest figure for any site in Spain) and there is a Cattle Egret colony in a stand of holm oaks near the reservoir bank.

PRACTICAL INFORMATION

- **Access:** Navalvillar de Pela and the area of Vegas Altas is accessible from Trujillo and Miajadas, both near the Autovía de Extremadura (N-V), or from the N-430 from Mérida.
- **Accommodation:** in Navalvillar de Pela and, somewhat further, Don Benito and Villanueva de La Serena. The hostal Acueducto Gran Ruta (tel. 924 825 018) in Acedera is often used by birdwatchers.
- **Visitor infrastructure:** an interpretation centre with tower hides for watching the movement of the cranes is to be opened soon near La Charca de Gorbea (Moheda Alta).
- **Visiting tips:** best from November to April and, above all, winter for the abundance of water birds and cranes. Summer is unbearably hot. Take care along dirt tracks after rain.
- **Recommended:** the NGO Naturex (tel. 924 866 711, naturex@terra.es) provides specialist guides and organises ornithological work camps and courses.
- **Further information:** maps 1:50,000, n°. 731, 732, 754 and 755 (SGE).

EXTREMADURA

Return again to the EX-355 and after just a few metres, turn left to Campo Lugar (9 km) or continue along the EX-355 to Zorita (11 km). Both options are good for Great Bustard and possibly also for Little Bustard and Black-bellied Sandgrouse; Montagu's and Marsh Harriers, here sharing habitat in the cereal fields, are regularly seen flying over the area.

MOST REPRESENTATIVE SPECIES

Residents: Cattle and Little Egrets, Black-winged Kite, Marsh Harrier, Little and Great Bustards, Stone Curlew, Black-bellied Sandgrouse, Calandra Lark, Crag Martin, Red Avadavat, Common Waxbill.
Summer visitors: Little Bittern, Montagu's Harrier, Lesser Kestrel, Black-winged Stilt, Collared Pratincole, Gull-billed Tern, Roller, Short-toed Lark, Black-eared Wheatear.
Winter visitors: Black-necked Grebe, Great Cormorant, Greylag Goose, Red-crested Pochard and other duck,

Hen Harrier, Common Crane, European Golden Plover, Northern Lapwing, Black-headed and Lesser Black-backed Gulls.
On migration: Black Stork, Osprey, Avocet, Black-tailed Godwit, Eurasian Curlew and other waders.

83. MÉRIDA

The river Guadiana passes through the city of Mérida and its gently flowing waters and islands covered with verdant woodland add a touch of charm to the city centre. Downstream beyond the confluence with the river Aljucén, the Guadiana is controlled by the small reservoir of Montijo, from where issue a number of canals that water the irrigation projects of Las Vegas Bajas. In the surrounding area, industry and food-processing plants are widespread, although there are also interesting natural sites such as the reservoir of Los Canchales and the Cornalvo Natural Park, both classified as SPAs. The rest of the area, however, lacks any legal protection.

La Oliva de Mérida © Eduardo de Juana

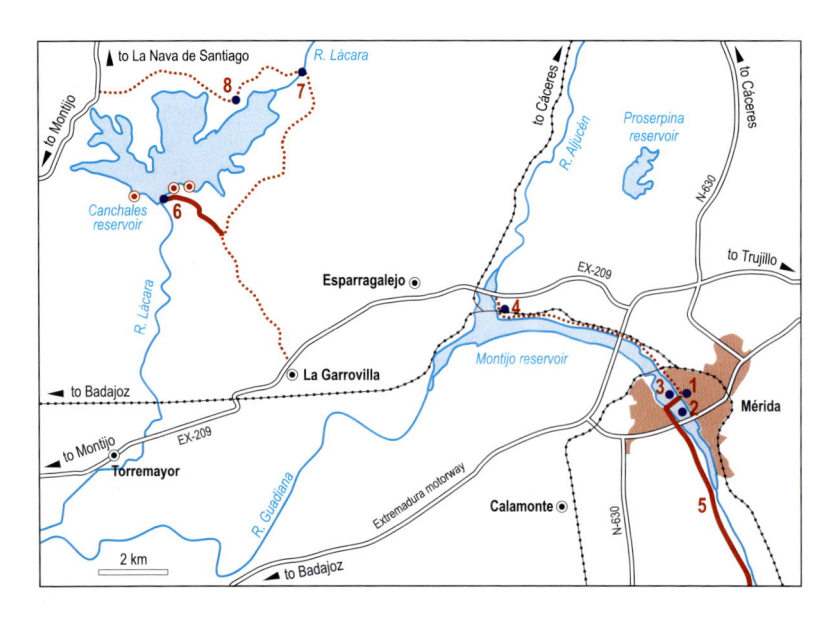

ORNITHOLOGICAL INTEREST

The river Guadiana harbours a number of colonies of Cattle Egret, which include a few Night Herons and Little Egrets, while at the mouth of the river Aljucén there is 20-pair strong Purple Heron colony. Other breeders along the river include Little Bittern, escaped passerines such as Red Avadavat and Common Waxbill and warblers. The reservoir of Los Canchales has colonies of Collared Pratincole and Gull-billed Tern, Eurasian Spoonbill is regular on passage and has even bred, and Greylag Goose and Common Crane both winter. Mérida city is full of White Storks and Lesser Kestrels and the surrounding plains are good for Black-winged Kite.

ITINERARIES

1. 20 km

In Mérida you can walk along either bank of the Guadiana from the Roman Bridge (1). From the bridge itself – home to breeding Common, Pallid and Alpine Swifts – look upstream (2) to an islet with a mixed heronry dominated by Cattle Egrets. In winter the island is used by the egrets and vast numbers of starlings as a roost site. On the other side of the bridge, another island (3) is frequented in winter by Great Cormorants.

Downstream from the bridge a track runs between the river and the railway and reaches Aljucén railway station (4). From here continue up the river Aljucén as far as the road to Montijo and look for Marsh Harrier, Night and Purple Herons and Cattle and Little Egrets.

All along the bank of the Guadiana the stands of fluvial woodland (*sotos*) are good for breeding Golden Oriole and Black Kite, while the river banks and stands of bulrush have Little Bittern, Reed and Great Reed Warblers in spring and summer, and Bluethroat, Penduline Tit and Reed Bunting in winter. Throughout the year Kingfishers are seen, as are Red Avadavat and Common Waxbill. The same species can be observed from the asphalted pedestrian walk (5)

EXTREMADURA

heading upstream from the Roman Bridge along the left-bank of the river.

2. 20 km 🚗 ☀ ➡

From the village of La Garrovilla, a track (asphalted at the end) heads for the reservoir of Los Canchales. The asphalt ends at the dam (6), where three hides have been built on the nearby shore. Nevertheless, the point at which the river Lácara enters the reservoir in the north is the best spot for birds. Pick up a path off to the right from just before the dam and head north with the reservoir on your left.

After the bridge (7) across an interesting stretch of river, the path reaches a good point (8) for viewing the wader and tern colonies on the artificial islands installed by the Confederación Hidrográfica del Guadiana with expert assessment from the University of Extremadura. Return along the same path or continue to the road from La Nava de Santiago to Montijo.

Groups of Eurasian Spoonbill pass through Los Canchales on migration, along with waders such as Little Stint, Dunlin, Ruff, Black-tailed Godwit, Common Redshank and Greenshank. Less regular are Great Egret and Osprey. Large numbers of White Stork build up as the breeding season finishes and winter sees the arrival of Great Cormorants, duck and gulls and the establishment of a Common Crane roost. Black-winged Kite and Montagu's Harrier are seen in the area around the reservoir.

MOST REPRESENTATIVE SPECIES

Residents: Cattle and Little Egrets, Black-winged Kite, Marsh Harrier, Stone Curlew, Kingfisher, Penduline Tit, Red Avadavat, Common Waxbill, Cirl Bunting.

Summer visitors: Little Bittern, Night and Purple Herons, White Stork, Black Kite, Montagu's Harrier, Lesser Kestrel, Black-winged Stilt, Collared Pratincole, Little and Gull-billed Terns, Alpine and Pallid Swifts, Reed and Great Reed Warblers, Golden Oriole.

Winter visitors: Great Cormorant, Greylag Goose and other wildfowl, Red Kite, Common Crane, Black-headed and Lesser Black-backed Gulls, Bluethroat, Common Starling, Reed Bunting.

On migration: Great Egret, Eurasian Spoonbill, Osprey, waders, Black Tern.

PRACTICAL INFORMATION

- **Access:** Mérida lies at the junction of the Autovía de Extremadura (N-V) and the N-630, the road that runs north-south through western Spain. From km point 341 of the Autovía or from Mérida itself, take the EX-209 towards Montijo for La Garrovilla (itinerary 2).
- **Accommodation:** lots of hotels in Mérida.
- **Visitor infrastructure:** except for the hides at Los Canchales, no visitor centre or infrastructure for birdwatchers.
- **Visiting tips:** all year, but avoid summer (too hot).
- **Recommended:** 15 km north of Mérida lies the Natural Park that protects the Roman reservoir of Cornalvo and the surrounding *dehesas*. The information centre is in the village of Trujillanos and there are a number of marked itineraries.
- **Further information:** maps 1:50,000, n°. 776 and 777 (SGE).

GALICIA

84. Ría de Ortigueira
85. Costa da Morte
86. Ría de Arousa
87. Islas Atlánticas

Galicia at its wildest and most natural is best sought along its long convoluted granite coastline, frequently battered by the worst Atlantic gales and periodically scoured by ecological disasters such as the Prestige oil spill of November 2002.

A succession of cliffs and sandy beaches with many small groups of offshore islands represents the typical Galician seascape, which may remind some more of Ireland or Scotland than the Iberian Peninsula. The most singular geomorphological features are the *rías*, valleys of tectonic origin that have been drowned by a rise in sea-level and in some cases moulded by the action of the rivers.

Inland Galicia consists of flat land separated by low, rounded mountains; only in the south-east of this autonomous community on the frontier with León and Portugal in Los Ancares do mountain ranges reach a certain height. Population densities are still relatively high as many people still live in the many small villages scattered throughout the region. The climate is humid and mild, and the vegetation is dominated by a mosaic of small-scale cultivation (crops and hay meadows) and pine and eucalyptus plantations, which have replaced the original forest cover and brought about a considerable impoverishment of bird populations.

The sites of greatest interest for birdwatchers are thus all along the coast. The sea-cliffs on islands and the mainland harbour good populations of seabirds, of which the most important are the only Spanish colonies of Kittiwake and Guillemot on Cabo Vilán and the Islas Sisargas.

The prominent position of the Galician coastline makes it an ideal place for seawatching for shearwaters, gannets, scoter, skuas, terns and auks passing close to the Galician mainland, above all in summer and autumn. The estuaries, on the other hand, in the depth of the *rías* are home to many birds in winter and provide shelter to nationally important numbers of waders such as Oystercatcher, Grey Plover, Sanderling, Bar-tailed Godwit, Eurasian Curlew, Common Redshank and Greenshank on passage.

Kittiwake   Carlos Sánchez © nayadefilms.com

A number of spots, including some of the beaches and coastal lagoons, have become famous for regularly playing home to groups of rare birds – mainly divers, sea ducks and gulls – from the north. As well, an unusually high number of North American vagrant birds such as Pectoral Sandpiper turns up, in the main blown onshore by Atlantic gales.

The Rías Altas, flanked by vertiginous cliffs, occupy the northern tip of the Galician coastline and the best sites here for birdwatchers are the intertidal flats of Ría de Ortigueira (84) and the nearby promontory of Estaca de Bares. Further west the rugged coastline of the grimly but accurately named Costa da Morte (85) also includes the gentler landscapes of the wetland at Baldaio, Ensenada de A Insua, Laguna de Traba, Cabo de Touriñán and Playa de Carnota, all musts for visiting birdwatchers.

The Rías Bajas lie between the provinces of A Coruña and Pontevedra and concentrate most of the Galician population and its industrial, commercial and tourist activity. Despite being the least well preserved part of the Galician coastline, many interesting habitats for birds still remain.

This is the case of the broad Ría de Arousa (86), where the *ensenada* (shallow bay) of O Grove has come to be recognised in recent years as the best site for waders in north-east Iberia. Other interesting sites near here include the Corrubedo Natural Park and, further south, the estuary of the river Miño on the border with Portugal.

A visit to Galicia in search of wildlife would be incomplete without an excursion to the Islas Cíes. Lying off the Ría de Vigo, these islands, the main group within a loose archipelago of islands known as the Islas Atlánticas (87), have been declared a national park. One of the main attraction for the naturalist here are the colonies of Shag and Yellow-legged Gull, which are some of the largest in the whole of Western Europe.

84. RÍA DE ORTIGUEIRA

In the extreme north of the province of A Coruña the estuary of the river Mera opens out into the broad Ría de Ortigueira, connected to the secondary Ría de Ladrido. This site contains some of the most important salt-marshes and mudflats in northwest Spain and has been declared a Ramsar Site and an SPA. Sandy beaches and dune systems surround the *ría* and within lie several islands. Further east, Estaca de Bares, the northernmost point of Spain, juts out into the Atlantic; this scrub-covered promontory is flanked by stunning sea-cliffs and has been declared a *Sitio de Interés Nacional*.

ORNITHOLOGICAL INTEREST

Famous for its water birds and, above all, waders on migration and in winter, the Ría de Ortigueira is with Santoña (Cantabria) the most important wintering site for Eurasian Curlew in Spain. There are also important build-ups of Oystercatcher, Bar-tailed Godwit, Common Redshank and Greenshank, as well considerable numbers of Eurasian Wigeon and Pintail in winter.

Many Balearic Shearwater, Northern Gannet, skuas, terns, auks and

Ría de Ortigueira © Fernando Barrio

GALICIA

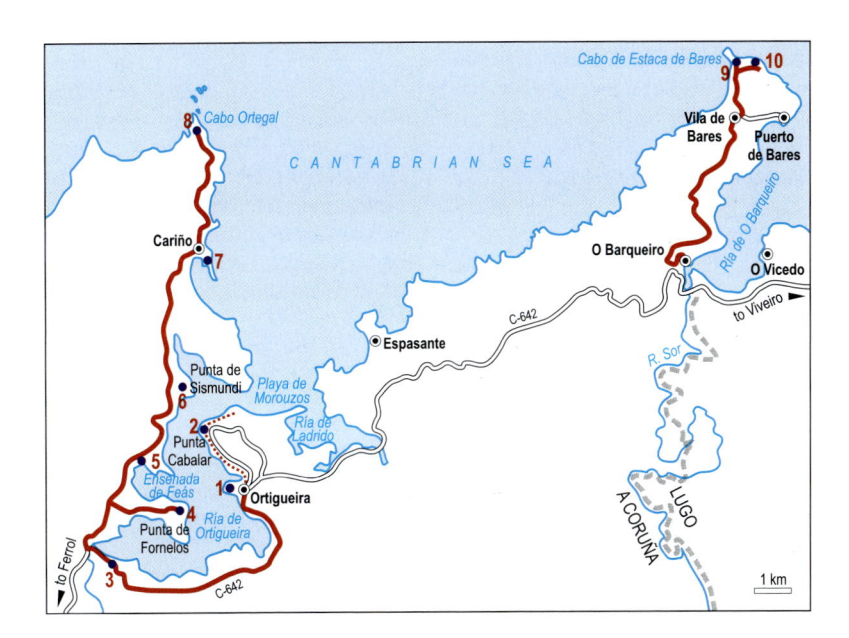

other seabirds are seen offshore during passage and the cliffs and islets hold breeding populations of European Storm-petrel, Shag and Yellow-legged and, occasionally, Lesser Black-backed Gulls.

ITINERARIES

1. 25 km 🚗 🚶 ☀ ▬

Good views of the whole *ría* can be had from the port of Ortigueira (1) and low-tide is ideal for looking for feeding waders. As well, winter brings herons, duck and the possibility of a Red-necked Grebe.

A 3-km long footpath runs along the eastern side of the *ría* from the town of Ortigueira to the Playa de Morouzos. As the tides rise and fall, Oystercatchers, Eurasian Curlews, godwits, plovers and other waders are in constant motion to and from their roosting sites. Divers, scoters, gulls and Eurasian Spoonbill on migration penetrate the estuary and can be seen from Punta Cabalar (2), halfway along the path and the best place to stop and observe the estuary.

Similar species can be observed by taking the road from Ortigueira towards Ferrol and turning right towards Cariño. Very soon, you should park opposite the hotel-discotheque La Ría (3) and scan the mudflats of the *ría* at low tide for Common Redshank, Greenshank, other waders and gulls. Other viewing points include Punta de Fornelos (4), accessible along an asphalted track, Ensenada de Feás (5), opposite the port of Ortigueira and Punta de Sismundi (6) to the north.

It is worth carrying on to Cariño and its port, beyond the *ría*, where many species of seabirds such as Great Northern and, less frequently, Red-throated Divers, Common Scoter, Common and Great Black-backed Gulls and Razorbill take refuge. The breakwater has regular Purple Sandpiper.

From Cariño a splendid asphalted track continues along the coast to Cabo Ortegal and its lighthouse (8). Aside from being a good seawatching site, the Peregrine Falcons that feed in the *ría*, breed on the cliffs here and there are colonies of Shag and Yellow-legged Gull on the offshore islands.

2. 8 km

Almost any promontory on the northern Galician coast is good for seawatching in autumn, especially after westerly or north-westerly gales. Nevertheless, during the season be prepared for the good days with strong passage of seabirds, as well as the birdless days in which the seas seem all but devoid of birdlife.

Bearing this in mind, Estaca de Bares is *the* seawatching site in Spain. Take the minor road north from O Barqueiro towards the port of Bares and turn left after 5 km just after the village of Vila de Bares and head for the lighthouse (9). Park and head on foot to the point itself.

Nevertheless, those who know the area best prefer a site less popular with tourists: just before the lighthouse, turn right and park on the far side of an abandoned American communication base. Walk down to the small building (10) that operates as a hide (normally closed) and view from wherever you please.

The Northern Gannet is the commonest species on passage and daily totals can be spectacular. From August to October the maximum numbers of shearwaters, scoter, skuas and terns are recorded; from November onwards groups of Little Gulls and Kittiwakes, Guillemots, Razorbill and even Puffins appear.

MOST REPRESENTATIVE SPECIES

Residents: Shag, Peregrine Falcon, Oystercatcher, Lesser Black-backed and Yellow-legged Gulls.
Winter visitors: Red-throated and Great Northern Divers, Red-necked Grebe, Little Egret, Grey Heron, Eurasian Wigeon, Common Teal, Pintail, Northern Shoveler, Common Scoter, Grey Plover, Knot, Sanderling, Purple Sandpiper, Dunlin, Black-tailed and Bar-tailed Godwits, Eurasian Curlew, Whimbrel, Common Redshank, Greenshank, Common and Great Black-backed Gulls, Sandwich Tern.
On migration: Sooty, Manx and Balearic Shearwaters, Northern Gannet, Eurasian Spoonbill, Pomarine, Arctic and Great Skuas, Little Gull, Kittiwake, Common Tern, Razorbill, Guillemot, Puffin.

PRACTICAL INFORMATION

- **Access:** The C-642 from Ferrol reaches Ortigueira in just over 50 km and passes near Espasante (itinerary 1) and O Barqueiro (itinerary 2).
- **Accommodation:** in Ortigueira, O Barqueiro and O Vicedo.
- **Visitor infrastructure:** the hide at Estaca de Bares is for scientific use only.
- **Visiting tips:** best between October and March for wintering and passage birds; from August to January seabirds on migration. Warm and wind-proof clothing is required, even at times in summer.
- **Recommended:** La Sociedade Galega de Historia Natural (tel. 981 584 426, sghn@sghn.org) and La Sociedade Galega de Ornitología (sgosgo@sgosgo.es) organises birdwatching trips to various sites in Galicia.
- **Further information:** maps 1:50,000, n°. 1, 2, 7 and 8 (SGE).

GALICIA

Punta Roncudo, Costa da Morte © Pedro Retamar

85. COSTA DA MORTE

This abrupt coastline (80 km) in the west of A Coruña province, exposed to the full fury of the Atlantic gales, consists of a continuous line of cliffs with a number of offshore islets such as Las Islas Sisargas. The cliffs are broken up by a number of beaches, often backed up by dune cordons protecting saline lagoons and small estuaries. Most of the coast between Baldaio and Cabo Vilán (the latter also a *Sitio de Interés Nacional*) and including Las Islas Sisargas are an SPA.

ORNITHOLOGICAL INTEREST

Las Islas Sisargas and Cabo Vilán hold Spain's only Kittiwakes (about 20 breeding pairs) and Guillemots (all but extinct). European Storm-petrel and Shag breed on cliffs and islands and other seabirds pass by on passage. A surprisingly good number of waders winter on or migrate through the beaches and estuaries and nowhere in Spain can beat this site's records of Nearctic rarities blown onshore by Atlantic gales. Playa de Carnota holds the best Galician population of Kentish Plover.

ITINERARIES

1. [1 km]

Halfway between A Coruña and Malpica lies Baldaio (1), a small wetland behind a large beach with a notable dune system. This is a good site for migrant waders and, to a lesser extent, seabirds: Knot and Black-tailed Godwit in spring and wintering groups of Great Northern Diver offshore. To reach this protected area, head for the nearby beach of Pedra do Sal from the villages of Arteixo or Carballo.

2. [30 km]

Between the towns of Corme and Laxe the small but wide estuary of the river Anllóns (also known as Ensenada de A Insua) harbours wintering groups of Oystercatcher and other waders, which feed on mud and sand exposed at low tide. Autumn storms bring vagrant American birds to the site, the most common of which is the Pectoral Sandpiper.

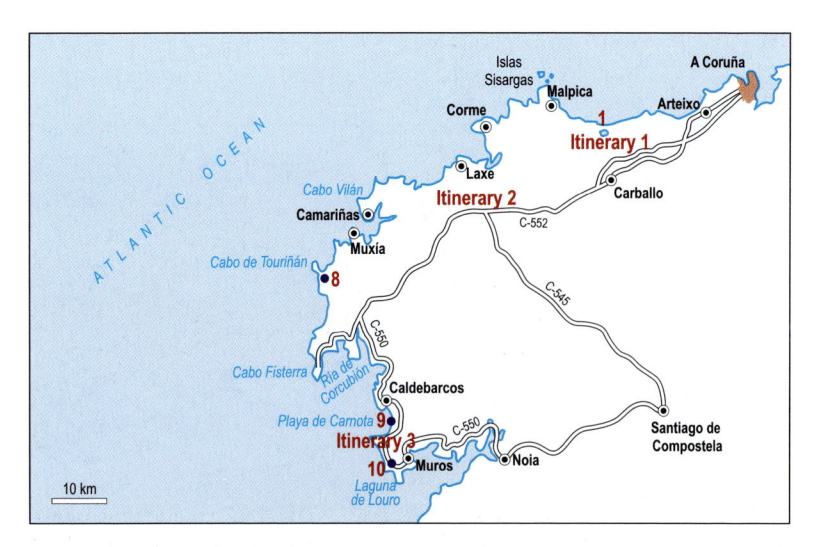

On the road from Ponteceso to Corme near the village of Cospindo (1 km) an asphalted track (2) heads off left to the village of Currás and drops down to the northern shore (3) of the *ensenada*. It is best to walk this route, stopping at your leisure.

The southern shore of A Insua can be reached from the road between Ponteceso and Laxe. A first good spot is the salt-marsh (4) behind the bridge over the river Anllóns. Two kilometres further on, a turn-off to the football pitch of As Redondas takes you down to a small beach (5), from where there are good views over the central part of the estuary. Two more kilometres further on, stop at the As Grelas viewpoint (6), an ideal site for scanning the tip of the sand-bar for birds at high-tide.

Laguna de Traba is another site well-known to ornithologists for its rarities. After passing near Laxe on the road to Camariñas, about 5 km further on the right-turn (7) towards the hamlet of Mórdomo passes near this wetland. Water Rail and Reed Bunting breed; winter sees the lagoon full of diving duck and waders and gulls in the outlet of the lagoon into the sea. Richard's Pipits have been recorded from the surrounding area.

Another worthwhile site for sea-watching and one of the most pleasant such spots in Galicia is Cabo de Touriñán (8), south of Muxía.

3. 3 km 🚗 🚶 ☀ ➡

The largest beach (7 km) in Galicia – Carnota (9) – lies in the extreme

GALICIA

> ## PRACTICAL INFORMATION
>
> - **Access:** the C-552 between A Coruña and Fisterra, joined by the C-545 from Santiago de Compostela, is the main communication axis to the site and provides access to Baldaio and to the Ría de Corme-Laxe. The C-550 follows the coast from Fisterra to La Guardia and goes near the Playa de Carnota; alternatively, from Santiago take the C-543 to Noia.
> - **Accommodation:** from north to south: A Coruña, Malpica, Ponteceso, Laxe, Camariñas, Muros and other towns in the area.
> - **Visitor infrastructure:** no visitor centre or infrastructure for birdwatchers.
> - **Visiting tips:** best from September to March, wintering and migrant seabirds and water birds; Atlantic storms blow rarities on to the coast. Wellington boots and wind-proof clothing can come in handy.
> - **Further information:** maps 1:50,000, n°. 43, 44, 67, 68 and 119 (SGE).

south of the Costa da Morte facing the Ría de Corcubión. Behind its dune cordon sits a complex of lagoons and tidal flats divided in two by a natural outlet to the sea. The northern sector, accessible from the village of Caldebarcos, is the most attractive for birdwatchers and produces a wide range of waders at low tide in winter and during migration periods.

The offshore waters are frequented by Great Northern and Black-necked Divers, Shag and, on occasions, large numbers of Common Scoter. A walk along the beach will give good views of Oystercatcher, Sanderling and Sandwich Tern.

South of Carnota towards Muros, the C-550 passes next to the Laguna de Louro (10) just before the village of the same name. Separated from the sea by a dune cordon and surrounded by reed- and rush-beds, this is a good site for diving duck, as well as being a place of great beauty.

MOST REPRESENTATIVE SPECIES

Residents: Shag, Peregrine Falcon, Water Rail, Kentish Plover, Lesser Black-backed and Yellow-legged Gulls, Reed Bunting.

Winter visitors: Black-throated and Great Northern Divers, Great Cormorant, Common Scoter, Common Pochard, Tufted Duck, Oystercatcher, Grey Plover, Knot, Sanderling, Dunlin, Black-tailed and Bar-tailed Godwits, Eurasian Curlew, Whimbrel, Greenshank, Turnstone, Sandwich Tern, Razorbill, Richard's Pipit, Snow Bunting.

On migration: Sooty, Manx and Balearic Shearwaters, Northern Gannet, Eurasian Spoonbill, Pectoral Sandpiper and other waders, skuas, Common Tern, Aquatic Warbler.

86. RÍA DE AROUSA

The main *ría* in Galicia is Arousa, on the border between A Coruña and Pontevedra provinces. The most interesting sites for birdwatchers in the southern sector are grouped around Ensenada de O Grove, a shallow bay separated from the sea by the isthmus of La Lanzada that links the small O Grove peninsula to the mainland. The Isla de La Toja lies within the bay. This is a Ramsar Site and an SPA, as is the Corrubedo Natural Park in the north-west on the A Coruña side of the *ría*.

Ría de Arousa   © José Manuel Reyero

ORNITHOLOGICAL INTEREST

This complex of salt-marshes, saline pools and beaches around the Ensenada de O Grove is one of the main sites in Galicia for seabirds and water birds, both on migration and in winter. Overall, it is unparalleled in the region in terms of waders, with the best populations in Spain of Oystercatcher, Grey Plover, Sanderling and Bar-tailed Godwit. The coastal sands are breeding grounds for Kentish Plover and, at Corrubedo, for Stone Curlew. Good numbers of Eurasian Spoonbill pass through and some stay all winter; Great Northern Diver and Balearic Shearwater frequent the open sea.

ITINERARIES

1. 25 km 🚗 🚶 ☀️ ➡️

Ensenada de O Grove is the best site in the Ría de Arousa for migrant and wintering water birds. After passing through Vilagarcía de Arousa and Cambados, the C-550 provides good access to the south-eastern sector of the site. At the village of Vilalonga (1), turn right opposite a petrol station on a road that splits after one kilometre. Take the right fork to the hamlet of Fianteira (2), whose small jetty provides good views at low tide over mudflats with feeding waders and ducks.

The left fork takes you to the cement factory at Arnosa (3). Park at the entrance and walk along an asphalted track westwards past some abandoned gravel pits (Barreiras de Rouxique; kaolin extraction) frequented by Common Pochard and Tufted Duck. Just offshore are the islands of Os Leiros, covered in low vegetation and magnificent roosting sites for large numbers of Dunlin, as well as Grey Plover, both godwits and Whimbrel. Further out to sea the islets of Tourís and Marma rise out of the bay and provide refuge at high

GALICIA

tide for Great Cormorants, herons, Eurasian Spoonbill (often in good-sized groups), duck and waders, including Oystercatcher.

Continuing along the C-550, you cross the isthmus that links the O Grove peninsula to the mainland. On the eastern side of this narrow spit of land scan the salt-marshes of O Bao (4), good for Eurasian Wigeon and waders at low tide, from the two hides on the right of the road.

Between the two hides there is a roundabout, from where a turn-off heads left to the beach of La Lanzada, on the west of the isthmus. Check the sea for Great Northern and possibly Black-throated Divers, as well as Common Scoter, Sandwich Tern and Razorbill. Northern Gannet and Balearic Shearwaters can be seen along with other seabirds on passage.

The last part of the C-550 before the village of O Grove provides further views over the O Bao salt-marshes from a hide at Punta Surmuiño (6).

2. 20 km

Near Ribeira (Santa Uxía) in the northern sector of the *ría* at the end of the VRG-11 a road leads off to the village of Aguiño and the beach of Vilar, passing through the Corrubedo Natural Park and past its visitor centre (7). A marked itinerary heads southwards to the freshwater Laguna de Vixán, a good site for Great Cormorants, Eurasian Coot and gulls. In winter Marsh Harriers and Reed Buntings frequent the reed-beds and Richard's Pipits have been recorded from the surrounding pastures and beaches.

An itinerary heading north takes you to the Carregal salt-marshes, connected to the sea by a natural channel. Waders including Eurasian Curlew and Oystercatcher move around the marshes and the beach and in spring passage Bar-tailed Godwit and Whimbrel are regular. Stone Curlew breed in the dunes, although they are easier to find in

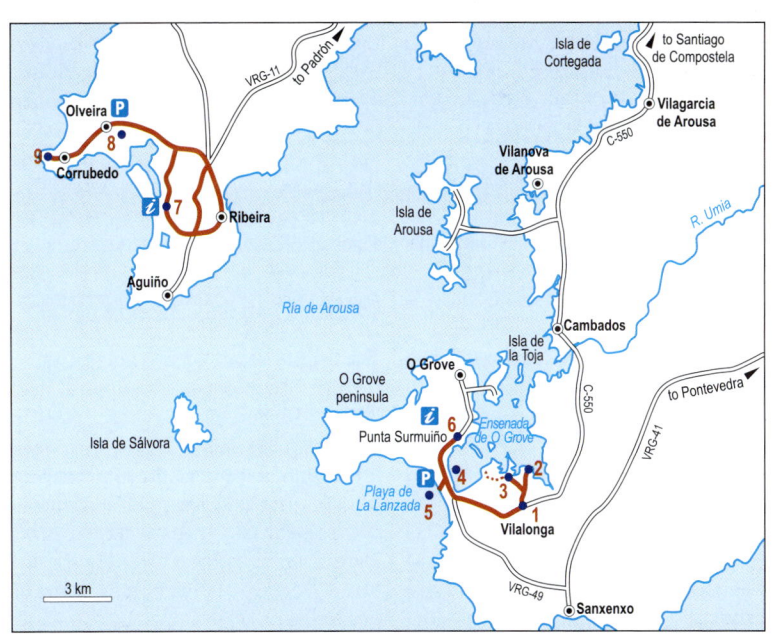

> **PRACTICAL INFORMATION**
>
> - **Access:** from Pontevedra via Sanxenxo; from Santiago de Compostela, the C-550 goes to O Grove (itinerary 1); from Padrón, the VRG-11 reaches Ribeira (itinerary 2).
> - **Accommodation:** many places to stay in O Grove. For Corrubedo, try Ribeira or further away, Noia, Rianxo or even Santiago de Compostela.
> - **Visitor infrastructure:** the Siradella interpretation centre (tel. 986 805 469) near O Grove provides information about the local environment. In summer open every day, morning and afternoon; the rest of the year, Monday to Friday, morning only. The Corrubedo Natural Park has a visitor centre (tel. 981 878 532) known as Casa da Costa: open every day, morning and afternoon.
> - **Visiting tips:** best from September to March. The tides greatly affect where birds will be. Wellington boots may be necessary in some areas.
> - **Recommended:** from the port of O Grove, boat trips visit the Ría de Arousa.
> - **Further information:** maps 1:50,000, n°. 151, 152, 184 and 185 (SGE).

winter. Off shore, Great Northern Diver, Northern Gannet and Razorbill are not rare.

From the road between Ribeira and Corrubedo a left turn just before the village of Olveira takes you to a car-park (8) next to the enormous mobile dune that closes off the northern part of the park. In autumn, it is worth continuing on to Corrubedo and climbing up to the nearby lighthouse (9) to seawatch: best with winds from the west.

MOST REPRESENTATIVE SPECIES

Residents: Little Grebe, Gadwall, Water Rail, Stone Curlew.
Summer visitors: Kentish Plover, Great Reed Warbler.
Winter visitors: Black-throated and Great Northern Divers, Great Cormorant, Shag, Eurasian Wigeon, Common Teal, Common Pochard, Tufted Duck, Greater Scaup, Common Scoter, Oystercatcher, Grey Plover, Knot, Sanderling, Dunlin, Black-tailed and Bar-tailed Godwits, Eurasian Curlew, Whimbrel, Common Redshank, Greenshank, Turnstone, Sandwich Tern, Razorbill, Richard's Pipit, Reed Bunting.
On migration: Cory's, Manx and Balearic Shearwaters, Northern Gannet, Eurasian Spoonbill.

87. ISLAS ATLÁNTICAS

The three Islas Cíes, characterised by imposing west-facing cliffs and gentler beaches and dune systems on their eastern shores, lie at the mouth of the Ría de Vigo. Along with the Isla de Ons in the mouth of the Ría de Pontevedra, these islands constitute the main attractions of the Islas Atlánticas National Park; they are also an SPA.

ORNITHOLOGICAL INTEREST

The Islas Cíes hold the largest colonies of Shag (over 1,000 pairs) and Yellow-legged Gulls (over 20,000 pairs) in Western Europe; lesser numbers also breed on Isla de Ons. European Storm-petrel and Lesser-black Backed Gull breed in small numbers; the few Guillemots that once bred have recently disappeared.

GALICIA

Islas Cíes, Islas Atlánticas de Galicia National Park © Pedro Retamar

ITINERARIES

1. 4 km

One of the most attractive excursions in and around the Islas Cíes is the 6-nautical mile boat trip from the Ría de Vigo, during which it is easy to see Balearic Shearwater and Razorbill (above all on post-breeding migration) and other seabirds such as Northern Gannet, Skuas and terns. Large groups of Shag fish in the waters around the islands.

A number of walking routes start at the small harbour at the north of Playa de Rodas, the sandy beach that links the islands of Faro and Monte Agudo. The most interesting for birdwatchers is the 4-km uphill walk to the top of Monte Agudo along the abrupt wave-battered western coast. After passing by the Muxieiro dune system (2) and climbing slopes covered in scrub and eucalyptus plantations, you reach a junction (3). The path to the right heads to the lighthouse; the left-hand path is better for birds and takes you to a hide (4) that provides views of part of the Shag and Yellow-legged Gull colony. Adult and young (from June onwards) birds perch on the rocks next to the hide.

2. 60 km

If you have two or more days at your disposal, combine a visit to the Islas Cíes with a tour of the southern-most stretch of the Galician coast. From Vigo, take the C-550 through Baiona and onto Cabo Silleiro: the lighthouse here is a good seawatching spot for birds such as Balearic Shearwater and Northern Gannet almost all year round. Look out for Kittiwakes and other seabirds accompanying the fishing boats entering the Ría de Vigo.

Cabo Silleiro (5) also offers the chance to spot waders such as Bar-tailed Godwit and Whimbrel on spring passage. The rocky islands offshore are frequented by Shag and in winter are good sites for Purple Sandpiper and Turnstone.

Another alternative is to head for the Miño estuary on the border be-to look for Black-necked Grebe and Red-breasted Merganser. Take the road towards A Guarda and after 300 m a right-turn takes you to two hides (10), one right next to a camp-site and the other past the camp-site, overlooking the *ría*. These hides are excellent points for finding wintering birds such as divers, herons, duck, waders, gulls and, with luck, Osprey.

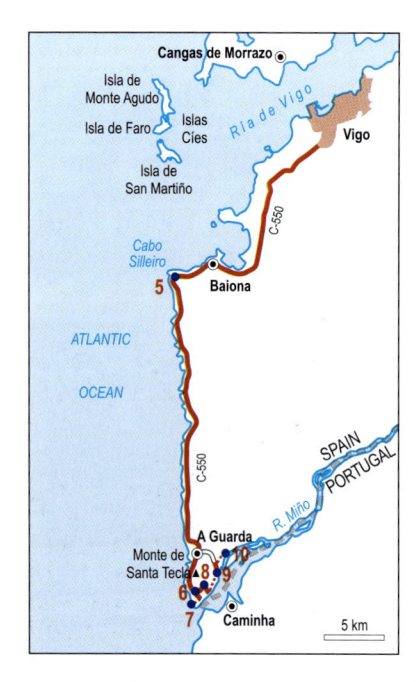

The reed-beds near the second hide are hunted over by Mash Harrier and have breeding Great Reed Warbler and Reed Bunting, as well as Bluethroat on passage. As a curiosity, this reed-bed has been colonised by two exotic species, Common Waxbill and Yellow-crowned Bishop.

tween Spain and Portugal. Thirty kilometres south of Baiona, head for A Guarda, from where a road heads around the western, seaward side of Monte de Santa Tecla. In 3 km, turn right (6) towards the beach of Camposancos: however, before reaching the beach, turn right again along a track that will take you to a rocky headland to the south (7). Here you should find Oystercatcher, Purple Sandpiper, Turnstone and Sandwich Tern in winter, as well as Great Cormorants moving along the coast and other seabirds on migration.

Return to the road and continue in the same direction as before as far as a turn-off (8) in 200 m to Playa de Armona, from where a track towards Pasaxe takes you along the shore of the *ensenada*. At low tide the mudflats hold herons, duck, gulls and waders such as Sanderling, Knot and Curlew Sandpiper.

The quaysides at Pasaxe (9) provide good views of the estuary and in winter are a good point from which

MOST REPRESENTATIVE SPECIES

Residents: Shag, Goshawk, Kentish Plover, Lesser Black-backed and Yellow-legged Gulls.

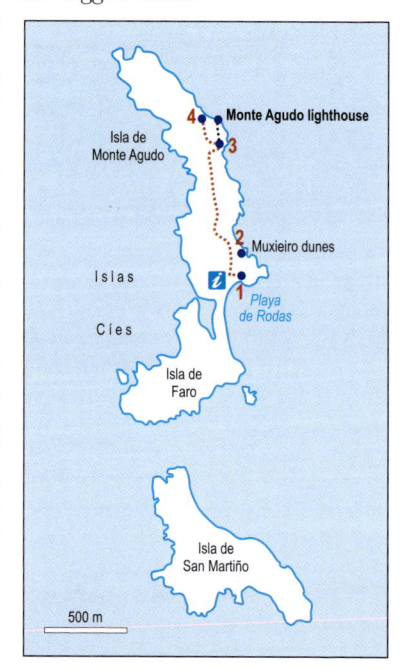

GALICIA

Winter visitors: Black-necked Grebe, Great Cormorant, Eurasian Wigeon, Gadwall, Common Teal, Tufted Duck, Red-breasted Merganser, Osprey, Oystercatcher, Grey Plover, Sanderling, Purple Sandpiper, Dunlin, Bar-tailed Godwit, Whimbrel, Eurasian Curlew, Greenshank, Turnstone, Great Black-backed Gull, Sandwich Tern.
On migration: Cory's, Manx and Balearic Shearwaters, Northern Gannet, Knot, Curlew Sandpiper, Great Skua, Kittiwake, Arctic and Common Terns, Guillemot, Razorbill, Richard's Pipit.

PRACTICAL INFORMATION

- **Access:** The Islas Cíes are only accessible by boat: Naviera Mar de Ons (tel. 986 225 272) provides a service from Vigo, Baiona and Cangas de Morrazo in summer, at Easter and on spring weekends. From Vigo via Baiona the C-550 follows the coast to A Guarda on the mouth of the river Miño.
- **Accommodation:** in Vigo, Baiona and A Guarda. On the Islas Cíes there is a camp-site (986 438 358), open at Easter and in summer. Reservation required.
- **Visitor facilities:** the National Park has a visitor centre on the islands that opens when the tourist boats arrive.
- **Visiting tips:** best in spring when the Shag and gull colonies are at their busiest. From August to March the estuary of the Miño and other coastal sites are good for wintering and migrating birds.
- **Further information:** maps 1:50,000, n°. 222, 223, 261 and 299 (SGE). Get in touch with the park offices (tel. 986 858 593) in Pontevedra. Open Monday to Thursday, morning and afternoon; Fridays only mornings. Also www.mma.es/parques/lared/islas_atlan.

LA RIOJA

| 88. Sierra de La Demanda | 89. Iregua, Leza and Cidacos |

La Rioja, lying at the intersection of the regions of Navarre, Castille and Aragon, is the smallest autonomous community in mainland Spain (a little over 5,000 km^2) and can be clearly divided into two discrete geographical regions of approximately the same size.

The northern and much flatter half of La Rioja is watered by the river Ebro and is home to most of the region's major towns – Haro, Logroño, Calahorra and Alfaro. Agriculture in the form of cereal cultivation, olive groves, irrigated crops and, naturally, the vineyards that make La Rioja famous world over predominates.

In this very humanised area of La Rioja only small patches of natural habitat remain, principally along the river Ebro; the sotos near the town of Alfaro are one of the best examples. Artificial habitats including reservoirs (La Grajera near Logroño, for example) offer added interest for birdwatchers, although in both cases similar habitats can be found in nearby Navarre and are dealt with in the chapter on that autonomous community.

The northern-most sector of the Sistema Ibérico lies in the southern half of La Rioja and consists of a well-wooded mountainous region – above all in the west – with a cool, moist climate. Its rural economy is based on extensive stock-raising, forestry and, more recently, tourism in the shape of the ski-station at Valdezcaray.

These mountains are the most attractive areas for the birdwatcher to explore. Ranges such as Los Picos de Urbión, the Sierra de Cebollera and, above all, the Sierra de La Demanda (88), all peaking at over 2,000 m, are true Atlantic islands in a sea of Mediterranean habitats, as the presence of beech forests near the southern edge of their European range demonstrates. Birds more commonly associated with northern Spain such as the Grey Partridge, here in one of its few breeding sites outside its Pyrenean and Cantabrian bastions, and Woodcock, Eurasian Treecreeper, Marsh Tit, Red-backed Shrike, Bullfinch and Yellowhammer can all be found here.

The river valleys hurrying down northwards to the river Ebro provide an interesting contrast to the mountains to the south. Atlantic and Mediterranean influences mix and aridity increases the further east you go. Many limestone and conglomerate cliffs line the banks of these rivers and provide habitat for Griffon and Egyptian Vultures, Golden Eagle and Peregrine Falcon, as well as some of the last Bonelli's Eagles in the north of Spain. The Iregua, Leza and Cidacos valleys (89) offer a good selection of easily watchable cliff habitats.

Griffon Vulture © Jordi Bas

88. SIERRA DE LA DEMANDA

In the south-west of La Rioja (and also partly in Burgos province) rises the Sierra de la Demanda, the highest range in the northern part of the Sistema Ibérico. Peaking at over 2,000 m, these mountains constitute an island of Atlantic climate sandwiched between the continental climes of the Meseta to the south and the Mediterranean Ebro valley to the north. Some beech forests survive – here all but at the southern limit of the European distribution of this essentially Atlantic tree species – amidst the otherwise predominant Pyrenean oak and, at higher altitudes, pine plantations. Intensive grazing has transformed much of the forests into pasture lands that, nevertheless, are being invaded by scrub. On the southern slopes of La Demanda holm oak forests appear in valleys such as that of Najerilla. The whole area is partially protected by an SPA of almost 140,000 ha.

ORNITHOLOGICAL INTEREST

The most outstanding of the high-altitude bird communities of the site is the Grey Partridge, here in one of

Sierra de La Demanda © Pedro Retamar

its few breeding sites outside its Pyrenean and Cantabrian bastions. La Demanda is also home to species all but non-existent further south such as Woodcock, Eurasian Treecreeper, Marsh Tit and Bullfinch, all generally closely tied to beech forests, and Red-backed Shrike and Yellowhammer in more open areas. Raptor communities include Honey Buzzard in forested areas, Hen Harrier in the high scrub and Griffon Vulture, Golden Eagle and Peregrine Falcon on cliffs; Red Kites and Egyptian Vultures both breed but are rather scarce. Mountain passes sees a massive passage of Wood Pigeons and other migratory birds.

ITINERARIES

1. 40 km

The road from the town of Ezcaray climbs to the Valdezcaray ski-station and provides easy access to the north-facing slopes of La Demanda. The road rises through mixed forests with Atlantic forest bird species such as Marsh Tit and Bullfinch, as well as Honey Buzzard, Short-toed and Booted Eagles and Goshawk. After parking at the ski-station (1), walk up any of the tracks that cross the area: a good option is to follow the track that reaches the base of Pico de San Lorenzo (2), passing through pastures and scrub with Water Pipit, Northern Wheatear, Red-backed Shrike, Rock Bunting and Yellowhammer. Wallcreepers frequent rocky outcrops in winter and patches of woodland hold Citril Finch. Autumn sees large movements of Wood Pigeons heading south through the passes here and elsewhere in the area.

A similar but rather less transited option is to drive up the valley of the river Oja. The final stretch passes through lush beechwoods before the tarmac finishes at El Collado de la Cruz de la Demanda (3): from here continue on foot and look for the scarce Grey Partridge in the scrub. This is also the best site for Wood

LA RIOJA

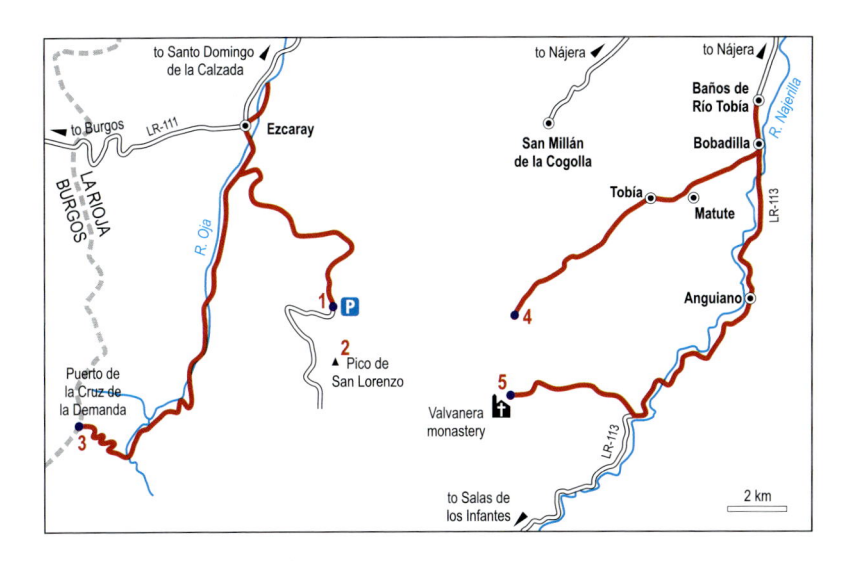

Pigeon passage in autumn, although birdwatchers have to compete with the many hunters who position themselves here. November is more peaceful and sees groups of Common Crane and Greylag Geese heading south.

2. 40 km 🚗 🚶 ☀ ━

The Tobía valley contains one of the best mature beech woods in the area and is an ideal place to look for Eurasian Treecreeper. Drive up the Najerilla valley along the LR-113 as far as Baños de Río Tobía. Two kilometres later, cross the river Bobadilla and head right along the road to the village of Tobía and its nearby conglomerate cliffs, home to Griffon and the here rare Egyptian Vultures, Golden Eagle, Peregrine Falcon and Eagle Owl. From here an 8-km asphalted track takes you to the picnic site at El Rajao (4), from where you can explore the surrounding beech forest.

Return to the LR-113 and continue through Anguiano (check the cliffs here) and 9 km later turn right along the road to the monastery of Valvanera, which passes through beech on north-facing and Pyrenean oak on south-facing slopes. From the monastery (5), you can walk the tracks along the mountain sides through patches of beech and Scots pine plantations with forest avifauna.

MOST REPRESENTATIVE SPECIES

Residents: Red Kite, Griffon Vulture, Hen Harrier, Goshawk, Golden Eagle, Grey Partridge, Woodcock, Eagle Owl, Crag Martin, Water Pipit, Dipper, Alpine Accentor, Song Thrush, Goldcrest, Firecrest, Eurasian Treecreeper, Marsh Tit, Red-billed Chough, Citril Finch, Common Crossbill, Bullfinch, Yellowhammer, Rock Bunting.

Summer visitors: Honey Buzzard, Short-toed and Booted Eagles, Tree Pipit, Northern Wheatear, Rufous-tailed Rock Thrush, Bonelli's Warbler, Pied Flycatcher, Red-backed Shrike.

Winter visitors: Fieldfare, Redwing, Brambling, Hawfinch.

On migration: Greylag Goose, Common Crane, Wood Pigeon.

PRACTICAL INFORMATION

- **Access:** between Burgos and La Rioja, the N-120 passes through Santo Domingo de la Calzada; from here a minor road leads in 14 km to Ezcaray. Continuing along the N-120, in Nájera take the LR-113 to the Tobía and Valvanera valleys.
- **Accommodation:** in Ezcaray, Santo Domingo de la Calzada, San Millán de la Cogolla and other towns in the area.
- **Visitor infrastructure:** no visitor centre or infrastructure for birdwatchers.
- **Visiting tips:** best in late spring and in autumn for passage migrants. Take care in high areas not to get lost in the fog and at Collado de la Cruz de la Demanda and other mountain passes when hunters are shooting Wood Pigeon.
- **Recommended:** in the south of La Rioja lies the Sierra Cebollera Natural Park: similar birds but with the addition of a few Bluethroat and Ring Ouzel. Access from the village of Najerilla via the village of Viniegra de Abajo along the LR-332 or from Soria and Logroño via the N-111 over the Puerto de Piqueras. There is an information centre in Villoslada de Cameros (tel. 941 468 216, sierra.cebollera@larioja.es). More information on www.larioja.org/cebollera.
- **Further information:** maps 1:50,000, n°. 240 and 241 (SGE). Also, try the La Rioja Department of Tourism, Environment and Territorial Planning (tel. 941 291 100, fauna.flora@larioja.org).

89. IREGUA, LEZA AND CIDACOS

A series of secondary valleys, cleaving open the northern slopes of the Sistema Ibérico as they drop down to the great river Ebro, are protected by two SPAs: the Iregua, Leza and Jubera valleys to the west and the Mediterranean and even partially semi-arid Cidacos and Alhama valleys further east.

A Biosphere Reserve of almost 120,000 ha has been declared to protect these fluvial corridors and the limestone and conglomerate cliffs that line stretches of the rivers. The vegetation consists predominantly of box scrub in the more Atlantic valleys, grading into genista, rock-rose and rosemary scrub in drier areas.

ORNITHOLOGICAL INTEREST

These valleys hold the best raptor populations in La Rioja and include important populations of Griffon Vulture, as well as Egyptian Vulture, Golden Eagle, Peregrine Falcon and Eagle Owl; one of the few remaining pairs of Bonelli's Eagle in La Rioja breeds in the Cidacos valley. Furthermore, there are occasional sightings of Lammergeiers and Black Vultures straying from their breeding populations in, respectively, the Pyrenees and central Spain. The warmer valleys contain appropriate habitats for Mediterranean passerine communities that include a good diversity of wheatears, warblers and buntings.

ITINERARIES

1. 12 km

Head for the village of Islallana along the N-111; from the bridge (1) over the river Iregua, walk along the right-bank of the river through a few houses and allotments in the shadow of the conglomerate cliffs of Peña La Cruz and Peña Sayón.

After a short while, the route leaves the river valley and climbs gently

LA RIOJA

Cidacos river in Iregua, Leza and Cidacos © Pedro Retamar

amongst the rock outcrops: look out for Griffon and Egyptian Vultures, Crag Martin, Blue Rock Thrush, Red-billed Chough and Rock Sparrow; in winter Alpine Accentor and Wallcreeper turn up. After 3 km you come to Viguera where the viewpoint of Peñueco (2) in the outskirts of the village is a good spot for Alpine Swift and for listening for Eagle Owl.

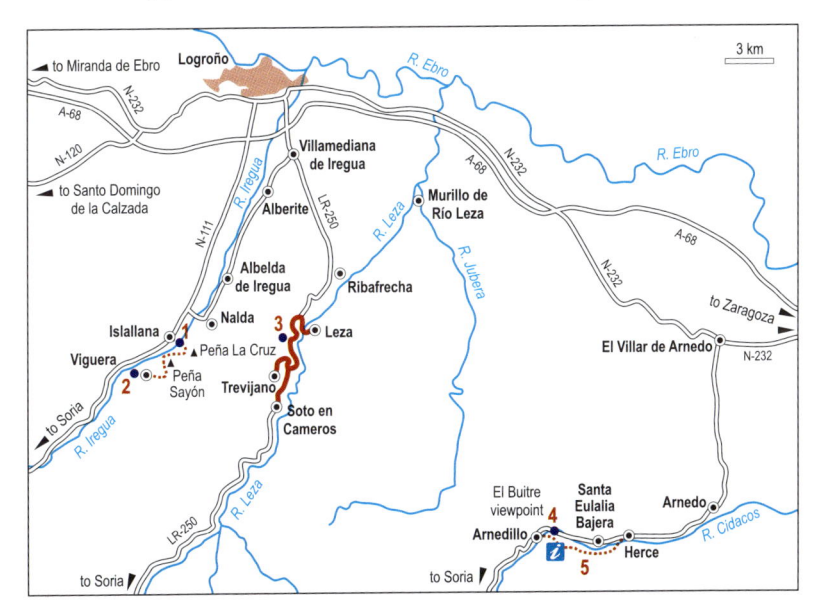

The Leza valley is also of interest: either return to Logroño along the N-111 or take minor roads through Nalda, Albelda de Iregua, Alberite and Ribafrecha to reach the road (3) between the villages of Leza and Soto en Cameros that passes through a spectacular limestone gorge. Just before the turn-off to Trevijano and two kilometres further on you can view the gorge from beside the road: look for raptors and Rufous-tailed Rock Thrush in summer.

2. 10 km 🚶 🌅 ➡

The Cidacos valley is at its most abrupt near the village of Arnedillo and can be viewed from the Mirador del Buitre, just 800 m away. From Arnedillo you can walk up to this viewpoint in an abandoned quarry or park just before the entrance to the village (coming from Arnedo) in a lay-by (4). Visit also the interpretation centre with a live video feed from a camera installed in the nearby Griffon Vulture colony (100+ nests).

The Mirador del Buitre lies on the Vía Verde del Cidacos, a bike lane (that can be walked) that follows the former course (5) of a mine railway. Walking towards the station of the village of Herce (8 km), the river and its fluvial woodland lies on your right and vineyards and olive and almond groves mixed in with patches of scrub on your right. Look out for Thekla Lark, Tawny Pipit, Black-eared Wheatear, Dartford and Sardinian Warblers, Rock Bunting and, with luck, Red-backed Shrike.

MOST REPRESENTATIVE SPECIES

Residents: Griffon Vulture, Golden and Bonelli's Eagles, Common Kestrel, Peregrine Falcon, Eagle Owl, Thekla Lark, Crag Martin, Dipper, Blue Rock Thrush, Black Wheatear, Dartford and Sardinian Warblers, Red-billed Chough, Rock Sparrow, Rock and Cirl Buntings.

Summer visitors: Egyptian Vulture, Short-toed and Booted Eagles, Alpine Swift, Red-rumped Swallow, Tawny Pipit, Black-eared Wheatear, European Bee-eater, Rufous-tailed Rock Thrush, Spectacled and Orphean Warblers, Golden Oriole, Red-backed Shrike.

Winter visitors: Alpine Accentor, Wallcreeper.

PRACTICAL INFORMATION

- **Access:** from Logroño the N-111 towards Soria reaches Islallana in the Iregua valley in 15 km. The LR-250 enters the Leza valley from the village of Ribafrecha. The N-232 towards Zaragoza passes through El Villar de Arnedo, where you turn off to Arnedo and Arnedillo in the Cidacos valley.
- **Accommodation:** in Logroño and in Arnedo and Arnedillo for the Cidacos valley.
- **Visitor infrastructure:** the Mirador del Buitre (tel. 941 394 226, oficina@valcidacos.es), run by the Arnedillo town council, has an interpretation centre devoted to cliff-loving birds and a specialised guiding service.
- **Visiting tips:** best in spring. Take great care not to disturb birds breeding on the cliffs.
- **Recommended:** many dinosaur footprints have been found in the valleys described in this chapter. The Cidacos valley has the most and in Enciso, 10 km from Arnedillo, there is a wonderful palaeontology centre (tel. 941 396 093) that provides information and guided visits.
- **Further information:** maps 1:50,000, n°. 204, 241, 242, and 243 (SGE). Also, try the La Rioja Department of Tourism, Environment and Territorial Planning (tel. 941 291 100, fauna.flora@larioja.org).

MADRID

90. Valle del Lozoya
91. Alto Manzanares
92. South-west Madrid
93. Valle del Jarama
94. South-eastern wetlands

Located in the centre of the peninsula, this single-province autonomous community is centred around the Spanish capital Madrid. Despite this exaggerated concentration of most of the regional population in the large Madrid conurbation (almost 5 million inhabitants) and the pressure it exerts, there are still plenty of well-preserved birdwatching habitats in the region, in part thanks to the survival until fairly recently of an emphasis on traditional land-uses such as stock-raising, cereal cultivation and forestry.

Madrid province ranges from peaks of over 2,000 m in the north to the lowlands of the Tajo valley in the south. In between a whole series of altitudinal gradients, combined with different relief-types, climates and land-uses, make for a notable diversity of habitats, thereby granting Madrid the privilege of possessing a small-scale compendium of most of the Iberian Peninsula's natural inland habitats.

Guadarrama and Somosierra are the main mountain chains in the region and provide Madrid, respectively, with its western and northern borders. Despite the urban development that has heavily transformed these ranges, the higher slopes still harbour excellent Pyrenean and Scots pine forests. The forests of the Lozoya valley (90), with its large Black Vulture colony that extends into neighbouring Segovia province, are probably the most attractive for the birdwatcher.

High altitude birds can be found around Peñalara which, at around 2,400 m, is the highest point of the province. The headwaters of this valley are rainy and are more reminiscent of the north of Spain, as is the bird life, with northern species such as Tree Pipit, Whinchat and Red-backed Shrike breeding here.

Another of the great natural wildlife corridors of the region is the upper Manzanares valley (91), rather warmer than Lozoya and with its Mediterranean influences patent in the excellent holm oak forests of its piedmont that extend down to Monte El Pardo, a former royal hunting estate. Higher up, the rock chaos of La Pedriza is home to a notable Griffon Vulture colony and the Santillana reservoir is a good site for wintering water birds.

The mountains south-west of Madrid (92) connect the Guadarrama and Gredos mountain chains. Plains

Spanish Imperial Eagle © Juan Martín Simón

covered with pine and holm oak forests, *dehesas* and scrub formations provide excellent habitats for typically Mediterranean bird communities, especially rich in raptors. This area is one of the most important in the country for the highly threatened Spanish Imperial Eagle.

Very different is the eastern part of the province around the south of the Jarama valley (93), where habitats otherwise non-existent in Madrid such as limestone cliffs, Lusitanian oak forests and dry cereal cultivation are the order of the day. Great Bustard and other steppe birds breed here and many large raptors come to feed.

Downstream the Jarama is joined by its tributaries the Henares, Manzanares and Tajuña and flows between, in places, impressive gypsum cliffs that provide home for a number of cliff-loving species. The attraction of the site for ornithologists is increased by the presence of a number of lagoons – generally abandoned gravel pits – such as those near the towns of Rivas-Vaciamadrid and Velilla de San Antonio that are all part of a site known as the South-east Wetlands (94), home to breeding Purple Heron, Marsh Harrier and Purple Swamp-hen and a large winter gull roost.

90. VALLE DEL LOZOYA

The eastern sector of the Sierra de Guadarrama peaks at Peñalara (2,430 m), the highest point in the province of Madrid and part of a Natural Park (almost 800 ha) that also protects a series of glacial lakes. With its dense forests of Pyrenean Oak and Scots pine and, above 1,800 m, its bare rocks, pastures and broom scrub, the Lozoya valley has a truly northern feel to it. At lower altitude an attractive mosaic of hedgerows, hay meadows, ash groves and well-preserved stretches of fluvial woodland takes over. The whole valley is protected by an SPA of over 7,800 m.

MADRID

Peñalara, Lozoya valley © Pedro Retamar

ORNITHOLOGICAL INTEREST

Lozoya is well-known for its important Black Vulture colony of over 60 pairs that breeds on Scots pines and other forest raptors such as Honey Buzzard and Red Kite. Furthermore, passerine communities are well represented by Rufous-tailed Rock Thrush and Bluethroat at altitude and by Tree Pipit, Whinchat, Song Thrush, Goldcrest and Red-backed Shrikes – all species here near the southern limit of their Atlantic distributions – in the wood-pasture-meadow mosaics at lower altitudes. Wood Pigeons and other migrants cross the mountain passes during migration periods.

ITINERARY

[40 km] 🚗 🚶 ☀ ➔

Puerto de Cotos (1) offers birdwatchers the chance to see numerous upland birds. One possibility is to walk along the track (2) to the Lagunas de Peñalara. First you enter a Scots pine forest that gradually opens out and grades into meadows and broom scrub. Once at the Arroyo de Peñalara, begin to climb this stream in the midst of a delightful alpine landscape up to the glacial corrie occupied by Laguna Grande. Alternatively, drive or walk up to the Valdesquí ski-station (3) along a 2-km road passing through pines and then broom scrub.

Along both routes, the edges of the pine forests are frequented by Citril Finch, while the scrub holds Dunnock, Northern Wheatear and Rock and Ortolan Buntings. Bluethroats are harder to find: search in areas of tall broom scrub near areas of damp meadows or pools (with abundant Water Pipits). The cliffs of Peñalara resound to the calls of Red-billed Chough and any rock outcrop or stone building – above all in Valdesquí – can have Rufous-tailed Rock Thrush. In winter, Alpine Accentors frequent the area.

From Cotos, the M-604 drops down towards Rascafría through magnificent pine forests and a few patches of Pyrenean oak. After about 10 km, between km points 32 and 31, turn left along an asphalted track to the viewpoint of Los Robledos (4). Park and enjoy the unforgettable views of Peñalara and the valley's pine forests, home to the Black Vulture colony.

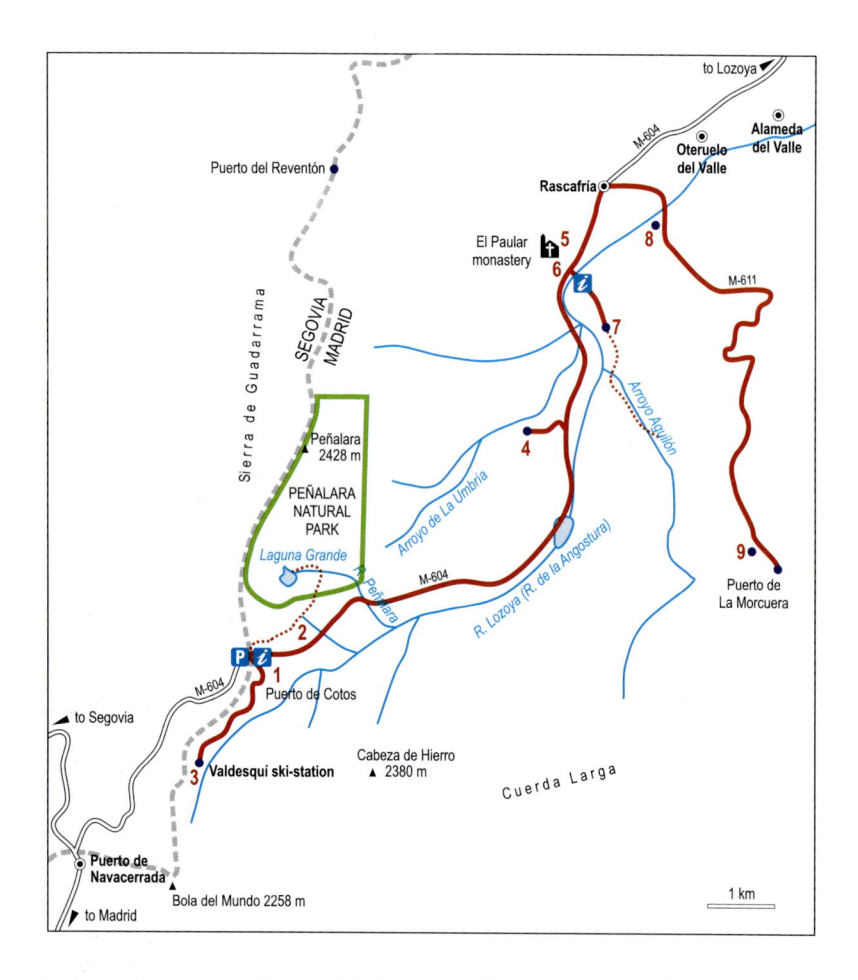

From this viewpoint it is easy to follow the daily movements of these large raptors and to spot other forest raptors such as Honey Buzzard and Booted Eagle. A walk in spring through the surrounding pine forest will be rewarded by sightings of Mistle Thrush, Firecrest, Goldcrest, Crested Tit and Common Crossbill.

Return to the M-604 and after 3.5 km opposite the monastery of El Paular (5), turn right along an asphalted track that takes you immediately to El Puente del Perdón (6), a bridge over the river Lozoya. Look for Dipper from the bridge and as you continue on through the oak forest as far as the Las Presillas (7) picnic site listen out for Tree Pipit, Bonelli's Warbler and Common Whitethroat. From the picnic site pick up a path that climbs up towards Puerto de La Morcuera along Arroyo Aguilón. The juniper bushes in between the oaks at the beginning of the climb are popular with thrushes, including Fieldfare and Redwing in winter.

Two kilometres on from El Paular you come to the village of Rascafría, where you should investigate the

MADRID

PRACTICAL INFORMATION

- **Access:** leave Madrid on the Autovía de A Coruña (A-6) and at Villalba, take the M-601 to Puerto de Navacerrada. From here the M-604 leads to Puerto de Cotos.
- **Accommodation:** in the Lozoya valley there is a good variety of rural accommodation, as well as the luxury hotel near the monastery of El Paular. Both Segovia and Madrid are relatively nearby.
- **Visitor facilities:** At Puerto de Cotos the Peñalara Natural Park has set up an interpretation centre, while at Puente del Perdón there is an environmental education centre (tel. 918 691 757) dedicated to the Lozoya valley that organises excursions and workshops. Open every day, morning and afternoon.
- **Visiting tips:** best once the snow has gone in late spring. Summer is also good for birds despite the massive human presence in the area. Ice and snow in winter make conditions tricky in higher areas.
- **Further information:** maps 1:50,000, nº. 483, 484, 508 and 509 (SGE).

mosaic of hay meadows, hedgerows and copses around the river Lozoya for Red-backed Shrike and Whinchat: try the fields and hedges (8) on the right-hand side of the road that climbs up to Puerto de La Morcuera.

Just before reaching this mountain pass, on the left-hand side of the road opposite a hostal (9) there is a large expanse of broom scrub with Bluethroat, as well as Tawny Pipit, Northern Wheatear, Rufous-tailed Rock Thrush and Ortolan Bunting. Both Griffon and Black Vulture and Golden Eagle are often seen flying over the area.

MOST REPRESENTATIVE SPECIES

Residents: Red Kite, Griffon and Black Vultures, Goshawk, Golden Eagle, Peregrine Falcon, Dipper, Dunnock, Song and Mistle Thrushes, Goldcrest, Firecrest, Crested Tit, Eurasian Jay, Red-billed Chough, Citril Finch, Common Crossbill, Cirl and Rock Buntings.
Summer visitors: Honey Buzzard, Black Kite, Short-toed and Booted Eagles, Wryneck, Skylark, Tawny, Tree and Water Pipits, Bluethroat, Whinchat, Northern Wheatear, Rufous-tailed Rock Thrush, Common Whitethroat, Bonelli's Warbler, Pied Flycatcher, Red-backed Shrike, Ortolan Bunting.
Winter visitors: Alpine Accentor, Redwing, Fieldfare.
On migration: Wood Pigeon, passerines.

91. ALTO MANZANARES

The river Manzanares drains a large sector of the south-facing slopes of the Sierra de Guadarrama, including the wooded plains of El Pardo near to the city of Madrid itself. The broad altitudinal range provides for a succession of habitats, starting with the high rock outcrops of the spectacular granite rock chaos of La Pedriza and continuing through Scots pine and Pyrenean oak forests, Mediterranean scrub, fluvial and ash woodland, holm oak forests and *dehesas* down to rough grazing and the Santillana reservoir.

A large percentage of the area is protected by the Cuenca Alta del Manzanares Regional Park (50,000 ha), a Biosphere Reserve that is also the largest protected area in Madrid prov-

La Pedriza, Alto Manzanares © Eduardo de Juana

ince. In the southern-most part of the river basin lie the SPAs of El Pardo and Soto de Viñuelas.

ORNITHOLOGICAL INTEREST

The largest Griffon Vulture colony in Madrid breeds on the cliffs of La Pedriza (around 100 pairs) and there are also interesting communities of forest raptors that include Red Kite. The holm oak woods in the south of the site (El Pardo and surrounding areas) hold a few pairs of Spanish Imperial Eagle and there are also large numbers of breeding White Stork. The Santillana reservoir holds one of the largest winter roosts of Black-headed and Lesser Black-backed Gulls in the interior of the Iberian Peninsula.

ITINERARIES

1. 15 km

The Santillana reservoir is one of the most important of all wetlands in Madrid if its rather variable water levels are appropriate. Best for birds is half-full and if so, stop next to the roundabout (1) opposite a camp-site halfway between Soto del Real and Manzanares.

On the other side of the road, an ash grove (2) is home to a large colony (100+ pairs) of White Storks and a number of nests can be seen from outside the fenced-off estate. Large numbers of storks – including the occasional Black Stork – gather on the reservoir shores after the breeding season is over.

Winter brings Black-necked Grebe and Greylag Goose to this section of the reservoir in winter, as well as abundant Great Cormorant, duck, Eurasian Coot and gulls. Migration time sees the arrival of good numbers of waders and the occasional Eurasian Spoonbill, Osprey and Common Crane.

Similar species can be seen from the hide placed just before the entrance into Manzanares El Real next to the bus-stop in the Peña el Gato (3) housing estate. From here, walk along the edge of the reservoir past a number of good waders pools as far

MADRID

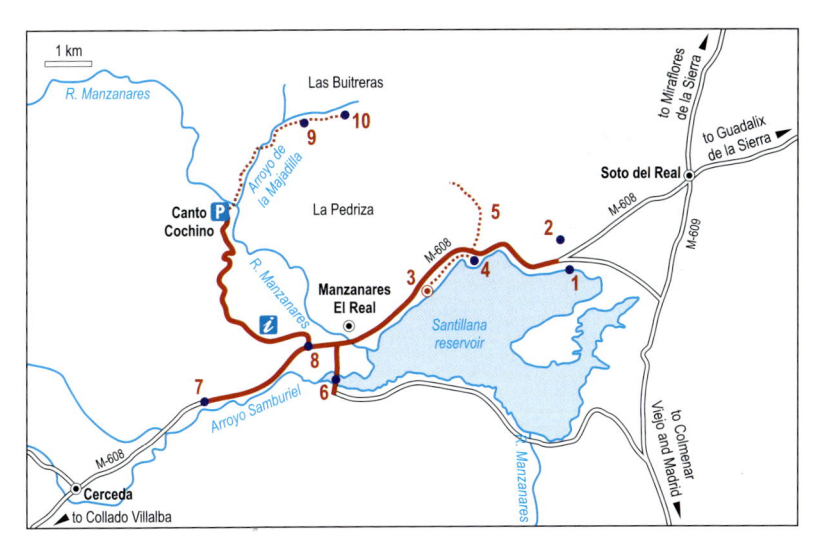

as a sewage station (4). On the opposite side of the road a path (5) heads across a publicly-owned *dehesa* towards the base of the south-face of La Pedriza with its Griffon Vulture colony and other cliff-loving birds (see itinerary 2).

Another option is to stop next to the bridge over the Arroyo Samburiel (6), west of Manzanares El Real on the south side of the reservoir. This is a good place for seeing grebes and other water birds close up. Upstream, there is an area of fluvial woodland (7) with a picnic site, which can be reached via the M-608 towards Cerceda. Aside from Golden Oriole and other birds typical of these habitats, Lesser Spotted Woodpecker has been recorded from this site.

2. 4 km

The interpretation centre of the Regional Park can be reached via a road (8) that heads off right from the M-608 towards Cerceda just 800 m after Manzanares El Real. After passing through a barrier, in 8 km you come to a car-park at Canto Cochino, from where you can walk along a well-marked track running parallel to Arroyo de La Majadilla, a stream that holds Dipper.

Begin to climb gently towards the mountain hut of Giner de los Ríos (9). From here the path then begins to climb more steeply in between maritime pine plantations, which are gradually substituted by rock-rose and other scrub-forming plants. Look out for Dunnock, Dartford and Sardinian Warblers and Azure-winged Magpie. The path finishes at the base of El Tolmo (10), a huge granite boulder from where there are good views of the vulture colony on the cliffs of Las Buitreras away up to the left. Here there are dozens of pairs of Griffon Vulture, along with Peregrine Falcon, Golden Eagle, Crag Martin, Blue Rock Thrush and Red-billed Chough.

MOST REPRESENTATIVE SPECIES

Residents: Cattle and Little Egrets, White Stork, Red Kite, Griffon Vulture, Golden Eagle, Peregrine Falcon, Eagle Owl, Kingfisher, Lesser Spotted Woodpecker, Thekla Lark, Crag

> ### PRACTICAL INFORMATION
>
> - **Access:** from Colmenar Viejo (35 km north of Madrid), take the M-609 and M-608 (via a short connecting road) to Santillana and Manzanares El Real.
> - **Accommodation:** the city of Madrid is nearby, although there are various places to stay in Manzanares El Real.
> - **Visitor infrastructure:** the Cuenca Alta del Manzanares Regional Park has an interpretation centre (tel. 918 539 978) at the main entrance to La Pedriza that organises excursions and workshop. Open every day, morning and afternoon.
> - **Visiting tips:** best between September and May; the summer heat is best avoided. Only 300 cars at a time are allowed along the road to Canto Cochino and so at weekends there can be problems. In this case, another option is to walk from Manzanares El Real or to make use of the special bus service.
> - **Recommended:** the town council of Colmenar Viejo has carried out a successful reintroduction of Lesser Kestrel (12 pairs) in the parish church.
> - **Further information:** maps 1:50,000, n°. 508 and 509 (SGE).

Martin, Dipper, Blue Rock Thrush, Dartford and Sardinian Warblers, Azure-winged Magpie, Red-billed Chough, Rock Bunting.
Summer visitors: Black Kite, Short-toed and Booted Eagles, Red-rumped Swallow, Black-eared Wheatear, Subalpine Warbler.
Winter visitors: Little, Black-necked and Great Crested Grebes, Great Cormorant, Grey Heron, Greylag Goose and other wildfowl, Black-headed and Lesser Black-backed Gulls, Water Pipit, Fieldfare, Redwing.
On migration: Black Stork, Eurasian Spoonbill, Osprey, Common Crane, waders.

92. SOUTH-WEST MADRID

On the border of the provinces of Ávila and Toledo a series of mountains drained by the river Alberche and its tributaries the Cofio and Perales act as transition zone between the two imposing ridges of Sierra de Guadarrama and Sierra de Gredos. Lying south-west of the city of Madrid, the higher northern half of this site is abrupt and covered by maritime pine plantations; further south, the landscape softens and patches of holm oak and stone pine *dehesa* and pastureland predominate. Most of the site is protected by an SPA of over 80,000 ha.

ORNITHOLOGICAL INTEREST

This site is home to the best Spanish Imperial Eagle population (10 pairs) in Madrid; as well, there are a few small colonies of Griffon (often breeding in trees) and Black Vultures, the latter all but extinct in the area. The area can also boast good numbers of forest raptors, Black-winged Kite (a recent coloniser) and 12 pairs of Black Stork.

ITINERARIES

1. 7 km

Three kilometres from Aldea del Fresno along the M-507 towards Villa del Prado, take an asphalted track (1) that drops down into the valley of the river Alberche and reaches the reservoir of Picadas. Park after crossing the dam and walk up a broad flat

South-west Madrid © Rubén Moreno-Opo

track (2) along the eastern bank of the reservoir.

This is in fact the course of a railway line that, despite its tunnels and bridges, was never finished; halfway along the walk, a bridge (3) takes you over the river to where the mountain sides are covered by dense stands of stone pine and patches of holm oak and scrub. The track eventually takes you out to the M-501 near Puente de San Juan (4), a bridge over the river Alberche; return from here (or whenever you want) to your starting point.

This is an excellent route for seeing Spanish Imperial Eagle in flight or perched on a holm oak. As well, you should look out for Black Kite, Short-toed and Booted Eagles, Azure-winged Magpie and a good variety of Mediterranean passerines in the scrub and woodland. The cliffs near the dam hold Eagle Owl and all rock outcrops should be checked for Crag Martin, Blue Rock Thrush and Red-rumped Swallow.

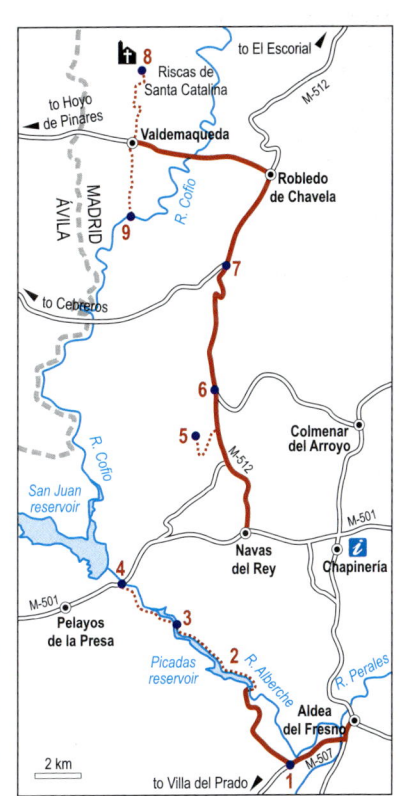

2. (30 km)

From Navas del Rey, the M-512 towards El Escorial passes through patches of holm oak and stone pine woodland, substituted in places by scrub and pastures, frequented by Spanish Imperial Eagle.

A good viewpoint is the trig point of Monteagudillo (5) that can be reached along a dirt track off to the left just after the turn-off to San Martín de Valdeiglesias. Park (the steep track is chained off) and walk up through an area of Mediterranean scrub with Red-necked Nightjar and Mediterranean passerines such as Subalpine and Orphean Warblers to the summit: the views from the top are more than just reward for the effort involved.

A more comfortable option is to continue north along the road and stop at the junctions (6,7) with the roads to Colmenar del Arroyo (right) and Cebreros (left). In more open areas look for Thekla Lark and the scarce Spectacled Warbler.

Once at Robledo de Chavela, head for Valdemaqueda, where you walk for 3 km along a well-marked path uphill (northwards) towards Prados del Hoyo (8), a small chapel surrounded by maritime pine woods lying in the shadow of the cliffs of Santa Catalina. A number of pairs of Griffon Vultures and rather fewer of Black Vulture breed here, along with Golden Eagle and Peregrine Falcon. The woods here hold Goshawk, Booted Eagle, Mistle Thrush, Common Crossbill and Hawfinch.

Another interesting option is to descend southwards from Valdemaqueda along a comfortable track through stone pine forests to Puente Mocha (9), a magnificent medieval bridge over the river Cofio. The area around the river is used by both Griffon and Black Vultures and Golden and Spanish Imperial Eagles; Black Stork are not uncommon and in winter Eagle Owl can be heard calling.

MOST REPRESENTATIVE SPECIES

Residents: Griffon and Black Vultures, Goshawk, Golden and Spanish Imperial Eagles, Peregrine Falcon, Eagle Owl, Kingfisher, Thekla Lark, Crag Martin, Blue Rock Thrush, Mistle Thrush, Dartford and Sardinian Warblers, Crested and Coal Tits, Azure-winged Magpie, Rock Sparrow, Common Crossbill, Hawfinch, Rock Bunting.

PRACTICAL INFORMATION

- **Access:** from Madrid take the Autovía de Extremadura (N-V) and in Navalcarnero turn off along the M-507 towards Aldea del Fresno (itinerary 1). Another possible access is the M-501 to Navas del Rey (itinerary 2).
- **Accommodation:** being so close, Madrid is the best option.
- **Visitor infrastructure:** El Águila (tel. 918 652 098), an environmental education centre in Chapinería devoted to Mediterranean forests and run by the Madrid Department of the Environment, organises walks and workshops. Open every day, morning and afternoon.
- **Visiting tips:** spring is ideal for breeding birds, although take care not to disturb nest sites. Avoid the summer heat and do not leave tracks and paths as most of the site consists of privately owned hunting estates.
- **Further information:** maps 1:50,000, n°. 532, 557 and 580 (SGE).

Summer visitors: Black and White Storks, Black Kite, Short-toed and Booted Eagles, Great Spotted Cuckoo, Red-necked Nightjar, Red-rumped Swallow, Black-eared Wheatear, Subalpine, Spectacled and Orphean Warblers.
Winter visitors: Great Cormorant, Grey Heron, Merlin.
On migration: Osprey, Common Crane.

93. VALLE DEL JARAMA

The eastern part of the Autonomous Community of Madrid is drained by the river Jarama, which, beyond its confluence with the river Lozoya in the northern sector of this site, runs through the foothills of the Sierra de Ayllón in a landscape of slates, schists and a single limestone ridge, one of the few such outcrops in the Madrid area. Most of the area is covered by rock-rose scrub with a few scattered patches of Mediterranean forest; in the south, the scrub gives way to a succession of gullies, low hills and cereal cultivation with a few copses of Lusitanian oak and Mediterranean scrub. Along the river Jarama and its tributaries a few *sotos* (fluvial woodland) remain. Part of the site is included in an SPA of over 30,000 ha.

ORNITHOLOGICAL INTEREST

The cereal plains of the mid-course of the river Jarama can boast one of the best populations of Great Bustards and other steppe birds in central Spain. Black Vultures and Golden Eagles, the latter with a number of territories in the mountainous terrain in the upper course of the Jarama and Lozoya, often fly over the area. The limestone cliffs are complemented by a varied mosaic of copses, scrub, dry farming and *sotos*, all with a broad diversity of bird communities.

ITINERARIES

1. 5 km

From Torrelaguna take the M-102 to Patones de Abajo (6 km), from where

Valle del Jarama © Pedro Retamar

a road climbs up to the picturesque village of Patones de Arriba. Alternatively, walk up a steepish, 2-km footpath (1) alongside the *arroyo* (stream) that joins the two villages. The nearby limestone cliffs hold Black Wheatear, Blue Rock Thrush, Crag Martin and Red-billed Chough. Once at Patones de Arriba, explore the extensive tracts of rock-rose scrub for Thekla Lark, Black-eared Wheatear, Dartford and Sardinian Warblers and Rock Bunting.

Return to Patones de Abajo and continue east along the M-102 and in 5 km, park on the left of the road. Walk down to the right of the road to the reservoir of El Pontón de la Oliva (2), an attractive reservoir on the river Lozoya near its confluence with the Jarama, the latter here flowing through limestone cliffs which was once home to a pair of Bonelli's Eagles. A walk along either side of the river upstream from the dam will take you through an interesting mix of rock outcrops, scrub and pastures with correspondingly diverse bird communities, including hunting Golden Eagles.

2. 50 km 🚶 ☀ ➡

The river Jarama flows through a broad valley flanked by a succession of interesting limestone mountains with Lusitanian oak woodland – look here for Bonelli's and Orphean Warblers and Hawfinch – in areas such as the last part of the N-320 (3) from Torrelaguna on towards the Autovía de Burgos. The minor roads between Torrelaguna and the villages of El Vellón (4) and El Espartal (5) pass through aromatic scrub with Thekla Lark and various warblers including Spectacled Warbler.

Many *sotos* line the river banks and some are home to colonies of Spanish Sparrow: check out the poplar woods and the *soto* (6) on the river Jarama near the town of Talamanca de Jarama and its impressive Roman bridge over the river for Kingfisher, Wryneck, Great Spotted Woodpecker, Penduline Tit and Golden Oriole.

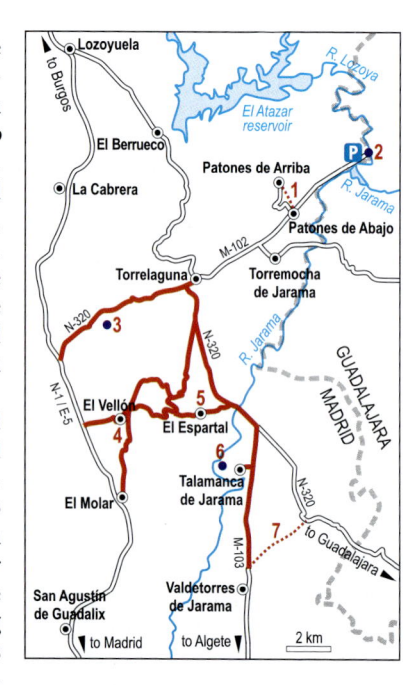

Continue south on the M-103 from Talamanca de Jarama as far as a roundabout just before the industrial estate north of Valdetorres de Jarama: turn left along a track across open cereal fields (best on foot since the track gets worse as it progresses). Look out for groups of male Great Bustard in March and April, as well as Little Bustard, Stone Curlew, Black-bellied Sandgrouse and Calandra and Short-toed Larks.

These arable plains have surprisingly good raptor populations: Marsh, Montagu's and Hen Harriers and Common and Lesser Kestrels all breed, and Merlins appear in winter. Other large raptors such as Black Vulture and Golden Eagle also feed here.

MADRID

PRACTICAL INFORMATION

- **Access:** from Madrid, take the Autovía de Burgos (N-I / E-5) along the western edge of the mid-course of the Jarama valley and then the N-320 across the valley itself. The town of Torrelaguna is the best starting place for the routes described.
- **Accommodation:** being so close, Madrid is the best option.
- **Visitor infrastructure:** no visitor centre or infrastructure for birdwatchers.
- **Visiting tips:** any time of year except high summer (too hot). Early spring is the best time of year to see steppe species and many of the other birds of the area.
- **Further information:** maps 1:50,000, n°. 485, 509 and 510 (SGE).

MOST REPRESENTATIVE SPECIES

Residents: Red Kite, Black Vulture, Marsh and Hen Harriers, Golden Eagle, Peregrine Falcon, Great and Little Bustards, Stone Curlew, Northern Lapwing, Black-bellied Sandgrouse, Kingfisher, Calandra Lark, Thekla Lark, Crag Martin, Dunnock, Blue Rock Thrush, Dartford and Sardinian Warblers, Penduline Tit, Red-billed Chough, Spanish Sparrow, Hawfinch, Rock Bunting.
Summer visitors: White Stork, Black Kite, Short-toed Eagle and Booted Eagles, Montagu's Harrier, Lesser Kestrel, Wryneck, Short-toed Lark, Red-rumped Swallow, Black-eared Wheatear, Bonelli's, Subalpine, Spectacled and Orphean Warblers, Golden Oriole.
Winter visitors: Merlin, European Golden Plover, Common Snipe.

94. SOUTH-EASTERN WETLANDS

Not far from the city of Madrid, the Sureste Regional Park (30,000+ ha; also SPA) extends along the lower courses of the rivers Manzanares and Jarama through flood plains occupied by numerous reed-fringed gravel pits and stands of tamarisk, willow, poplar and ash woodland. Some parts of the valleys are flanked by imposing gypsum cliffs that plunge down from the plains occupied by holm oak and pine woodland, holly oak scrub and cereal cultivation.

ORNITHOLOGICAL INTEREST

Home to some of Madrid's most important heronries, this site can boast breeding Purple (50 pairs) and Grey Herons, Little Bittern, Black-necked Grebe, Gadwall, Red-crested and Common Pochards, Marsh Harrier and Black-winged Stilt. White Stork also breed and many winter, while Purple Swamp-hen have become much commoner in recent years. Winters sees the largest gull (Black-headed and Lesser Black-backed) roost in central Spain form and the arrival of numerous duck. The nearby gypsum cliffs hold Peregrine Falcon, Black Wheatear and Red-billed Chough.

ITINERARIES

1. 8 km

Lying next to the town of Rivas-Vaciamadrid, the lagoon of El Campillo is one of the largest and best-known in the Regional Park. From the town, take the road alongside the railway line as far as a barrier (1),

Cantiles de El Piul, south-eastern wetlands © José Manuel Reyero

where you should park. From here, walk the track along the southern bank of the abandoned gravel pit and the river Jarama. Grebes, Great Cormorants and gulls congregate on the open water, along with many duck (above all, Northern Shoveler) in winter. Common Crane can, with luck, be seen in migration.

The track reaches the park information centre (2), which also has viewing facilities. The full itinerary (5 km) can be completed by walking around the northern bank of the lagoon past a marshy area (3) of reeds and bulrushes with Little Bittern, Purple Swamp-hen and Purple Herons, which come here to fish. The last part of the route passes by the foot of the gypsum cliffs with Common Kestrel, Black Wheatear and Red-billed Chough; Eagle Owl call here in winter.

Of similar interest are the lagoons close to Velilla de San Antonio, upstream along the Jarama. Near to this town, there is a picnic site and car-park from where you can walk along the banks of the lagoons of El Raso (4) and El Picón de los Conejos (5) for 2 km as far as a fence (6). On your left rise the cliffs of El Piul – with Peregrine Falcon and the other cliff-loving species of the region – and

MADRID

the river Jarama, with Black Kite in the *sotos* and a number of Cattle Egret roosts.

2. 2 km 🚶 ☀ ⭕

In the lower course of the river Tajuña, just before its confluence with the Jarama, lies the Laguna de San Juan (just south of the Regional Park). The best access is from Titulcia along the M-404 towards Chinchón: halfway between both towns, turn right along a track (7) just after crossing the river Tajuña, continue along the base of the cliffs and park in 2 km in a small car-park.

The itinerary follows the southern bank of the lagoon, hidden from view by a tall reed-bed that is home to an important colony of Purple Heron. Here too there are enormous post-breeding Sand Martin, Barn Swallow and Spotless Starling roosts and Penduline Tit and Reed Bunting in winter. The track passes by a small hide giving views over the open water: wait patiently for Little Bittern, Water Rail and Purple Swamp-hen to appear. Winter brings Great Cormorants, Grey Heron and many duck, including Red-crested Pochard.

Some 300 m further on a path off to the left climbs steeply to the surrounding plains (8) through the gypsum cliffs that here hold Eagle Owl, Thekla Lark, Black Wheatear and Rock Sparrow. From the edge of the cliff there are excellent views of Purple Heron and Marsh Harrier in flight.

MOST REPRESENTATIVE SPECIES

Residents: Little and Great Crested Grebes, Grey Heron, White Stork, Red-crested Pochard, Marsh Harrier, Common Kestrel, Peregrine Falcon, Water Rail, Purple Swamp-hen, Stone Curlew, Eagle Owl, Kingfisher, Thekla Lark, Black Wheatear, Penduline Tit,

PRACTICAL INFORMATION

- **Access:** leave the Autovía de Valencia (A-3) at junction 17 to Rivas-Vaciamadrid or at junction 22 to Campo Real and then in 3 km head north towards Velilla de San Antonio (itinerary 1). For itinerary 2, leave the Autovía de Andalucía (E-5) at junction 29 and pass through Ciempozuelos and on to Titulcia.
- **Accommodation:** in Madrid and nearby towns.
- **Visitor infrastructure:** the Sureste Regional Park has an interpretation centre (tel. 600 508 638) next to Laguna del Campillo. Open every day, morning and afternoon.
- **Visiting tips:** the greatest diversity of birdlife occurs in winter.
- **Recommended:** SEO-Monticola (seo-monticola@seo.org) has a ringing station in the Sureste Regional Park and organises bird-related activities.
- **Further information:** maps 1:50,000, n°. 583 and 606 (SGE). Also, www.elsoto.org.

Red-billed Chough, Jackdaw, Spotless Starling, Rock Sparrow.
Summer visitors: Little Bittern, Purple Heron, Black Kite, European Bee-eater, Sand Martin, Barn Swallow, Reed and Great Reed Warblers, Golden Oriole.

Winter visitors: Great Cormorant, Cattle Egret, Northern Shoveler, Common Pochard, Tufted Duck and other duck, Black-headed and Lesser Black-backed Gulls, Reed Bunting.
On migration: Osprey, Common Crane.

MURCIA

95. Mar Menor

This small, single-province autonomous community, with a relatively low population density (just three major towns or cities – Murcia, Cartagena and Lorca), is squeezed into a space between the Comunitat Valenciana and Andalusia and occupies the heart of the so-called 'Sudeste Ibérico', the most arid part of Europe. Despite this aridity, the economy of Murcia is based on intensive market gardening whose life-line is water pumped in from river basins outside of Murcia (for example, the canal linking the Tajo and Segura rivers).

Thus, visitors can expect to see tracts of green agricultural land backing on to the extremely contrasting semi-arid and almost desert-like habitats that are the typical of landscapes of the region. High temperatures and low rainfall – which when it occurs tends to be torrential and potentially very erosive – provide for landscapes that have a lot in common with parts of North Africa.

There are few permanent water courses and the only one of any size, the river Segura, is very contaminated and over-exploited in the main; in its upper course, however, there are a number of interesting stretches of fluvial woodland and populations of Otters. The typical rivers of the region are thus seasonal (*ramblas*) and only carry water after heavy storms.

The birdlife of the region is probably not as varied as that of some other parts of Spain, although the fact that many species are well adapted to arid habitats all but unknown in other parts of Europe, makes the birds of the region especially relevant in a European context. The abrupt mountains of the eastern ranges of Las Sierras Béticas and the vast pine forests that cover them are good areas for raptors such as Griffon Vulture, which has recently colonised the region, and Golden Eagle.

Bonelli's Eagles prefer the lower coastal mountains such as the heavily mined ridges behind Cartagena whose semi-arid scrub habitats are home to the only European population of Barbary arbor-vitae, a rare North African shrub. Trumpeter Finches, likewise a denizen of desert and semi-arid North African habitats, have recently colonised the area.

Another of the main relief units in Murcia consists of flattish inland areas, either depressions or raised plateaux. The plains around Yecla in the extreme north of the region are a continuation of the great plains of

Audouin's Gull   Carlos Sánchez © nayadefilms.com

Castilla-La Mancha and their cereal fields and fallows hold Little Bustard and Black-bellied Sandgrouse. Other such areas include Los Saladares del Guadalentín, an inland area of saline steppe with an important population of Rollers, and the thyme and needle-grass steppe of Los Llanos de las Cabras with a small population of Dupont's Lark.

A few sections of semi-virgin habitat between Cabo Cope and Las Puntas de Calnegre in the south of Murcia still remain along the region's almost 200 km of coastline, much of which is rocky. Offshore a number of islands hold seabird colonies: a good example is the Audouin's Gull colony on Isla Grossa.

Wherever the coastline is lower and gentler with broad sandy beaches, the existence of a tourist trade that has transformed much of the Mediterranean coastline becomes more patent. A good example is the urban sprawl stretched out along the sand-spit of La Manga and around the large coastal lagoon of the Mar Menor it protects (95). This hypersaline lagoon and the nearby salt-pans of San Pedro del Pinatar are, nevertheless, the main site in Murcia for water birds and seabirds.

95. MAR MENOR

The Mar Menor is the largest saline coastal lagoon in Spain (13,000+ ha) and is all but cut off from the Mediterranean by La Manga, a heavily built-up sand-spit and dune cordon of over 20 km. Despite the intense pressure from tourist infrastructures, good areas of saline steppe, salt-marsh, reed-bed and dune, mostly protected by a Paisaje Protegido, still exist. The working salt-pans of San Pedro del Pinatar have been declared a Regional Park; both areas have been declared conjointly a Ramsar Site and separately SPAs.

ORNITHOLOGICAL INTEREST

Las Salinas de San Pedro del Pinatar are the most important wetland in Murcia and harbour breeding Shelduck, Black-winged Stilt, Avocet, Kentish Plover, Black-headed Gull

MURCIA

Calblanque, Mar Menor © Pedro Retamar

and Gull-billed, Common and Little Terns. Despite being seen throughout the year, Greater Flamingos do not breed: likewise, the Audouin's Gulls that are frequent, breed not here but on the nearby Isla Grossa. The lagoon of the Mar Menor is one of the best places in Spain in winter for Black-necked Grebe and Red-breasted Merganser.

ITINERARIES

1. 4 km

At the entrance to the Regional Park by the Las Salinas information centre, a path (1) alongside the road to the port runs through the northern part of the salt-pans and provides good views of Shelduck, Black-winged Stilt, Avocet and Kentish Plover. Greater Flamingo are abundant and in winter and during migration periods Audouin's and Slender-billed Gulls are common.

The salt-pans give way to a large sandy area (2), partially covered by Aleppo pines: turn left to a hide (3) with excellent views over the salt-pans. Return to the path and continue between the pines and salt-pans as far as Charca del Coterillo (4), a recently restored wetland with a hide giving good views of the Avocet, Black-headed Gull and Common and Little Tern colonies.

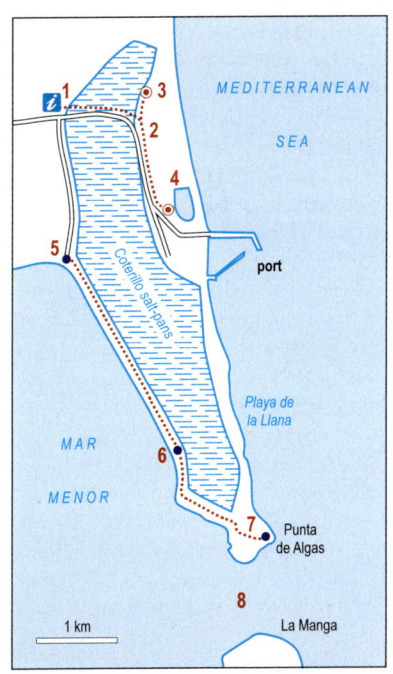

2. 50 km 🚗 🚶 ☀️ →

From the entrance to the Regional Park a road runs along the western side of the salt-pans and then along the northern shores of the Mar Menor. From the windmill of Quintín (5), a 3-km track, asphalted at first, runs between the large pools of the salt-pans and the lagoon: look out for Black-necked Grebe – often here in their hundreds – in winter.

At a second windmill, La Calcetera (6), park and walk towards the beach of La Llana and Punta de Algas (7), 2 km away. Southwards, scan Las Encañizadas (8), a shallow channel connecting the Mar Menor to the sea that is full of islands and shallow bays and an important feeding ground for herons, Great Cormorant, gulls, terns and waders. In winter, Eurasian Spoonbill are easy to see and, with a little bit more luck, you should also find Great Egret and Red-breasted Merganser.

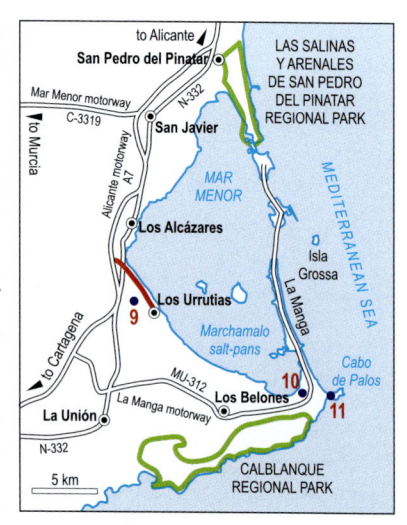

As for the rest of the Mar Menor, try the Marina del Carmolí (9), a saline steppe. Take the local road off

PRACTICAL INFORMATION

- **Access:** from the city of Murcia, take the Autovía del Mar Menor (C-3319) to San Javier and from here to San Pedro del Pinatar. The road to this town's port passes through the Regional Park that protects the salt-pans. Another possibility is the Autovía de Alicante (A-7) that heads south near the western side of the Mar Menor and connects to the Autovía de La Manga (MU-312) on the southern side of the lagoon and provides access to La Manga and Cabo de Palos.
- **Accommodation:** plenty of accommodation in and around the Mar Menor and La Manga, as well as in Cartagena and Murcia.
- **Visitor infrastructure:** the Centro de Investigación y Conservación de Humedales Las Salinas (tel. 968 178 139, salinas.sanpedro@cotambiental.com) acts as a information centre for the Regional Park. Open all year, from Tuesday to Sunday, morning and afternoon (at weekends, mornings only).
- **Visiting tips:** any time of year, although summer sees too many tourists in the area. Do not leave the recommended itineraries so as not to disturb birds, above all during the breeding season.
- **Recommended:** south of the Mar Menor, the Calblanque Regional Park protects a sector of coastal mountains with breeding Golden and Bonelli's Eagles, Peregrine Falcon, Eagle Owl and other cliff-loving species. The information centre is in the village of Cobaticas, near Los Belones.
- **Further information:** maps 1:50,000, n°. 956 and 978 (SGE). *Guía de aves acuáticas del Mar Menor*, published by the Murcia Department of Agriculture, Water and the Environment (2000), provides useful information.

MURCIA

the Autovía de Alicante (A-7) towards the village of Los Urrutias and drive along the western border of the Mar Menor with the steppe flooded by irrigation run-off inland. Any of these small pools may hold Shelduck, Black-winged Stilt, Kentish Plover and other waders and, in drier areas, look for Stone Curlew and Calandra and Lesser Short-toed Larks.

In the southern-most part of the Mar Menor, the salt-pans of Marchamalo (10) just off the Autovía de La Manga (MU-312) are also worth a visit. They are frequented by Greater Flamingo, Audouin's and Slender-billed Gulls and waders. In winter Black-necked Grebe and Red-breasted Merganser swim on the Mar Menor near here. Make a final stop at Cabo de Palos (11), a good place in autumn for seawatching for shearwaters and Northern Gannet.

MOST REPRESENTATIVE SPECIES

Residents: Greater Flamingo, Shelduck, Black-winged Stilt, Avocet, Stone Curlew, Kentish Plover, Black-headed, Audouin's and Slender-billed Gulls, Calandra and Lesser Short-toed Larks.

Summer visitors: Gull-billed, Common and Little Terns.

Winter visitors: Great Crested and Black-necked Grebes, Great Cormorant, Northern Gannet, Great Egret, Eurasian Spoonbill, Red-breasted Merganser, Osprey, Grey Plover, Sanderling, Little Stint, Dunlin, Black-tailed Godwit, Whimbrel, Eurasian Curlew, Turnstone, Sandwich Tern.

On migration: Cory's and Balearic Shearwaters, Northern Gannet, waders, Black and Whiskered Terns.

NAVARRE

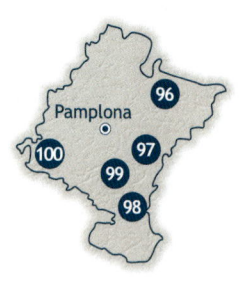

96. Roncesvalles and Irati
97. Lumbier and Arbayún
98. Bardenas Reales
99. Laguna de Pitillas
100. Embalse de Las Cañas

Navarre is probably the most diverse single-province autonomous community in Spain, with natural sites as different as the high Pyrenees and the Ebro depression all within 100 km of each other. It only lacks the sea, although its north-eastern corner all but reaches the Cantabrian coastline: this is consequently the most oceanic, verdant and mildest part of Navarre and its pleasant landscapes of copses, hedgerows and small-holdings are very reminiscent of the neighbouring Basque Country.

Nevertheless, the Pyrenees have always been the great attraction in Navarre for naturalists. Lower than the Catalan and Aragonese Pyrenees and with more open and accessible valleys, the Navarrese Pyrenees is blessed with a relatively respectful tourism, as the lack of ski-stations indicates. On the other hand, stock-raising is well developed and has left its mark on the upland pastures and scrub formations. As well, Navarre can proudly point to its upland forests, some of the best in Spain.

The beech and European silver-fir forests of Roncesvalles and Irati (96) on the border with France are home to an extremely rich forest avifauna, including the Peninsula's best populations of White-backed Woodpecker. Above the forests, the mountain passes are used by countless migrating birds and autumn sees spectacular waves of raptors, storks, cranes, pigeons and passerines heading south into the Peninsula. The highest peaks in the Navarrese Pyrenees border on Aragon and their heights provide habitat for species such as Ptarmigan, here at the south-west limit of its range.

Southwards the landscape evolves into mid-mountain habitats of east-west-facing limestone cliffs of great biogeographical interest. Mediterranean and Atlantic influences meet, as the juxtaposition of beech and broad-leaved oak forests with holm and holly oak forests and Mediterranean scrub formations demonstrates. The erosive power of the rivers running down from the Pyrenees has carved out a number of *foces* (97) (the local word for gorge), of which Lumbier is the most accessible and Arbayún the most spectacular. Large colonies of Griffon Vultures pack these cliff faces, sharing breeding habitat with a pair of Lammergeiers.

Black Woodpecker
Carlos Sánchez © nayadefilms.com

The landscape changes radically as you approach the decidedly Mediterranean Ebro depression. The all but completely treeless land rolls away gently before your eyes and land-use is predominantly agricultural (irrigated and non-irrigated). The driest areas are semi-arid and the pseudo-steppes of the Bardenas Reales (98) are famous for their startlingly eroded landscapes. Steppe-bird populations are good and there are also a fair number of raptors.

Habitat diversity in the south of Navarre is enriched by good stands of fluvial woodland (*sotos*) along the river Ebro and its main tributaries, areas of rice-paddies and a number of natural and artificial wetlands. The largest wetland is the Laguna de Pitillas (99), surrounded by an enormous reed-bed with various breeding herons including the scarce Great Bittern, and good numbers of Marsh Harrier. In the east of Navarre near the city of Logroño (La Rioja province), the Embalse de Las Cañas (100) provides further opportunities for finding water birds.

96. RONCESVALLES AND IRATI

The slopes of this sector of the Pyrenees in northern Navarre bordering on France are covered by some of the best-preserved broad-leaved woodland in Spain. The grandiose Selva de Irati – 13,000 ha of mature beech forest mixed with European silver-fir at its western-most point of its European distribution – lies at the heart of the site; large areas of grazing and heather scrub extend above the forests. Both the forests and grazing are intensively exploited commercially on a communal basis. An SPA of over 18,000 ha and three Reservas Naturales – Lizardoya, Mendilaz and Tristuibartea – protect part of the area.

ORNITHOLOGICAL INTEREST

Forest birds are of obvious interest and practically all Iberian woodpeckers including the main Spanish population of White-backed Woodpecker, a bird of mature beech forests, breed here. Black Woodpeckers are abundant and there have even been a few records of Middle-spotted Woodpecker from the oak forests.

Many raptors hunt and feed in the area: look out for the one pair of Lammergeiers and the small populations of Egyptian and Griffon Vultures and Golden Eagle. More important is the migration over the mountain passes, especially spectacular after the breeding season, when large numbers of Honey Buz-

Selva de Irati © Pedro Retamar

zard, Black and Red Kites and other raptors head south into the Peninsula. Massive numbers of Wood Pigeon, Black Stork, Greylag Goose, Common Crane and many passerines also pass through.

ITINERARIES

1. 40 km

From Roncesvalles, take the N-135 towards France as far as the pass of Ibañeta (1). From here take a track 3 km up to Collado de Lindux (2) on the border between France and Spain. This is one of the best places from August onwards to watch for migrating soaring birds (above all raptors), although even from the end of July there are groups of Black Kite beginning to pass through. Hunting starts in October and birdwatching becomes difficult and rather fraught.

White-backed and Black Woodpeckers live in the surrounding beechwoods and Citril Finches frequent the forest edges. Breeding raptors such as Red Kite, Egyptian and Griffon Vulture, Golden Eagle and, with luck, Lammergeier fly over the forests.

Return to Roncesvalles and continue along the N-135 towards Pamplona. Turn left after Burguete along the NA-140 and then between Garralda and Aribe (7 km) on the right-hand side of the road stop at the lookout point of Ariztokia (3): view north to the cliffs with Griffon and possibly Egyptian Vultures. The pedunculate oak forest of Olaldea with interesting forest birds is also viewable from here.

From the village of Aribe, a road takes you in 6 km to Orbaiceta and then a little further on to the ruins of the Orbaiceta arms factory (4). A track from here heads north to the French border at Collado de Azpegi (5), a good site for Water Pipit, Northern Wheatear, Mistle Thrush and Alpine Chough in the pastures, forest birds in the surrounding woods and raptors (breeding and on passage).

NAVARRE

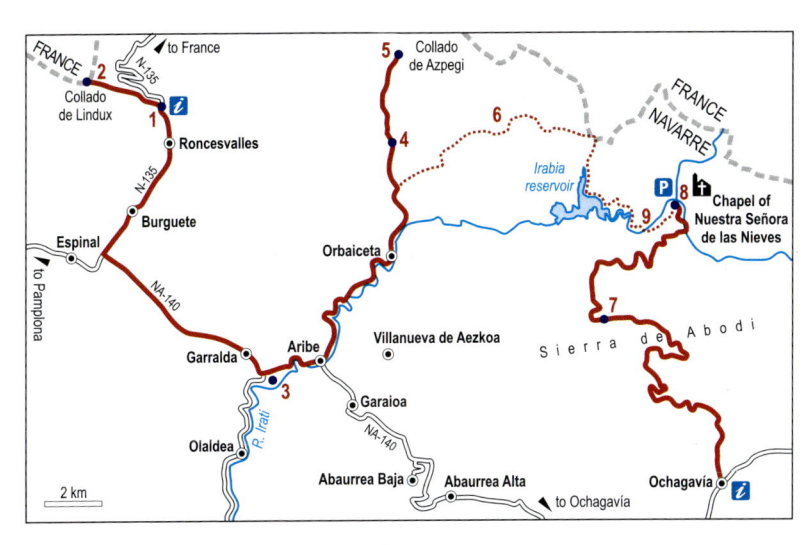

2. 8 km 🚶 ☀️ ➡️

Two km before the arms factory, a long cement track (6) heads into the western sector of the Irati forest. Nevertheless, given that it is best to explore this imposing forest on foot, we recommend a shorter option in the east of the forest. Head north from Ochagavía across the Sierra de Abodi and El Paso de Tapla (7) with scrub bird communities and a few Grey Partridge.

After 20 km you reach the car-park near the chapel of Nuestra Señora de las Nieves (8). Begin walking along a track (9) running parallel to the river Irati and then alongside the Irabia

PRACTICAL INFORMATION

- **Access:** from Pamplona the N-135 takes you to Roncesvalles. After Espinal you can take the NA-140 along the south edge of this sector of the Pyrenees to Ochagavía.
- **Accommodation:** in Roncesvalles, Espinal, Burguete, Ochagavía and other towns and villages.
- **Visitor infrastructure:** Gurelur (tel. 948 151 077, gurelur@bme.es) run a centre dedicated to bird migration at Collado de Ibañeta. Open every day, morning and afternoon between July and November during post-breeding migration. In Ochagavía there is an interpretation centre (tel. 948 890 641, oit.ochagavia@cfnavarra.es) devoted to the wildlife of the western Pyrenees. Open every day in the morning and some afternoons.
- **Visiting tips:** late spring is best and, for migration, July onwards. From October onwards Pyrenean passes are best avoided owing to the numbers of hunters.
- **Recommended:** Organbidexka Col Libre (tel. 0033 559 256 203, ocl@wanadoo.fr), with its HQ in Bayona (France) carries out a post-breeding survey (Transpyr) of birds passing through Pyrenean passes such as Lindux. Volunteers welcome. More information: www.organbidexka.org.
- **Further information:** maps 1:50,000, n°. 91, 91 bis, 116 and 117 (SGE). The best reference book is *El Parque Natural Pirenaico en Navarra II. Irati-Ibañeta* by Jesús Elósegui *et al.* (Gobierno de Navarra, 1989).

reservoir. Look out for Black Woodpecker, Firecrest, Marsh Tit, Eurasian Treecreeper and Bullfinch in the beech, while the stands of fir also hold Goldcrest and Common Crossbill. The jewel in Irati's crown – the White-backed Woodpecker – breeds here but is hard to find.

MOST REPRESENTATIVE SPECIES

Residents: Red Kite, Lammergeier, Griffon Vulture, Goshawk, Grey Partridge, Black and White-backed Woodpeckers, Dipper, Mistle Thrush, Goldcrest, Firecrest, Marsh, Crested and Coal Tits, Eurasian Treecreeper, Red-billed and Alpine Choughs, Citril Finch, Common Crossbill, Bullfinch, Yellowhammer, Rock Bunting.
Summer visitors: Egyptian Vulture, Water Pipit, Dunnock, Northern Wheatear, Red-backed Shrike.
On migration: Black Stork, Greylag Goose, European Bee-eater, Black Kite, Short-toed and Booted Eagles, Marsh and Montagu's Harriers, Osprey, Common Crane, Wood Pigeon, Stock Dove.

97. LUMBIER AND ARBAYÚN

In the east of Navarre the rivers Irati and Salazar have sliced open the pre-Pyrenean Sierra de Leyre and

Foz de Arbayún © Pedro Retamar

NAVARRE

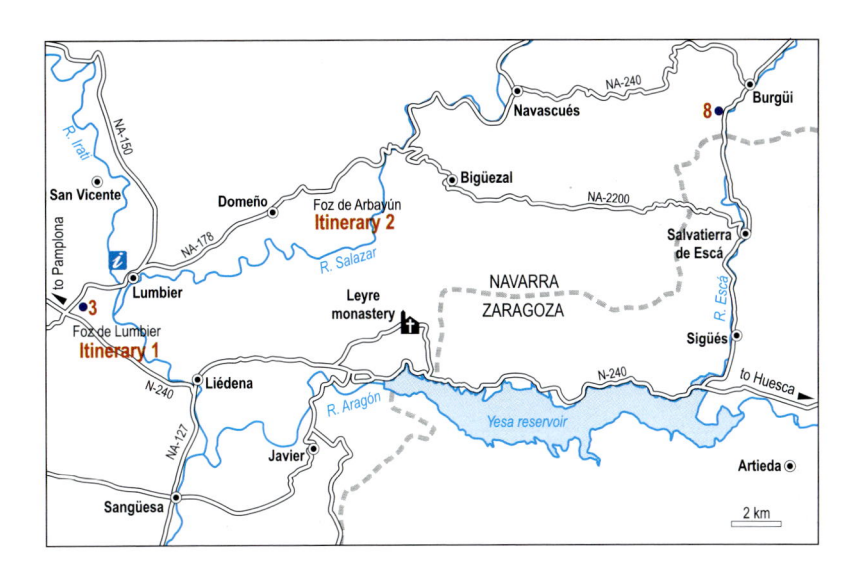

created two deep limestone gorges (*foces*). Lumbier, the smaller and more open, is much more Mediterranean in feel and is populated by holm and holly oaks and Mediterranean scrub; Arbayún, on the other hand, is the jewel in the crown of the Navarrese *foces* and the northern-facing base of its 200-m high cliffs are covered with beech and other deciduous tree species. Both *foces* are part of a *reserva natural* and an SPA.

ORNITHOLOGICAL INTEREST

This site holds Navarre's best Griffon Vulture colonies (well over 300 birds) as well as a single pair of Lammergeiers, Egyptian Vulture, Golden Eagle, Peregrine Falcon and Eagle Owl. Until the 1990s there was even a few pairs of Bonelli's Eagle. Forest raptors include the here scarce Honey Buzzard, Red and Black Kites and Short-toed and Booted Eagles. The rest of the bird communities reflect the Mediterranean, Pyrenean and Atlantic influences that converge here and are thus fairly diverse.

ITINERARIES

1. 6 km

From the town of Lumbier a road takes you in 2 km to the car-park at the entrance to the Foz de Lumbier. An easy-to-walk track (1) parallel to the river Irati takes you into the canyon along the 1-km route of an old railway line, complete with tunnels. Watch out for typical rock-loving species and in winter for Eagle Owl and Wallcreeper.

The passerine communities - Northern Wheatear, Subalpine, Sardinian and Orphean Warblers and Cirl, Rock and Ortolan Buntings – are of interest and are best searched for in the second half of the walk (2) as you return to the car-park via the scrubby slopes above the gorge. Keep an eye open for Lammergeier and other raptors.

Coming from Pamplona, just after turning along the NA-178 towards Lumbier, park at a viewpoint (3) next to a small chapel (see main map). This spot gives good views of a vulture feeding station set up to provide Griffon Vultures with food and at times over 100 birds congregate here. Black and Red Kites and Egyptian Vultures also come down to feed.

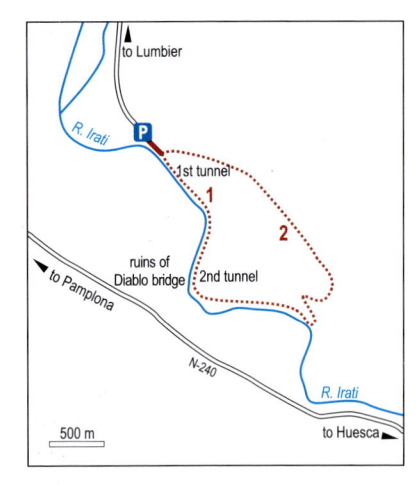

2. 40 km 🚗 🚶 ☀️ →

The larger and deeper Foz de Arbayún is accessible from Domeño, 6 km from Lumbier on the NA-178. Take a minor road to the village of Usún and then pick up a track that takes you to a open area, where you should park. From here, a path (4) crosses the river Salazar and heads upstream along its left-bank: look out for Black Woodpecker and other forest birds. Nevertheless, this path is somewhat dangerous as it includes two vertiginous stretches near the edge of the cliff: do not attempt if you have no head for heights.

A better option for getting a view of the *foz* is to head for the viewpoint of Iso (5) on the right of the NA-178, 6 km from Domeño. This is a wonderful place to sit and watch the comings and goings of the here abundant Griffon Vultures: Lammergeiers are seen reasonable frequently and there are also Black and Red Kites, Egyptian Vulture, Golden Eagle, Peregrine Falcon and Alpine Swift to look out for.

Continuing along the NA-178, after crossing a bridge over the river Salazar, turn right to Bigüezal (6). Just past this village, take a steep narrow road off to the right that climbs up to the TV masts on Alto de Arangoiti (7), the summit of the Sierra de Leyre. From here there are marvellous views of the Navarrese Pyrenees and raptors in flight are easily spotted. The road passes through Scots pine forests dotted here and there with beech and birch: forest birds here include Honey Buzzard, Goshawk, Black Woodpecker, Citril Finch and Common Crossbill.

Back on the NA-178, continue north to Navascués (5 km), where you should turn right to Burgui. Between this village and Salvatierra de Escá, the road passes through the Foz de Burgüi (8) on the frontier between Navarre and Aragon (see main map). A number of view points along the road allow you to park and examine the sheer cliffs of the *foz*.

MOST REPRESENTATIVE SPECIES

Residents: Red Kite, Griffon Vulture, Lammergeier, Goshawk, Golden Eagle, Peregrine Falcon, Eagle Owl, Kingfisher, Black Woodpecker, Crag Martin, Dipper, Dunnock, Blue Rock

NAVARRE

PRACTICAL INFORMATION

- **Access:** from Pamplona, take the N-240 towards Jaca (Aragon). In 36 km, turn left for the Salazar valley, the town of Lumbier and the other itineraries described here.
- **Accommodation:** best in Lumbier and Sangüesa.
- **Visitor infrastructure:** The Foces interpretation centre in Lumbier (tel. 948 880 874, cinlumbi@cfnavarra.es) is open every day in the morning and sometimes in the afternoon. Closed on Monday.
- **Visiting tips:** all year, but best in spring and summer.
- **Further information:** maps 1:50,000, n°. 142, 143 and 174 (SGE).

Thrush, Sardinian Warbler, Firecrest, Red-billed Chough, Rock Sparrow, Citril Finch, Common Crossbill, Bullfinch, Cirl and Rock Buntings.
Summer visitors: Honey Buzzard, Black Kite, Egyptian Vulture, Short-toed and Booted Eagles, Alpine Swift, Northern Wheatear, Subalpine and Orphean Warblers, Red-backed Shrike, Ortolan Bunting.
Winter visitors: Alpine Accentor, Wallcreeper.

98. BARDENAS REALES

In south-east Navarre on the border with Aragon lies the remarkable uninhabited steppe of Las Bardenas Reales, almost 50 km from north to south and covering over 40,000 km. A succession of broad tablelands form gigantic steps dropping down to the river Ebro and are deeply scarred by highly seasonal valleys, made all the more interesting by the alternating mudstone, gypsum, clay and sandstone strata that have been eroded into evocative shapes and buttresses.

More than half of the site has been transformed into cereal steppe with a few areas of irrigated land here and there. There are still vast areas of natural vegetation, including gypsum and saline scrub and a few patches of Aleppo pine and holm oak woodland. Protected by a Natural Park – that excludes the bombing range in the middle – and a Biosphere Reserve, there are also three *reservas naturales* in the area: Rincón del Bu, Caídas de la Negra and, slightly further away, Vedado de Eguaras.

ORNITHOLOGICAL INTEREST

Las Bardenas is best known for its steppe birds, including Stone Curlew, Pin-tailed and Black-bellied Sandgrouse and Dupont's and Lesser Short-toed Larks. Over 30 pairs of Egyptian Vulture breed on the cliffs, along with Griffon Vulture, Golden Eagle, Peregrine Falcon and Eagle Owl. The pinewoods are a good place to search for Red-necked Nightjar, rare so far north, and some of the irrigation pools in the area have breeding Great Bittern and good collections of wintering ducks.

ITINERARIES

1. 50 km

Just over 1 km south of Arguedas along the NA-134 towards Tudela, turn left along a narrow road heading into Las Bardenas. After 12 km at Alto de Aguilares (1) you enter Las Bardenas proper and then soon come to the entrance to the military (2) base in the heart of La Blanca, home to perhaps the most impressive and solitary landscapes in the

Cabezo de Castildetierra, Las Bardenas Reales © José Manuel Reyero

region. The bombing range is circumnavigated by a track (25 km in all) that can be driven and used to explore the most characteristic habitats of Las Bardenas.

If you turn right and head anti-clockwise, after 4 km you pass near the Balsa de Zapata (3), a small lagoon within the bombing range with breeding Purple Heron and Marsh Harrier and good numbers of wintering duck. A few kilometres further on, just after the track turns north, you approach the cliffs and buttresses (4) of Sanchicorrota, El Rallón, La Ralla and Pisquerra where Griffon and Egyptian Vultures breed and with luck you may get a view of Golden Eagle. The track passes through a mosaic of natural steppe (5) with low-growing scrub, cereal fields and rocky outcrops. Here you should look for Stone Curlew, both sandgrouse and most of the larks – Dupont's, Calandra, Thekla, Crested, Short-toed and Lesser Short-toed – present in the site. Here too you may be lucky to find all three Iberian wheatears (Black, Black-eared and Northern) and Spectacled Warbler. Once at the curiously shaped buttress of Castildetierra (6), you can take a track directly back to the entrance to Las Bardenas (7).

2. 15 km

The NA-134 between Arguedas and Tudela runs through an area of rice paddies (8) of great ornithological interest: explore along any of the agricultural tracks entering from the road. Aside from the good population of breeding Black-winged Stilt, it is not difficult to spot Purple Heron, White Stork and Marsh Harrier in summer, Reed Bunting in winter and Common Crane and waders on passage.

The western sector of the rice paddies extends as far as the river Ebro and a well-preserved area of fluvial woodland. One of the most interesting such woodlands is Soto de Vergara (9), accessible from the NA-134 south from Arguedas and then along one of the tracks that heads right through the rice paddies. A walk along the river inside the *soto* should give you views of herons and raptors such Black Kite and Booted Eagle. European Bee-eaters and Sand Martins

NAVARRE

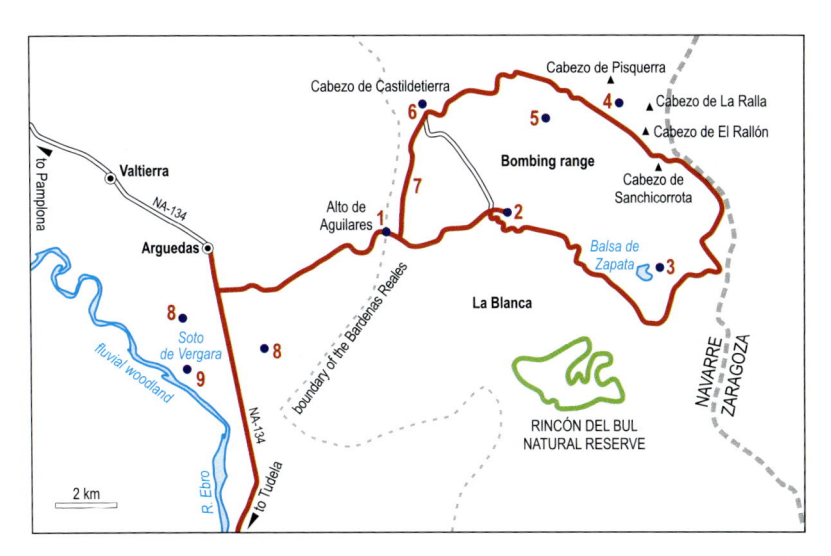

form mixed colonies, Kingfishers speed up and down the river and the trees of the *soto* are home to Wryneck, Penduline Tit and Golden Oriole.

MOST REPRESENTATIVE SPECIES

Residents: Great Bittern, Grey Heron, Griffon Vulture, Marsh Harrier, Golden Eagle, Stone Curlew, Black-bellied and Pin-tailed Sandgrouse, Eagle Owl, Dupont's, Calandra, Lesser Short-toed and Thekla Larks, Black Wheatear, Blue Rock Thrush, Dartford and Sardinian Warblers, Penduline Tit, Red-billed Chough, Rock Sparrow.

Summer visitors: Purple Heron, Black Kite, Egyptian Vulture, Short-toed and Booted Eagles, Montagu's Harrier, White Stork, Alpine Swift, Wryneck, Sand Martin, Short-toed

PRACTICAL INFORMATION

- **Access:** 65 km south of Pamplona the N-121 heads towards Arguedas.
- **Accommodation:** in Tudela.
- **Visitor infrastructure:** near Arguedas on the way to Tudela the NGO Gurelur (tel. 948 151 077, gurelur@bme.es) has set up a biological station that also works as an information and environmental education centre.
- **Visiting tips:** all year except summer (too hot). Access to the Bardenas Reales Natural Park is only allowed in daylight and along certain tracks. After rain, take care not to get stuck in muddy tracks.
- **Recommended:** the nearby town (in La Rioja) of Alfaro has an interpretation centre dedicated to the Ebro *sotos*. The White Stork colony on the Colegiata de San Miguel is the largest on a single urban building anywhere.
- **Further information:** maps 1:50,000, n°. 207, 244, 245, 282 and 283 (SGE). The Natural Park is run (and public access organised) by the Comunidad de Bardenas Reales de Navarra (tel. 948 820 020, junta@bardenasreales.es) in Tudela. The book *Las Bardenas Reales* by Jesús Elósegui and Carmen Ursúa (Gobierno de Navarra, 1994) is very informative.

Lark, Tawny Pipit, Yellow Wagtail, Northern and Black-eared Wheatears, Spectacled Warbler, Golden Oriole.
Winter visitors: Great Cormorant, Merlin, Reed Bunting.
On migration: Common Crane, Osprey, waders.

99. LAGUNA DE PITILLAS

This endorheic (inwardly draining) steppe lagoon, the largest natural wetland in Navarre (200 ha), has been made more permanent by the building of a dyke and can thus provide a more regular supply of water for irrigation. Even so, given that it depends purely on rainfall, its water levels vary from year to year and even dries out in especially dry summers. A large reed-bed surrounds and partially invades the lagoon and gives way in places to rush-beds, saline pastures and salt-marsh. The surrounding land is largely devoted to cereal fields and vineyards, although there are some patches of rosemary, albardine and other types of Mediterranean scrub in the vicinity. Pitillas is a *reserva natural* (run by the Navarre government), a Ramsar Site and an SPA.

ORNITHOLOGICAL INTEREST

Along with the reservoir of Las Cañas, Pitillas is the most important wetland in Navarre, as well as one of the best – up to 4 males in some years – sites in Spain for Great Bittern. Large numbers of Grey Heron (200+ pairs) and rather fewer of Purple Heron breed in the reeds, as do around 40 pairs of Marsh Harrier, the largest colony in the Ebro valley. Winter also sees an important Marsh Harrier roost form.

Other notable breeding birds include Black-necked Grebe, Black-winged Stilt and Bearded Tit; Purple Swamp-hen have begun to be seen in recent years and, if confirmed, this would be the first breeding site for the species in Navarre. Winter and passage periods brings Greylag Geese and waders.

Laguna de Pitillas © Miguel A. Sánchez, Foto-Ardeidas

NAVARRE

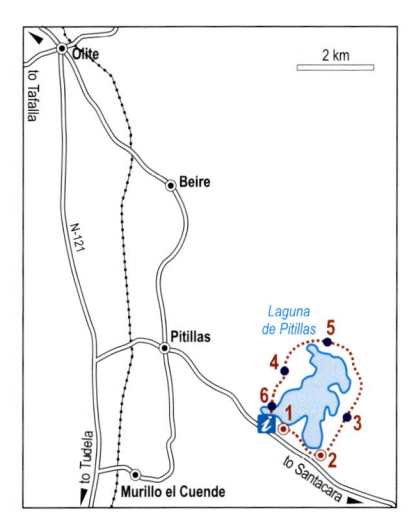

ITINERARY

| 12 km | 🚶 | ☀ | ⟳ |

Exactly what birds you will see at Pitillas depends a lot on water levels. Bearing this in mind, the main observatory (1), situated on a promontory, is the best place to begin a visit. This is one of the few hides in Spain from which, with luck, you may hear Great Bittern booming in March.

From the observatory there are good views across the reed-bed used by breeding herons to the patches of open water where, if water levels are adequate, Little, Black-necked and Great Crested Grebes – Spain's three breeding grebes – can all be seen. Red-crested Pochard also breed and winter sees large concentrations of Common Teal, Mallard, Northern Shoveler and Common Pochard.

During the breeding season it is best not to approach the lagoon edges: if in doubt, ask for information at the observation point that doubles as an information centre. The best idea is to continue on to the small wooden hide (2) at the extreme southern end of the lagoon, looking out for Purple Heron, Black-winged Stilt and waders during passage on the exposed mud.

Outside of the breeding season, you can completely circumnavigate the lagoon using the fence as a reference. The slightly raised scrubby slopes (3) to the east of the lagoon and arable fields to the west (4) give panoramic views over the lagoon and its immense reed-bed and are good sites to watch dozens of Marsh Harriers come into roost at dusk in winter.

The rush-beds and saline habitats to the north, as well as the nearby dry arable fields, are frequented by

PRACTICAL INFORMATION

- **Access:** from Pamplona, take the N-121 towards Tudela. In the town of Olite or 6 km beyond, there are turn-offs to the village of Pitillas. Two kilometres from the village on the road to Santacara, a track heads left to the car-park at the entrance to the lagoon.
- **Accommodation:** in Tafalla, Olite and Beire.
- **Visitor infrastructure:** the main observatory at the lagoon is also an information point (tel. 619 463 450). Open weekends and public holidays, morning and afternoon; in summer, open every day except Monday morning.
- **Visiting tips:** all year except summer, when bird activity is at its lowest point. The greatest variety of birds is in winter, although you should wrap up well as strong winds can seriously hamper birdwatching. When it is hot, mosquitoes can be a problem. Avoid disturbance to birds during the breeding season.
- **Further information:** maps 1:50,000, n°. 206 (SGE).

Greylag Geese on migration (a few stay to winter) and a few Stone Curlew and Hen Harrier. Just before completing the circuit, the lagoon dyke (6) provides a good viewing platform and is the best spot from which to look for Bearded Tit and, in winter, Reed Bunting.

MOST REPRESENTATIVE SPECIES

Residents: Little, Black-necked and Great Crested Grebes, Great Bittern, Grey Heron, Mallard, Red-crested Pochard, Griffon Vulture, Marsh and Hen Harriers, Water Rail, Bearded and Penduline Tits, Reed Bunting.
Summer visitors: Purple Heron, Black-winged Stilt, Little Ringed Plover, Reed and Great Reed Warblers.
Winter visitors: Greylag Goose, Eurasian Wigeon, Gadwall, Common Teal, Northern Shoveler, Common Pochard, Tufted and other ducks, European Golden Plover, Northern Lapwing, Black-headed Gull.
On migration: Little Egret, Black Stork, Eurasian Spoonbill, Garganey, Common Crane, Avocet and other waders, Whiskered and Black Terns.

100. EMBALSE DE LAS CAÑAS

The Las Cañas reservoir (*embalse*) lies near the city of Logroño (La Rioja province) in the extreme south-west of Navarre and close to the borders with the provinces of La Rioja and Álava. It was once an endorheic lagoon but since it was dammed a number of decades ago by two dykes, it now irrigates nearby agricultural areas. It covers 100 ha and much of its surface area ha been invaded by reeds and bulrushes; stands of tamarisk, rush-beds and damp meadows dot the perimeter of the reservoir.

ORNITHOLOGICAL INTEREST

Las Cañas holds an interesting breeding population of herons in its reed-beds that includes Cattle and Little Egrets and Grey (50 pairs), Purple (20 pairs) and Night (20 pairs) Herons: of the latter, many pairs have deserted Las Cañas in recent years

Las Cañas reservoir © Joseba del Villar

NAVARRE

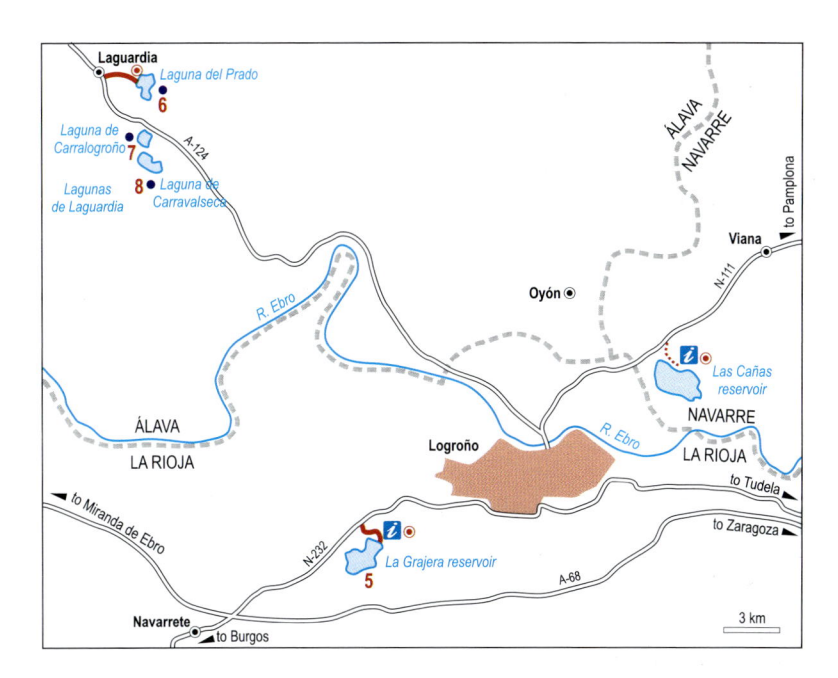

for other nearby wetlands. Water levels change from one year to another and the numbers and distribution of the colony varies accordingly. Great Bittern have not yet been confirmed as breeders, but are regularly seen on migration and in winter. However, Marsh Harrier, Savi's Warbler, Bearded Tit and Reed Bunting do all breed. In winter and on passage, a good numbers of birds and species pass through the site.

ITINERARY

5 km

Leave Logroño (La Rioja) towards Viana (Navarre) and turn right to the hide of El Bordón situated on a promontory on the northern side of the reservoir. Although a little distant from the reservoir, there are good panoramic views of the site from here.

From here you can walk all the way around the reservoir outside of the perimeter fence. Begin walking east (clockwise) and into an area of damp meadows (1) used during passage periods by Greylag Goose and groups of waders (Eurasian Curlew, Common Redshank, Greenshank and Ruff, for example).

A central dyke separates the two reservoirs: the eastern half is older and has large reed-beds (2) frequented by Marsh Harrier, Night, Purple and Grey Herons, egrets and passerines such as Penduline Tit, Reed Bunting and occasionally Bearded Tits. Hen Harriers hunt over the surrounded fields in winter. The southern bank of the reservoir has a number of promontories (3) that give good views over the open waters of the reservoir and the duck – above all Common Teal – that swim there in winter. The new dyke (4) is another good observation point.

Relatively near Las Cañas there are two other wetlands that are worth visiting. Both have heronries that include Purple Heron, Marsh Harrier, diverse communities of passerines and wintering duck.

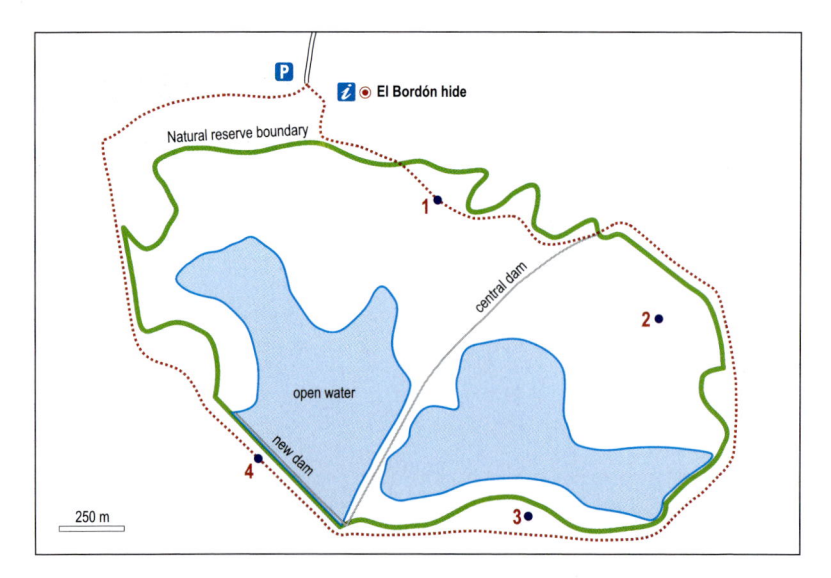

One of these sites, the reservoir of La Grajera (5) is in La Rioja and lies within an semi-urban park near the city of Logroño. Take the N-232 towards Burgos and turn left to the nature school, hide and 5-km walk through reed-beds, fluvial woodland and pine woodland.

The other site, in the province of Álava, consists of a singular group of endorheic lagoons near the town of Laguardia. They have been classified as a Biotopo Protegido – similar to a *reserva natural* – and are also protected as a Ramsar Site. From Laguardia itself, a good track takes you in 500 m to the artificially dammed Laguna del Prado (6). You can follow a circuit around the lagoon, which passes by a hide (ask for the key in the local tourist office). The main interest here is a large Night Heron colony of well over 100 pairs.

A little further away, although still part of the same group of lagoons, lie the lagoons of Carralogroño (7) and Carravalseca (8). Reach them along farm tracks off to the left of the road from Logroño just before you reach the town of Laguardia.

PRACTICAL INFORMATION

- **Access:** the access track to the reservoir turns off the N-111 halfway between Logroño (La Rioja) and Viana (Navarre).
- **Accommodation:** in Logroño.
- **Visitor infrastructure:** the information centre at El Bordón (tel. 696 830 898) is open at weekends and on public holidays, morning and afternoons: also every day in summer except Monday morning.
- **Visiting tips:** good all year given the site's interesting breeding, wintering and passage birds.
- **Further information:** maps 1:50,000, n°. 170 and 204 (SGE). The water and environment department of the Logroño City Council (tel. 941 277 000) and the tourist office in Laguardia (tel. 945 600 845, turismo@laguardia-alava.com) will provide information on, respectively, La Grajera reservoir and the Laguardia lagoons.

MOST REPRESENTATIVE SPECIES

Residents: Little and Great Crested Grebes, Grey Heron, Marsh Harrier, Water Rail, Bearded and Penduline Tits, Reed Bunting.

Summer visitors: Night and Purple Herons, Little and Cattle Egrets, Savi's, Reed and Great Reed Warblers.

Winter visitors: Great Cormorant, Great Bittern, Greylag Goose, Common Teal and other duck, Hen Harrier, Northern Lapwing, Jack Snipe, Black-headed Gull.

SPECIES INDEX

A

Accentor, Alpine 28, 33, 73, 76, 80, 83, 85, 103, 151, 199, 201, 204, 206, 208, 230, 232, 254, 259, 267, 293, 296, 301, 325
Avadavat, Red 172, 262, 273, 275
Avocet 35, 40, 48, 50, 53, 56, 58, 63, 66, 98, 154, 160, 163, 165, 168, 188, 190, 220, 240, 247, 273, 317, 330

B

Bee-eater, European 43, 68, 73, 77, 80, 83, 95, 103, 118, 196, 212, 232, 296, 312, 322
Bittern, Great 53, 92, 115, 172, 223, 227, 327, 330, 333
Bittern, Little 35, 40, 50, 53, 56, 60, 63, 93, 95, 115, 125, 160, 172, 220, 224, 227, 235, 240, 245, 247, 262, 270, 273, 275, 312
Blackbird 136, 140
Blackcap 25, 113, 136, 141
Bluethroat 40, 48, 51, 53, 56, 61, 63, 96, 103, 116, 128, 151, 160, 168, 191, 202, 208, 221, 224, 227, 241, 243, 245, 259, 262, 275, 301
Brambling 33, 93, 202, 208, 293
Bullfinch 28, 73, 77, 80, 83, 126, 127, 202, 204, 212, 230, 232, 256, 293, 322, 325
Bunting, Cirl 25, 85, 88, 174, 177, 204, 206, 259, 275, 296, 301, 325
Bunting, Corn 146, 180
Bunting, Ortolan 33, 103, 177, 202, 204, 206, 208, 214, 301, 325
Bunting, Reed 40, 48, 51, 53, 56, 61, 63, 93, 96, 108, 115, 126, 128, 130, 159, 163, 166, 168, 172, 191, 220, 227, 235, 241, 243, 245, 275, 283, 286, 312, 328, 330, 333
Bunting, Rock 25, 28, 31, 33, 77, 85, 88, 103, 177, 196, 199, 204, 206, 208, 214, 230, 254, 259, 293, 296, 301, 304, 306, 309, 322, 325
Bunting, Snow 105, 283
Bustard, Great 98, 163, 168, 172, 185, 187, 190, 193, 256, 264, 267, 270, 273, 309
Bustard, Houbara 143, 146
Bustard, Little 35, 87, 98, 159, 165, 168, 172, 180, 185, 187, 193, 235, 245, 256, 264, 267, 270, 273, 309
Buzzard, Common 33, 68, 113, 118, 136, 140, 143, 185, 245
Buzzard, Honey 28, 33, 68, 98, 113, 121, 151, 202, 204, 208, 232, 259, 293, 301, 325

C

Canary, Atlantic 136, 141, 143, 146
Capercaillie 73, 201, 229, 232
Chaffinch, Blue 136
Chaffinch, Common 113, 121, 136, 141
Chiffchaff 25, 33, 146
Chiffchaff, Canary Islands 136, 141
Chiffchaff, Iberian 103, 202, 204
Chough, Alpine 73, 76, 103, 151, 202, 204, 230, 232, 322
Chough, Red-billed 28, 31, 33, 73, 76, 80, 83, 85, 87, 103, 151, 177, 196, 199, 202, 204, 206, 208, 212, 214, 229, 232, 254, 259, 267, 270, 293, 296, 301, 304, 309, 312, 322, 325, 327
Cisticola, Zitting 125
Coot, Eurasian 38, 40, 56, 138, 143, 193
Coot, Red-knobbed 38, 40, 48, 53, 56, 221, 227, 243
Cormorant, Great 25, 38, 56, 58, 61, 93, 95, 105, 108, 113, 116, 126, 127, 130, 154, 160, 172, 174, 212, 214, 220, 224, 227, 235, 240, 245, 248, 262, 265, 267, 270, 273, 275, 283, 286, 289, 304, 307, 312, 317, 328, 333
Courser, Cream-coloured 143, 146
Crane, Common 36, 48, 51, 68, 85, 93, 98, 160, 163, 166, 172, 174, 180, 185, 188, 193, 262, 267, 270, 273, 275, 293, 304, 307, 312, 322, 328, 330
Crossbill, Common 28, 31, 33, 73, 77, 80, 83, 87, 113, 177, 208, 212, 230, 232, 293, 301, 306, 322, 325
Cuckoo, Great Spotted 25, 43, 80, 93, 168, 180, 235, 264, 267, 307
Curlew, Eurasian 58, 93, 98, 108, 128, 130, 154, 163, 221, 224, 273, 280, 283, 286, 289, 317
Curlew, Stone 35, 43, 48, 56, 63, 66, 87, 90, 93, 98, 118, 138, 143, 146, 168, 172, 174, 180, 185, 193, 204, 212, 214, 223, 235, 256, 264, 267, 270, 273, 275, 286, 309, 311, 317, 327

D

Dipper 31, 33, 73, 76, 80, 103, 127, 151, 177, 180, 204, 206, 208, 212, 214, 229, 232, 259, 293, 296, 301, 304, 322, 324
Diver, Black-throated 153, 224, 283, 286
Diver, Great Northern 108, 126, 127, 130, 153, 220, 280, 283, 286
Diver, Red-throated 153, 280
Dotterel, Eurasian 90, 224, 235, 245
Dove, Collared 138, 140, 146
Dove, Rock 43, 85, 113, 136, 138, 140, 146
Dove, Stock 322
Dove, Turtle 43, 136, 138, 141, 144, 146
Duck, Ferruginous 40, 48, 51, 126
Duck, Marbled 38, 40, 48, 50, 52, 63, 144, 240, 243, 245, 247
Duck, Ruddy 126
Duck, Tufted 38, 40, 48, 63, 125, 283, 286, 289, 312, 330
Duck, White-headed 35, 38, 40, 48, 50, 53, 60, 63, 115, 159, 163, 165, 168, 193, 245, 248
Dunlin 36, 58, 98, 108, 128, 130, 154, 163, 166, 168, 186, 191, 221, 280, 283, 286, 289, 317
Dunnock 25, 28, 33, 76, 121, 151, 201, 208, 230, 232, 301, 309, 322, 324

E

Eagle, Bonelli's 25, 28, 30, 43, 66, 80, 172, 174, 176, 203, 206, 214, 235, 253, 256, 259, 262, 267, 270, 296
Eagle, Booted 25, 28, 31, 33, 38, 48, 50, 61, 68, 80, 83, 85, 88, 103, 113, 118, 121, 174, 177, 180, 185, 196, 204, 208, 212, 245, 254, 256, 259, 262, 264, 270, 293, 296, 301, 304, 307, 309, 322, 325, 327
Eagle, Golden 25, 28, 30, 33, 73, 76, 80, 83, 85, 87, 90, 103, 151, 172, 176, 180, 196, 199, 201, 203, 206, 208, 211, 214, 229, 232, 235, 253, 256, 259, 267, 270, 293, 296, 301, 303, 306, 309, 324, 327
Eagle, Short-toed 25, 28, 31, 48, 50, 68, 77, 80, 83, 85, 88, 103, 151, 174, 177, 180, 196, 202, 208, 232, 254, 256, 259, 262, 264, 267, 270, 293, 296, 301, 304, 307, 309, 322, 325, 327
Eagle, Spanish Imperial 25, 48, 50, 172, 174, 180, 211, 253, 256, 262, 264, 306
Egret, Cattle 38, 40, 92, 95, 108, 115, 118, 172, 174, 243, 262, 264, 267, 270, 273, 275, 303, 312, 333
Egret, Great 48, 51, 53, 56, 93, 116, 220, 227, 240, 262, 275, 317
Egret, Little 40, 56, 58, 60, 92, 95, 108, 115, 118, 138, 141, 168, 172, 240, 243, 262, 273, 275, 280, 303, 330, 333
Eider, Common 154

F

Falcon, Barbary 138, 140, 143, 146
Falcon, Eleonora's 113, 116, 121, 146, 241
Falcon, Peregrine 25, 28, 30, 33, 43, 48, 51, 56, 60, 66, 73, 76, 80, 83, 85, 87, 90, 95, 103, 105, 108, 113, 116, 118, 121, 127, 130, 151, 153, 163, 166, 176, 180, 185, 188, 191, 193, 196, 199, 201, 204, 206, 208, 212, 221, 232, 235, 253, 259, 280, 283, 296, 301, 303, 306, 309, 311, 324
Falcon, Red-footed 118, 224, 235
Fieldfare 28, 85, 177, 199, 235, 254, 293, 301, 304
Finch, Citril 31, 73, 77, 80, 83, 103, 177, 202, 208, 212, 230, 232, 293, 301, 322, 325
Finch, Snow 73, 77, 103, 151, 232
Finch, Trumpeter 43, 66, 138, 143, 146
Firecrest 25, 31, 33, 83, 121, 208, 293, 301, 322, 325
Flamingo, Greater 35, 40, 47, 50, 53, 56, 58, 61, 63, 66, 116, 160, 163, 165, 168, 220, 223, 226, 243, 245, 247, 317
Flycatcher, Pied 106, 113, 121, 293, 301
Flycatcher, Spotted 121, 204

G

Gadwall 38, 98, 115, 125, 153, 187, 190, 220, 223, 226, 286, 289, 330
Gannet, Northern 48, 56, 61, 66, 68, 105, 113, 121, 128, 131, 154, 220, 224, 227, 240, 248, 280, 283, 286, 289, 317
Garganey 41, 61, 93, 98, 108, 116, 125, 160, 168, 188, 193, 221, 224, 235, 241, 243, 245, 248, 330
Godwit, Bar-tailed 58, 108, 154, 221, 280, 283, 286, 289
Godwit, Black-tailed 36, 53, 58, 63, 66, 93, 154, 163, 166, 191, 193, 221, 241, 245, 248, 273, 280, 283, 286, 317
Goldcrest 76, 83, 208, 212, 229, 293, 301, 322
Goldfinch 121

Goose, Barnacle 48, 154, 188, 191, 193
Goose, Greylag 40, 48, 51, 93, 98, 108, 116, 118, 126, 127, 131, 154, 160, 163, 166, 172, 174, 185, 188, 191, 193, 220, 224, 245, 273, 275, 293, 304, 322, 330, 333
Goose, Pink-footed 188
Goose, White-fronted 48, 188, 191, 193
Goshawk 25, 30, 73, 76, 83, 85, 103, 176, 203, 208, 229, 288, 293, 301, 306, 322, 324
Grebe, Black-necked 36, 38, 40, 47, 51, 56, 58, 61, 63, 66, 87, 98, 108, 126, 130, 154, 159, 163, 165, 168, 190, 193, 220, 224, 240, 243, 245, 248, 273, 289, 304, 317, 330
Grebe, Great Crested 38, 40, 59, 95, 125, 214, 220, 247, 304, 311, 317, 330, 333
Grebe, Little 31, 38, 40, 118, 125, 130, 262, 286, 304, 311, 330, 333
Grebe, Red-necked 153, 280
Greenfinch 121, 143
Greenshank 59, 108, 128, 163, 166, 280, 283, 286, 289
Guillemot 106, 108, 128, 130, 154, 280, 289
Gull, Audouin's 48, 56, 59, 61, 63, 66, 68, 113, 115, 118, 121, 220, 224, 227, 240, 243, 248, 317
Gull, Black-headed 35, 38, 58, 63, 92, 96, 98, 105, 108, 116, 128, 141, 159, 163, 165, 168, 172, 174, 187, 190, 212, 214, 235, 240, 247, 262, 265, 270, 273, 275, 304, 312, 317, 330, 333
Gull, Common 105, 128, 280
Gull, Great Black-backed 105, 108, 280, 289
Gull, Herring 105
Gull, Lesser Black-backed 36, 38, 53, 58, 105, 108, 141, 172, 174, 191, 220, 262, 265, 270, 273, 275, 280, 283, 288, 304, 312
Gull, Little 48, 56, 131, 227, 280
Gull, Mediterranean 56, 61, 63, 105, 108, 128, 131, 154, 221, 227, 243, 248
Gull, Slender-billed 35, 48, 50, 58, 63, 66, 116, 220, 224, 227, 240, 247, 317
Gull, Yellow-legged 58, 92, 105, 108, 113, 115, 118, 121, 127, 138, 140, 280, 283, 288

H

Harrier, Hen 33, 36, 53, 68, 88, 90, 95, 98, 103, 118, 160, 166, 168, 174, 180, 185, 188, 191, 193, 199, 201, 203, 232, 235, 265, 267, 273, 293, 309, 330, 333
Harrier, Marsh 36, 38, 40, 48, 50, 52, 56, 61, 63, 66, 68, 87, 90, 92, 95, 98, 108, 113, 115, 118, 121, 126, 154, 159, 163, 165, 168, 172, 187, 190, 193, 221, 223, 226, 232, 235, 240, 243, 245, 248, 262, 267, 273, 275, 309, 311, 322, 327, 330, 333
Harrier, Montagu's 35, 50, 68, 80, 88, 90, 98, 118, 160, 168, 172, 174, 180, 185, 188, 190, 193, 212, 214, 232, 235, 243, 245, 248, 264, 267, 270, 273, 275, 309, 322, 327
Hawfinch 25, 33, 80, 177, 204, 214, 235, 254, 293, 306, 309
Heron, Grey 25, 31, 36, 38, 56, 58, 115, 118, 125, 138, 141, 172, 174, 212, 214, 256, 262, 280, 304, 307, 311, 327, 330, 333
Heron, Night 47, 52, 56, 60, 63, 93, 95, 116, 160, 172, 220, 227, 235, 240, 245, 247, 262, 275, 333
Heron, Purple 40, 48, 50, 53, 56, 60, 63, 66, 93, 95, 116, 118, 125, 160, 172, 190, 220, 224, 227, 235, 240, 243, 245, 247, 262, 270, 275, 312, 327, 330, 333
Heron, Squacco 41, 48, 50, 53, 61, 63, 93, 116, 160, 220, 224, 227, 235, 240, 243, 245, 247, 262
Hobby 25, 48, 68, 185, 208
Hoopoe 138

I

Ibis, Glossy 47, 53, 61, 116, 160, 163, 166, 220, 224, 241, 243, 247

J

Jackdaw 312
Jay, Eurasian 208, 256, 259, 301

K

Kestrel, Common 30, 43, 68, 76, 90, 95, 118, 121, 136, 138, 140, 143, 146, 151, 185, 196, 253, 259, 264, 267, 296, 311
Kestrel, Lesser 28, 35, 68, 88, 90, 98, 160, 163, 165, 172, 174, 185, 188, 193, 212, 235, 256, 262, 264, 267, 270, 273, 275, 309
Kingfisher 25, 30, 60, 95, 108, 126, 127, 130, 177, 180, 199, 206, 212, 214, 220, 223, 227, 240, 253, 256, 259, 262, 264,

SPECIES INDEX

275, 303, 306, 309, 311, 324
Kinglet, Canary Islands 136, 141
Kite, Black 25, 28, 33, 48, 50, 68, 80, 85, 95, 103, 177, 180, 185, 188, 193, 196, 204, 206, 208, 212, 214, 232, 254, 256, 259, 267, 275, 301, 304, 307, 309, 312, 322, 325, 327
Kite, Black-winged 174, 180, 214, 256, 262, 264, 270, 273, 275
Kite, Red 48, 50, 68, 80, 85, 87, 113, 118, 174, 180, 185, 188, 191, 193, 196, 199, 206, 208, 211, 214, 232, 253, 256, 259, 264, 275, 293, 301, 303, 309, 322, 324
Kittiwake 106, 128, 131, 280, 289
Knot 59, 280, 283, 286, 289

L

Lammergeier 73, 76, 80, 83, 85, 229, 232, 322, 324
Lapwing, Northern 38, 48, 93, 98, 105, 108, 118, 159, 163, 165, 168, 174, 186, 187, 190, 193, 224, 243, 265, 267, 270, 273, 309, 330, 333
Lark, Calandra 35, 43, 85, 87, 90, 98, 163, 165, 168, 172, 180, 185, 188, 193, 199, 212, 214, 235, 264, 267, 270, 273, 309, 317, 327
Lark, Crested 90, 98, 256, 267,
Lark, Dupont's 66, 87, 90, 98, 163, 196, 199, 235, 327
Lark, Lesser Short-toed 50, 58, 63, 66, 87, 90, 143, 146, 163, 165, 168, 220, 235, 245, 247, 317, 327
Lark, Short-toed 35, 38, 43, 53, 61, 63, 66, 88, 90, 98, 118, 163, 165, 168, 172, 185, 188, 193, 196, 199, 212, 220, 235, 264, 267, 273, 309, 327
Lark, Thekla 28, 33, 43, 66, 87, 90, 95, 98, 118, 168, 180, 196, 199, 204, 214, 253, 256, 259, 267, 296, 303, 306, 309, 311, 327
Linnet 121, 136, 141, 143, 146

M

Magpie, Azure-winged 25, 174, 180, 196, 199, 206, 212, 254, 256, 259, 306
Mallard 38, 98, 330
Martin, Crag 28, 31, 77, 80, 83, 85, 95, 103, 127, 151, 177, 180, 196, 199, 201, 204, 206, 208, 212, 214, 232, 253, 259, 264, 267, 270, 273, 293, 296, 303, 306, 309, 324

Martin, Sand 95, 172, 193, 206, 243, 312, 327
Merganser, Red-breasted 56, 126, 128, 130, 154, 220, 289, 317
Merlin 36, 80, 88, 90, 98, 180, 185, 193, 267, 307, 309, 328
Moorhen 38

N

Nightjar, European 31, 212
Nightjar, Red-necked 43, 307
Nuthatch, Eurasian 31, 76, 83, 259

O

Oriole, Golden 25, 80, 95, 113, 121, 172, 196, 212, 254, 256, 259, 275, 296, 309, 312, 328
Osprey 28, 48, 51, 56, 58, 61, 63, 66, 68, 80, 93, 108, 113, 115, 118, 121, 126, 128, 131, 140, 146, 154, 160, 163, 166, 172, 174, 193, 221, 224, 227, 235, 240, 243, 245, 248, 262, 265, 273, 275, 289, 304, 307, 312, 317, 322, 328
Ouzel, Ring 28, 33, 73, 77, 80, 83, 85, 177, 201, 230, 232
Owl, Barn 146, 188
Owl, Eagle 25, 28, 30, 43, 66, 80, 87, 90, 95, 103, 172, 176, 180, 196, 199, 201, 204, 206, 208, 214, 232, 253, 256, 259, 267, 270, 293, 296, 303, 306, 311, 324, 327
Owl, Little 43, 87, 90, 95, 185, 267
Owl, Long-eared 176, 185
Owl, Short-eared 48, 51, 53, 56, 98, 186, 188, 193, 221
Owl, Tengmalm's 73, 229, 232
Oystercatcher 58, 63, 66, 106, 154, 220, 280, 283, 286, 289

P

Partridge, Barbary 136, 138, 143, 146
Partridge, Grey 73, 76, 103, 201, 229, 232, 293, 322
Petrel, Bulwer's 141
Phalarope, Red-necked 221
Pigeon, Bolle's 136, 140
Pigeon, Feral 138, 140, 146
Pigeon, Laurel 136, 140
Pigeon, Wood 256, 262, 293, 301, 322
Pintail, 38, 48, 53, 280
Pipit, Berthelot's 136, 138, 140, 143, 146

Pipit, Meadow 33, 106, 191, 193
Pipit, Red-throated 191, 224
Pipit, Richard's 105, 283, 286, 289
Pipit, Tawny 33, 88, 90, 98, 106, 118, 151, 185, 199, 208, 212, 296, 301, 328
Pipit, Tree 105, 151, 293, 301
Pipit, Water 28, 33, 73, 96, 103, 116, 151, 191, 201, 208, 214, 230, 232, 243, 293, 301, 304, 322
Plover, European Golden 51, 53, 61, 93, 105, 108, 154, 172, 174, 180, 185, 188, 191, 193, 224, 227, 235, 240, 243, 245, 248, 256, 265, 267, 270, 273, 309, 330
Plover, Grey 58, 108, 128, 130, 154, 163, 166, 221, 248, 280, 283, 286, 289, 317
Plover, Kentish 35, 40, 48, 50, 53, 56, 58, 60, 61, 63, 66, 98, 116, 118, 138, 143, 146, 163, 165, 168, 190, 193, 220, 224, 227, 240, 243, 247, 283, 286, 288, 317
Plover, Little Ringed 43, 93, 95, 116, 125, 130, 138, 143, 166, 168, 188, 190, 193, 227, 330
Plover, Ringed 58, 163, 166, 186
Pochard, Common 38, 60, 93, 125, 226, 245, 283, 286, 312, 330
Pochard, Red-crested 36, 38, 40, 48, 52, 63, 92, 115, 126, 159, 165, 168, 220, 240, 243, 245, 247, 273, 311, 330
Pratincole, Collared 35, 40, 48, 50, 53, 63, 160, 163, 165, 220, 227, 240, 243, 245, 248, 267, 273, 275
Ptarmigan 73, 229
Puffin 280

Q

Quail 118, 144, 146

R

Rail, Water 95, 118, 125, 130, 168, 262, 283, 286, 311, 330, 333
Raven 73, 76, 85, 113, 118, 136, 141, 143, 146
Razorbill 56, 61, 68, 106, 108, 128, 130, 154, 221, 224, 227, 280, 283, 286, 289
Redshank, Common 48, 58, 108, 154, 163, 166, 186, 187, 280, 286
Redshank, Spotted 116, 163, 166
Redstart, Black 33, 76, 85, 113, 121, 146, 199
Redstart, Common 28, 113, 121, 204, 206, 259
Redwing 25, 28, 33, 85, 177, 199, 202, 235, 254, 256, 293, 301, 304
Robin 25, 113, 121, 136, 140, 146, 259
Robin, Rufous Bush 40, 43, 66
Roller 25, 41, 43, 63, 66, 88, 174, 180, 185, 214, 224, 235, 256, 262, 264, 267, 270, 273
Ruff 93, 163, 166, 186

S

Sanderling 59, 221, 280, 283, 286, 289, 317
Sandgrouse, Black-bellied 43, 66, 87, 90, 98, 143, 159, 163, 168, 172, 185, 187, 190, 193, 199, 235, 264, 267, 270, 273, 309, 327
Sandgrouse, Pin-tailed 87, 90, 159, 163, 165, 168, 172, 185, 193, 235, 264, 267, 270, 327
Sandpiper, Common 95, 108, 163, 166, 186, 196, 214
Sandpiper, Curlew 59, 163, 168, 193, 248, 289
Sandpiper, Green 25, 96, 163, 166, 186, 199, 214
Sandpiper, Marsh 227
Sandpiper, Pectoral 283
Sandpiper, Purple 128, 280, 289
Sandpiper, Wood 163, 166
Scaup, Greater 286
Scoter, Common 48, 56, 106, 154, 243, 280, 283, 286
Scoter, Velvet 154
Shag 66, 105, 113, 118, 121, 127, 130, 153, 280, 283, 286, 288
Shearwater, Balearic 48, 61, 66, 105, 113, 118, 121, 128, 131, 220, 223, 227, 240, 243, 248, 280, 283, 286, 289, 317
Shearwater, Cory's 48, 61, 68, 105, 113, 118, 121, 138, 141, 144, 146, 220, 286, 289, 317
Shearwater, Great 105
Shearwater, Little 141
Shearwater, Manx 280, 283, 286, 289
Shearwater, Mediterranean 56
Shearwater, Sooty 105, 280, 283
Shelduck 36, 40, 47, 50, 56, 58, 66, 93, 98, 126, 130, 153, 163, 165, 168, 174, 188, 220, 243, 245, 247, 317
Shelduck, Ruddy 144
Shoveler, Northern 36, 38, 93, 125, 226, 240, 245, 280, 312, 330
Shrike, Great Grey 105
Shrike, Lesser Grey 224, 235
Shrike, Red-backed 73, 77, 80, 103, 105, 108, 127, 130, 151, 199, 202, 204, 206,

SPECIES INDEX

230, 232, 293, 296, 301, 322, 325
Shrike, Southern Grey 25, 28, 35, 40, 43, 136, 138, 143, 146, 180, 196, 199, 204, 206
Shrike, Woodchat 25, 33, 43, 85, 90, 180, 196, 199, 204, 206
Siskin 33, 177, 208, 212, 230
Skua, Arctic 61, 221, 280
Skua, Great 56, 61, 131, 221, 280, 289
Skua, Pomarine 280
Skylark 28, 33, 106, 121, 185, 208, 220, 301
Snipe, Common 61, 93, 98, 108, 116, 126, 130, 160, 174, 186, 191, 243, 309
Snipe, Jack 61, 243, 333
Sparrow, Rock 28, 31, 33, 43, 85, 95, 177, 196, 199, 206, 214, 254, 296, 306, 312, 325, 327
Sparrow, Spanish 136, 138, 141, 143, 146, 174, 180, 256, 262, 264, 267, 270, 309
Sparrowhawk 68, 83, 85, 113, 136, 140
Spoonbill, Eurasian 48, 50, 53, 56, 58, 61, 93, 108, 126, 128, 131, 153, 160, 172, 174, 188, 191, 193, 221, 224, 227, 243, 248, 262, 265, 275, 280, 283, 286, 304, 317, 330
Starling, Common 38, 116, 245, 275
Starling, Spotless 312
Stilt, Black-winged 35, 40, 48, 50, 53, 56, 58, 60, 63, 66, 93, 98, 116, 125, 144, 160, 163, 165, 168, 172, 185, 188, 190, 193, 220, 224, 227, 235, 240, 243, 245, 248, 270, 273, 275, 317, 330
Stint, Little 36, 59, 98, 163, 166, 168, 221, 317
Stint, Temminck's 53, 227, 241
Stonechat, Fuerteventura 143
Stork, Black 41, 48, 51, 53, 61, 66, 68, 98, 126, 131, 160, 172, 174, 180, 214, 254, 256, 259, 262, 264, 267, 270, 273, 304, 307, 322, 330
Stork, White 47, 53, 68, 103, 138, 160, 174, 180, 188, 212, 223, 256, 262, 264, 267, 270, 275, 303, 307, 309, 311, 327
Storm-petrel, European 48, 105, 113, 121, 141
Swallow, Barn 41, 193, 243, 312
Swallow, Red-rumped 25, 43, 80, 172, 206, 214, 254, 256, 259, 262, 264, 296, 304, 307, 309
Swamp-hen, Purple 35, 38, 40, 48, 50, 53, 56, 63, 92, 115, 159, 163, 172, 220, 223, 226, 240, 243, 245, 247, 262, 311
Swift, Alpine 28, 31, 33, 43, 66, 73, 77, 80, 83, 85, 93, 103, 113, 151, 199, 202, 204, 206, 214, 232, 254, 259, 264, 267,

275, 296, 325, 327
Swift, Common 212
Swift, Pallid 31, 33, 43, 118, 144, 146, 264, 275
Swift, Plain 136, 138, 141, 144, 146
Swift, White-rumped 28, 254

T

Teal, Common 36, 38, 48, 93, 95, 163, 166, 280, 286, 289, 330, 333
Tern, Arctic 106, 289
Tern, Black 48, 53, 61, 63, 66, 126, 131, 154, 160, 163, 166, 168, 191, 193, 227, 241, 245, 275, 317, 330
Tern, Caspian 48, 50, 56, 58, 128, 131, 154, 221, 227, 248
Tern, Common 61, 63, 66, 106, 108, 131, 220, 240, 243, 245, 248, 280, 283, 289, 317
Tern, Gull-billed 35, 48, 50, 53, 61, 98, 163, 165, 168, 188, 193, 220, 240, 262, 273, 275, 317
Tern, Lesser-crested 220
Tern, Little 50, 53, 56, 58, 61, 63, 66, 154, 160, 174, 220, 224, 240, 248, 275, 317
Tern, Sandwich 48, 50, 59, 63, 66, 106, 108, 116, 131, 141, 154, 220, 227, 240, 243, 248, 280, 283, 286, 289, 317
Tern, Whiskered 40, 48, 53, 61, 63, 66, 126, 160, 163, 165, 188, 190, 193, 220, 224, 227, 241, 243, 245, 248, 262, 317, 330
Tern, White-winged Black 48, 224
Thrush, Blue Rock 25, 28, 31, 33, 43, 66, 80, 85, 95, 103, 113, 118, 121, 127, 151, 153, 177, 196, 199, 201, 206, 208, 212, 214, 232, 235, 253, 256, 259, 267, 270, 296, 304, 306, 309, 324, 327
Thrush, Mistle 31, 76, 87, 199, 206, 208, 229, 301, 306, 322
Thrush, Rufous-tailed Rock 28, 31, 33, 77, 80, 83, 103, 113, 151, 177, 196, 199, 202, 204, 206, 208, 230, 232, 293, 296, 301
Thrush, Song 25, 113, 121, 146, 199, 206, 293, 301
Tit, Bearded 92, 159, 165, 190, 224, 240, 245, 330, 333
Tit, Blue 136, 141, 143, 146
Tit, Coal 28, 33, 83, 177, 208, 212, 232, 306, 322
Tit, Crested 25, 28, 31, 73, 83, 103, 151, 177, 202, 206, 208, 212, 229, 301, 306, 322

Tit, Marsh 73, 76, 103, 151, 201, 204, 229, 293, 322
Tit, Penduline 40, 48, 53, 56, 61, 63, 93, 95, 131, 159, 172, 191, 220, 223, 227, 241, 243, 245, 262, 275, 309, 311, 327, 330, 333
Treecreeper, Eurasian 76, 103, 151, 202, 229, 293, 322
Turnstone 59, 128, 283, 286, 289, 317

V

Vulture, Black 25, 113, 172, 174, 180, 208, 211, 253, 256, 259, 262, 264, 301, 306, 309
Vulture, Egyptian 25, 28, 31, 48, 68, 77, 80, 83, 85, 88, 90, 95, 103, 113, 118, 143, 146, 151, 177, 180, 196, 199, 202, 204, 206, 214, 254, 256, 259, 262, 264, 267, 270, 296, 322, 325, 327
Vulture, Griffon 25, 28, 30, 68, 73, 76, 80, 83, 85, 90, 103, 151, 174, 176, 180, 196, 199, 201, 203, 206, 208, 211, 214, 229, 232, 253, 256, 259, 262, 264, 267, 270, 293, 296, 301, 303, 306, 322, 324, 327, 330

W

Wagtail, Grey 31, 76, 136, 138, 140
Wagtail, White 146
Wagtail, Yellow 35, 53, 61, 98, 105, 108, 127, 168, 185, 188, 193, 220, 243, 245, 328
Wallcreeper 76, 80, 83, 85, 103, 151, 202, 204, 206, 229, 232, 296, 325
Warbler, Aquatic 126, 191, 243, 283
Warbler, Balearic 113, 121
Warbler, Bonelli's 25, 31, 33, 80, 83, 103, 151, 199, 202, 204, 208, 259, 293, 301, 309
Warbler, Common 38
Warbler, Dartford 25, 80, 85, 87, 90, 95, 118, 121, 127, 174, 177, 180, 196, 199, 204, 206, 212, 214, 235, 254, 296, 304, 306, 309, 327
Warbler, Garden 121, 196, 202
Warbler, Grasshopper 130, 191
Warbler, Great Reed 35, 38, 40, 63, 93, 125, 127, 168, 190, 235, 240, 243, 245, 262, 275, 286, 312, 330, 333
Warbler, Moustached 115, 159, 220, 223, 227, 240, 243, 245

Warbler, Olivaceous 43, 53
Warbler, Orphean 25, 31, 80, 85, 88, 174, 180, 204, 206, 212, 254, 256, 259, 296, 307, 309, 325
Warbler, Reed 35, 40, 61, 63, 93, 108, 127, 130, 168, 190, 235, 240, 243, 245, 262, 275, 312, 330, 333
Warbler, Sardinian 25, 35, 43, 80, 85, 95, 118, 121, 127, 136, 141, 143, 146, 153, 180, 199, 212, 214, 254, 296, 304, 306, 309, 325, 327
Warbler, Savi's 53, 160, 190, 240, 243, 245, 262, 333
Warbler, Sedge 53, 108, 131, 191, 262
Warbler, Spectacled 35, 43, 50, 66, 88, 90, 118, 138, 143, 146, 168, 174, 180, 196, 199, 204, 206, 214, 235, 267, 296, 307, 309, 328
Warbler, Subalpine 25, 31, 33, 80, 85, 113, 121, 177, 180, 185, 196, 199, 206, 212, 214, 232, 254, 256, 304, 307, 309, 325
Warbler, Willow 121
Waxbill, Common 273, 275
Wheatear, Black 28, 31, 33, 43, 66, 80, 85, 87, 90, 95, 196, 199, 214, 235, 253, 259, 267, 270, 296, 311, 327
Wheatear, Black-eared 25, 28, 31, 35, 40, 43, 66, 80, 85, 88, 90, 168, 174, 180, 196, 199, 204, 206, 208, 212, 214, 254, 264, 267, 273, 296, 304, 307, 309, 328
Wheatear, Northern 28, 31, 33, 73, 76, 88, 90, 98, 103, 106, 113, 151, 168, 177, 180, 185, 188, 196, 199, 202, 204, 206, 208, 212, 214, 230, 232, 293, 301, 322, 325, 328
Whimbrel 58, 154, 280, 283, 286, 289, 317
Whinchat 41, 73, 103, 106, 113, 206, 230, 301
Whitethroat, Common 33, 105, 121, 196, 202, 206, 301
Wigeon, Eurasian 36, 38, 48, 53, 128, 154, 212, 280, 286, 289, 330
Woodcock 73, 113, 121, 230, 232, 293
Woodpecker, Black 73, 76, 83, 103, 151, 201, 229, 232, 322, 324
Woodpecker, Great Spotted 30, 136
Woodpecker, Lesser Spotted 214, 253, 259, 303
Woodpecker, Middle Spotted 151, 201
Woodpecker, White-backed 322
Wryneck 28, 80, 95, 151, 196, 301, 309, 327

SPECIES INDEX

Y

Yellowhammer 73, 77, 103, 202, 204, 206, 230, 293, 322

Other titles available:

- *Atles dels ocells nidificants de Catalunya 1999-2002* (J. Estrada, V. Pedrocchi, Ll. Brotons & S. Herrando)
- *Fauna y Paisaje de los Pirineos en la Era Glaciar* (O. Arribas)
- *Grebes of our World* (A. Konter)
- *The Spanish Imperial Eagle* (M. Ferrer)
- *Ecology and Conservation of Steppe-land Birds* (G. Bota, M. B. Morales, S. Mañosa & J. Camprodon)
- *Las aves marinas de España y Portugal* (A. Paterson)
- *Guía de las cajas nido y comederos para aves y otros vertebrados* (J. Baucells, J. Camprodon, J. Cerdeira & P. Vila)
- *Parques Nacionales de España, 26 itinerarios para descubrirlos y conocerlos* (O. Alamany & E. Vicens)
- *Itineraris pels parcs naturals de Catalunya* (J. Bas, A. Curcó & J. Orta)
- *On observar ocells a Catalunya* (14 authors)
- *Catàleg dels ocells dels Països Catalans* (J. Clavell)
- *A Birdwatcher's Guide to Italy* (L. Ruggieri & I. Festari)
- *Birding in Venezuela* (M. L. Goodwin)
- *Aves de España* (E. de Juana & J. M. Varela)
- *Guía sonora de las aves de Europa* (10 CDs; J. Roché & J. Chevereau)
- *Guía de las aves de O Caurel* (J. Guitián, I. Munilla, M. González & M. Arias)
- *A Field Guide to the Birds of Peru* (J. F. Clements & N. Shany; colour drawings by D. Gardner & E. Barnes)
- *Birds of South Asia: the Ripley Guide* (P. C. Rasmussen & J. C. Anderton)
- *Annotated Checklist of the Birds of Argentina* (J. Mazar Barnett & M. Pearman)
- *Annotated Checklist of the Birds of Belize* (H. Lee Jones & A. C. Vallely)
- *Annotated Checklist of the Birds of Chile* (M. Marín)
- *Mamíferos de España* (F. J. Purroy & J. M. Varela)
- *Boscos vells. Artists for Nature als Pirineus catalans / Mature forests, Artists for Nature in the Catalan Pyrenees* (Artists for Nature Foundation & Fundació Territori i Paisatge de Caixa Catalunya)
- *Arte de pájaros / Art of Birds* (P. Neruda)
- *Threatened Birds of the World* (BirdLife International)
- *Curassows and Related Birds. Second Edition* (J. Delacour & D. Amadon)
- *Handbook of the Birds of the World* (J. del Hoyo, A. Elliott, D. Christie & J. Sargatal)
  Vol. 1: Ostrich to Ducks
  Vol. 2: New World Vultures to Guineafowl
  Vol. 3: Hoatzin to Auks
  Vol. 4: Sandgrouse to Cuckoos
  Vol. 5: Barn-owls to Hummingbirds
  Vol. 6: Mousebirds to Hornbills
  Vol. 7: Jacamars to Woodpeckers
  Vol. 8: Broadbills to Tapaculos
  Vol. 9: Cotingas to Pipits and Wagtails
  Vol 10: Cuckoo-shrikes to Thrushes
- *Wildlife Travel Maps of Spain. Catalonia (1/300,000)*
- *Wildlife Travel Maps of Spain. Balearic Islands (1/150,000)*

For more information, please visit our website:
www.hbw.com

Lynx

Montseny, 8, 08193 - Bellaterra, Barcelona (Spain)
Tel: (+34) 93 594 77 10 / Fax: (+34) 93 592 09 69
E-mail: lynx@hbw.com